KANT'S TRANSCENDENTAL DEDUCTION

VOLUME 222

ROBERT HOWELL

Department of Philosophy, State University of New York at Albany

KANT'S TRANSCENDENTAL DEDUCTION

An Analysis of Main Themes in His Critical Philosophy

KLUWER ACADEMIC PUBLISHERS
DORDRECHT / BOSTON / LONDON

Library of Congress Cataloging-in-Publication Data

Howell, Robert.
Kant's transcendental deduction : an analysis of main themes in his critical philosophy / by Robert Howell.
p. cm. -- (Synthese library ; v. 222)
ISBN 0-7923-1571-5 (alk. paper)
1. Knowledge, Theory of. 2. Kant, Immanuel, 1724-1804--Contributions in theory of knowledge. I. Title. II. Series.
B2799.K7H68 1992
121'.092--dc20 91-43991

ISBN 0-7923-1571-5

Published by Kluwer Academic Publishers,
P.O. Box 17, 3300 AA Dordrecht, The Netherlands.

Kluwer Academic Publishers incorporates
the publishing programmes of
D. Reidel, Martinus Nijhoff, Dr W. Junk and MTP Press.

Sold and distributed in the U.S.A. and Canada
by Kluwer Academic Publishers,
101 Philip Drive, Norwell, MA 02061, U.S.A.

In all other countries, sold and distributed
by Kluwer Academic Publishers Group,
P.O. Box 322, 3300 AH Dordrecht, The Netherlands.

Printed on acid-free paper

Printed in the Netherlands

In Memory of My Mother and Father,
Lorinda Katherine Cottingham Howell
Robert Donald Howell

Es ist mißlich, den Gedanken, der einem tiefdenkenden Manne obgeschwebt haben mag und den er sich selbst nicht recht klar machen konnte, zu erraten …

– Kant to Marcus Herz, May 26, 1789,
discussing Leibniz

Firm ground is not available ground.

– A. R. Ammons

TABLE OF CONTENTS

DISPLAYED SENTENCES REFERRED TO FREQUENTLY

ACKNOWLEDGMENTS

In my references in this book to Kant's work, 'Ak. 3, 45' refers to vol. 3, p. 45, of the Academy Edition of Kant's works. 'Schmidt, ed.' is Schmidt's edition of the *Kritik der reinen Vernunft*, as cited in the Bibliography. For Kant and other authors, I use the translations noted in the Bibliography, sometimes with alterations. Translations not otherwise identified are my own. As is standard, 'A4/B8' refers to p. 4 of the first edition and to p. 8 of the second edition of the first *Critique*, as noted in the margins of most editions, including Kemp Smith's translation, from which (with some changes) I quote. References to Locke and Berkeley and to Hume's *Treatise* are by book, chapter, and section ('IV.4.5'). Aristotle is cited by work, using the translations listed in the Bibliography. 'McKeon, ed.' refers to *The Basic Works of Aristotle* (1941), cited in the Bibliography.

Some parts of Chapters Two and Five through Eight are from two of my previous essays (1979 and 1981a, as noted in the Bibliography). I am grateful for permission to reuse the material.

PREFACE

The *Critique of Pure Reason* is one of the two or three supreme texts of Western philosophy and the most influential philosophical work of the last 250 years. The Transcendental Deduction of the Categories, in the *Critique*, is the central argument of Kant's theoretical work and underlies much subsequent philosophical investigation. The categories referred to are, according to Kant, the *a priori* concepts of our mind's faculty of understanding – concepts such as those of quantity, quality, substance, and cause and effect. The argument of the Transcendental Deduction answers a fundamental question of classical metaphysics and epistemology: Can we know, through such categories, substantive facts about objects *a priori*, in independence of the evidence of our senses?

As we will see in Chapter One, the opinion of rationalist metaphysicians was that we can. But the opposing empiricist tradition, culminating in the work of Hume, held that we cannot know facts about objects in such a way and, further, that concepts such as those of substance or of cause and effect are not *a priori* at all. In the Transcendental Deduction Kant tries to reconcile these two major positions and to settle once and for all the issue of the scope and limits of the categories.

Kant argued earlier in the *Critique of Pure Reason* that we can know objects only as they appear to us through our senses and not as they exist in themselves, in independence of our sense experience of them. He argued also that concepts such as those of substance or of cause and effect are *a priori*. In the Transcendental Deduction he now attempts to show that the categories do indeed apply to, and yield us knowledge of, objects, but he denies that they yield us knowledge of objects as they exist in themselves. Kant thereby vindicates the rationalist view that *a priori* knowledge of objects through the categories is possible. But at the same time he also curbs rationalism and vindicates empiricism by insisting that we have such *a priori* knowledge only of the objects of our sense experience.

In the Deduction, Kant's argument for his major conclusions is straightforward in its overall structure. Kant reasons, very roughly, that the mental representations, or intuitions, through which we know objects in sense experience are subject to what he calls unity of apperception: We can take all those representations, and their contained elements, to belong to ourself. From this fact Kant argues that, using our *a priori* concept of an object, we must judge the objects of such representations to be structured by what he calls the logical functions of thought in judgment. According to Kant, the result of this structuring is that those objects fall under the categories. So Kant proves the applicability of the categories to the objects that we know through sense experience. And he shows also that through the categories we can know only such objects and not objects as they exist in themselves.

Kant's argument – clear in outline but obscure in many of its details – is, I think, the deepest and most far-reaching in all philosophy. Besides the points already mentioned about the categories, it raises fundamental questions about the role of logic (or of the organization of concepts in judgment) in structuring the world that we know, the possibility of our knowing objects independently of such mind-imposed structures, the place of the self and its conceptual apparatus with respect to the world that it knows, the nature of self-knowledge, and the dependence of self-knowledge on knowledge of objects distinct from the self's mental states. These questions, and the points already mentioned, have had an immense influence on subsequent philosophy, and every theoretical philosopher, of whatever flavor, who engages with fundamental concerns is in one way or another working within, or reacting against, the intellectual situation into which Kant plunged us late in the eighteenth century.

The Transcendental Deduction has been the subject of much discussion from Kant's time on. But many of its particulars, including crucial points about unity of apperception, the concept of an object, judgment, the logical functions of judgment, and the categories, are still not well understood. And because the logical structure of Kant's inferences and the extensive role, in his reasoning, of the logical phenomenon of intensionality have not been widely grasped, even many of the points that have been worked through have not been fully comprehended.

Past approaches to Kant's theoretical philosophy have ranged from painstaking philological investigations of the texts to freewheeling 'reconstructions' of his theses that build up Kantian or semi-Kantian constructions out of contemporary philosophical ideas. The goal of some (but not all) parts of my previous essays on Kant, especially my 1973 *Noûs* paper (cited in the Bibliography), has been reconstructive – to show, in part, how various concerns from contemporary intensional logic and philosophy of language reflect, or can be used to illuminate, Kantian theoretical claims. But the aim of the present book is rather different. Although I make a number of reconstructive suggestions, I am not here attempting to re-create the Transcendental Deduction in terms of contemporary logic or to compare Kant's ideas in detail with all the recent logical views. Rather, I seek to use specific ideas from contemporary philosophy and intensional logic in order to understand Kant's own reasoning and views as he himself presents them in the Deduction. I intend my findings to throw considerable new light on crucial claims and arguments in Kant's theoretical philosophy and on many obscure details of the Deduction. Of course the ultimate justification for this undertaking can be seen only by studying its results. However, I can say now that the procedure is vindicated by the fact that many of Kant's concerns in the first *Critique* are reflected in current philosophical work, and the sharp tools that that work has forged make possible a finer, keener grasp of Kant's manifold subtleties than has hitherto been available – subtleties that, despite past scholarship, have kept much of the Deduction a mystery.

A brief outline of the course of my discussion may be useful here. In Chapters One to Five, I explain Kant's basic picture of knowledge and the ideas needed to understand the argument of the Deduction – for example, the notion of an intuition and its manifold of elements; the view that we know objects only as they appear to us through our senses; and the nature of a concept and of the elements of the manifold of intuition. I also examine the overall structure of the Transcendental Deduction and of its 1787 (or B-) version, which is (for reasons noted below and explained in Chapter Four) the especial focus of this book. And I discuss in detail the idea of intensionality and begin to show its central role in the Deduction.

In Chapters Six through Ten I then develop the main argument of the first half of the B-Deduction, which provides Kant's clearest, most developed, and philosophically most stimulating exposition of his main lines of thought. In Chapter Six I consider apperception and its unity. I complete that discussion in Chapter Seven and also there examine the idea – on which Kant places much emphasis – of the necessity of unity of apperception. In Chapter Eight I study Kant's deep and interesting claims about the union of the elements of intuition in the concept of an object. In doing so, I resolve an apparent fallacy at the heart of his reasoning from unity of apperception to that union. I then turn in Chapter Nine to Kant's account of judgment and of the logical functions of thought in judgment. In Chapter Ten I explain how, by appeal to that account and to his previous results, Kant finally infers category application to the objects that we know through sense experience. As I do so, I try to make clear, with greater exactness than has been previously attained, precisely how the structuring of judgments by the logical functions of thought leads to category application to the objects of those judgments. And I suggest an Aristotelian origin for some hitherto unobserved, puzzling features of Kant's claims about judgment and the categories.

Throughout this book I try to state Kant's views with logical precision. This project sometimes entails complexities, particularly in the discussion of Kant's central claims about apperception, objectivity, and the categories. But the difficulty and depth of the Deduction demand close attention from any serious reader, and I have tried to pursue Kant's views with the same sort of patience, clear-headed penetration, and use of the relevant tools that one would bring to the work of a contemporary philosopher for whom one has the greatest respect.

Because my goal in this book is to understand and evaluate Kant's own reasoning in the Deduction, I did not begin my work with any fixed idea of how far the Deduction might succeed. Given the importance of the issues that Kant raises, the depth to which he pursues them, and the influence of his reasoning and framework, it seemed sufficient to follow the argument of the Deduction wherever it led, without worrying about the outcome. As the reader will discover, the results of my study are negative. It turns out that there are serious problems with – or there is a serious lack of evidence for – each of Kant's main claims about unity of apperception and its necessity, about the unity of elements of intuition in the concept of an object, and about the logical functions and the categories. The argument of the Deduction, as Kant presents it, fails; and that fact has serious implications for Kant's own theoretical philosophy. As I urge in Chapter Ten, I think that this situation must be acknowledged and not pushed aside while the Deduction is wafted upward on clouds of unthinking piety.

A number of other commentators have thought the official argument of the Deduction a failure. (And certainly 200 years of investigation have failed to yield a clearly sound version of Kant's Deduction reasoning.) But no one, I believe, has pursued Kant's argument through every twist and turn, with the focus on exact statement and explicit inference, and the emphasis on intensionality and related matters, that I have tried to maintain here. Nor have earlier scholars treated apperception, the concept of an object, and the categories and logical functions in the manner and detail that I do. Nonetheless, some prospective readers may perhaps wonder what the value can be of a study whose main conclusions about the overall Deduction are negative. To my mind, there are two answers to this question.

First, and as I have emphasized, the Transcendental Deduction is unmatched in modern philosophy for importance, depth, and influence. The framework and ideas that Kant lays down in the Deduction are the basis from which all subsequent theoretical philosophy has proceeded, whether in agreement, rebellion, or bemusement. Each of us is gripped by, or is fighting the grip of, that framework today. One has to look no further than current disputes over realism and antirealism, the role of linguistic and conceptual frameworks in fixing our version of reality, and similar matters. It therefore is of great importance to realize that Kant's own fundamental claims and inferences in the Deduction are subject to the most serious objections. I think it is an open question whether, on some careful formulation, current and broadly Kantian ideas like those concerned with linguistic or conceptual frameworks and our version of reality will turn out to be correct. (Part of the effect of the Deduction is simply to have forced such questions on us.) But the fact that Kant's own claims in the Deduction fail means that those claims themselves cannot be appealed to in defense of the recent, broadly Kantian ideas. Nor can those claims support the many other views contemporary philosophers have tried to derive from the Deduction (for example, views about self-knowledge as requiring knowledge of objects distinct from one's experiences or views about the inevitability of our possessing the capacity for first-person ascription of our experiences). Whatever the fate of all the contemporary ideas whose ultimate origins lie in the Transcendental Deduction, my results show that that fate cannot be settled by appeal to the argument and claims of the Deduction itself.

Second, a cardinal task of scholarship is the dispassionate examination of the master texts of our tradition. In a famous essay, Kant praises enlightenment, which he describes as the release from a self-incurred tutelage in which one lacks the resolution and courage to use one's reason without direction from another. It is in the spirit of this ideal of enlightenment that I have tried to read the Deduction, endeavoring as far as possible to see for myself what actually lies in Kant's own reasoning in that text. My goal throughout is understanding, not cheerleading for Kant.

As I have noted, I concentrate on the first half of the 1787 version of the Transcendental Deduction as presented in the second (or B-) edition of the *Critique of Pure Reason*. I do so because of that version's clarity and philosophical interest and because it raises succinctly almost all the main issues about Kant's positive account of our knowledge. The B-Transcendental Deduction is Kant's final extended statement of the argument of the Deduction, and it explicitly introduces themes about judgment, the logical functions of thought, and the categories that the first (or A-) edition (1781) does not discuss. However, I comment on the second half of the B-Deduction as well as on the A-Deduction, and on various of Kant's remarks elsewhere about the Deduction. I also consider the relationship of Kant's ideas in the Deduction to other parts of his theoretical philosophy. Thus this book is a study of the main themes of the Transcendental Deduction and of Kant's overall theoretical philosophy, as well as of the actual B-Deduction itself. I try to take up most of the important disputed issues in the history of Deduction interpretation and to make a precise, reasoned contribution to the debate about them.

I have tried to push my examination of Kant's argument quite deep. The Deduction is, however, an enormously fertile piece of reasoning, and much more could be said, both philosophically and historically, than I have been able to say here. Thus Kant's views on self-knowledge, on the categories, and on the relationship of his claims in the Deduction to ideas that he presents in the Transcendental Dialectic deserve discussion beyond that which I can give them in this book. I also have had to truncate an account of Kant's historical background and of his position (in my opinion) as the supreme 'Aristotelian-Cartesian' philosopher – a philosopher who took into his hands, and reshaped, over two thousand years of thought. I hope to be able to present some of this additional material elsewhere.

This book has been long in the writing. I began it in the fall of 1972 and completed a first draft in August 1973. I presented some of my fundamental ideas (for example, about intuition, synthesis, individuation, intensionality, operator-shift fallacies, the concept of an object in general, and unity of apperception) in a series of papers published from 1973 to 1981. But most of those ideas are refined here, and there is much that is new. I also have tried to take into account relevant parts of the secondary literature, which is becoming enormous. In many respects, however, my approach differs from that of other authors, principally in its application of close logical analysis to Kant's argument, its emphasis on the role of intensionality, and its many specific points about the interpretation of apperception, unity of the object, the logical functions, and the categories. My stress on the importance of the Metaphysical Deduction of the Categories to the Transcendental Deduction – with which Kant would agree – also is unique to this work. A point of convergence exists with Guyer's criticisms, in his recent book, of Kant's treatment of necessity. Although my views (which go back to my 1973 manuscript) differ in detail from Guyer's, we both see Kant as falling victim to some of the same general kinds of errors about necessity. (So also does Harrison in a 1982 paper cited in the Bibliography.) I would like to think that this agreement is evidence for the general correctness of our discussions of Kant at this point.

I owe many debts. P. F. Strawson's lectures on Kant, which I heard at Oxford in 1965–66, have influenced my understanding of Kant's position, although I do not agree with everything that Strawson maintains in *The Bounds of Sense*. I have profited also from Jonathan Bennett's work on Kant. Jaakko Hintikka's well-known ideas about intensionality have been important to my view of the logical underpinnings of the Deduction, and Robert Paul Wolff's interpretation of the Transcendental Analytic both convinced me that Kant has one definite line of argument in that section of the *Critique* and helped me to see what it is. Both Hintikka and Wolff also gave me considerable, and much-appreciated, encouragement. I owe Wolff special thanks for persisting with detailed suggestions about substance and style while disagreeing strongly with many of my major claims. I found essays by Dieter Henrich, Charles Parsons, and Stephen Barker stimulating and far more helpful than many book-length works. Through his writings and in the few conversations I have had with him, Lewis White Beck has both clarified and altered my grasp of Kant's notion of the concept of an object in general. Early on, Konrad Marc-Wogau and Manley Thompson offered useful suggestions, as did Julius Moravcsik, who also improved my understanding of Aristotle. In addition, John Perry, Patrick Suppes, and Arthur Melnick deserve thanks for their support. In the SUNY Albany philosophy department, I want especially to thank Berel Lang and Robert G. Meyers for their efforts on my behalf. Further back, I owe much to my teachers V. C. Aldrich and Frithjof Bergmann. Thomas B. Kirsch was a source of unfailing insight. It goes without saying that none of these people should be held responsible for the opinions that I cxpress in this work.

Much of the final draft of this book was written while I was a Visiting Member at the Institute for Advanced Study in Princeton, and I am very much indebted to the Institute for having me and to Morton White for his help and encouragement while I was there. I must also thank the National Endowment for the Humanities for partial support at the Institute. I received funds for earlier essays related to this book from the American Council of Learned Societies, the SUNY Research Foundation, and from a National Endowment for the Humanities summer grant. I am grateful to SUNY Albany and its Faculty Research Awards Program (and to Francine Frank, dean of the College of Humanities and Fine Arts, and Jeanne E. Gullahorn, vice-president for research) for providing funds for manuscript preparation. And I appreciate the help of Donna Magee, who typed successive drafts with skill and dedication.

I want finally to thank my wife, Pam, for her patience and encouragement, and, as well, my children, Robbie and Kate, who were there, each day, messengers of hope as I scaled the sheer face of the Deduction.

CHAPTER ONE

KANT'S PICTURE OF KNOWLEDGE

1. KANT'S GOALS

Kant's goal in the Transcendental Deduction is to prove the objective validity of the categories. By the categories Kant means certain *a priori* concepts that are yielded to us by our understanding, or faculty of thought. These concepts include those of substance, cause and effect, and extensive spatial magnitude. Kant counts such concepts as *a priori* on the ground that they originate in operations or capacities of our mind that are independent of those mental operations involved in our having sense experience. Kant also counts these concepts as *a priori* because, as he sees it, we take them to apply with necessity and strict universality to all objects (or to all objects of a certain group). And he has, earlier, taken necessity and strict universality to be the marks of the *a priori* (both of judgments that are known *a priori* and, in a slightly different sense, of concepts that are possessed and utilized *a priori*). Yet – and here the question of the objective validity of the categories emerges – suppose that we regard the categories as *a priori* in this latter sense. Then it is still hardly obvious that the categories do apply with necessity and strict universality to all objects of the relevant group. Hence the problem arises of deducing the categories of the understanding – of justifying our right to employ those categories as though they did apply in that way to all such objects. And, to Kant's mind, the deduction of the categories is to be carried out by proving their objective validity – that is, by demonstrating, with respect to the relevant set of objects, that the categories in fact apply, with the proper sort of necessity and strict universality, to each object in that set.[1]

The demonstration of the objective validity of the categories, and thus the success of his deduction of them, is of great importance both to Kant and to us. For one thing, this demonstration provides a middle way between the competing views of rationalists and empiricists about such concepts. On the one hand, as Kant sees it, rationalist philosophers like Descartes and the Leibnizians have taken us to be able to demonstrate, via *a priori* metaphysical reasoning, the applicability of these *a priori* concepts to all actual and possible objects. And such philosophers have thought that by demonstrating this fact we enable ourselves to gain metaphysical knowledge, *a priori*, of substantive (or synthetic) truths that apply to all such objects, for example the truth that all objects are substances whose changes are caused, or the truth that all thoughts belong to persisting, immaterial spiritual substances. But, on the other hand, philosophers in the empiricist tradition, and particularly Hume, have argued that such supposedly *a priori* concepts are not *a priori* at all. As these philosophers see it, concepts like those of substance or of cause and effect have their origins in sense experience, not in the sense-experience-independent operations of our mind. Moreover, these philosophers, and especially Hume, believe that they can show that such concepts do *not* apply, with the relevant sort of necessity or strict universality, to *any* objects, let alone to all actual and possible objects. And such philosophers believe also that no *a priori* metaphysical knowledge of the sort sponsored by the rationalists is therefore possible.

One of Kant's intentions, in arguing for the objective validity of the categories, is to establish a position with regard to such concepts that falls between these rationalist and empiricist extremes. Kant agrees with the rationalists that, as we have noted, the categories are *a priori* concepts that have their origins independent of sense experience and apply with necessity and strict universality to all objects of a certain group. He also agrees that a certain form of philosophical reasoning can yield us knowledge, *a priori*, of various substantive, or synthetic, truths that apply to all such objects. But, contrary to the rationalists and in agreement with the empiricists, Kant denies that the categories can be shown to apply to all actual or possible objects whatsoever; and he denies that pure *a priori* metaphysical reasoning of the rationalist sort can demonstrate substantive, or synthetic, truths of the kind just mentioned. Rather, *a priori* concepts like the categories can be shown to apply, with necessity and strict universality, only to all those objects that we human knowers do or can know. Kant argues also that our own *a priori* philosophical reasoning can yield us knowledge, *a priori*, only of substantive, synthetic truths that concern just such objects and no others. In arguing in the Deduction for the objective validity of the categories he thus passes between the rationalist and empiricist camps by accepting that objective validity, but only with respect to a restricted set of objects – the set of objects that we do or can know, the set of objects of what he calls possible experience. In so proceeding, he therefore establishes the scope and limits of category application.

Another set of reasons for the importance of Kant's demonstration of the objective validity of the categories lies immediately within his theoretical philosophy. As we will see later, Kant supposes that, roughly, it is through the application of the *a priori* categories to the objects of our knowledge that we are first enabled to regard those objects as being things that go beyond simple, possibly disorganized collections of immediate sensory data. Through category application – and only through that application – we are first enabled to view such objects as being, in fact, members of causally organized collections of temporally persisting substances that stand in an objective time order distinct from the subjective time order that is established by our apprehension of those substances. But since the categories are *a priori* concepts not derived from our sense experience, it is not obvious to Kant why the individual objects of our knowledge – objects that surely are known by us only through our senses – should have to satisfy the categories at all. Why should not such objects simply *be* mere disorganized bits of sensory data? It is therefore of great importance to Kant, directly within his own theoretical philosophy, that he should be able to prove the objective validity of the categories with respect to all such objects, for by proving that point he is able to demonstrate that all the individual objects that we know are indeed something beyond such mere bits of data.

A third group of reasons for the importance of Kant's proof of the objective validity of the categories lies within contemporary philosophy. A good many philosophers would now want to abandon or at least to reformulate considerably Kant's sharp distinction of *a priori* from sense-experience-dependent, or *a posteriori*, concepts and judgments. Because of their strong doubts about his distinction between analytic and synthetic judgments, a good many philosophers also would now be quite uncomfortable about the Kantian position, noted above, concerning our knowledge (whether *a priori* or not) of various substantive, or synthetic, truths about objects. Numerous philosophers would in addition be dissatisfied either with the particular list of concepts that Kant presents as *the* fundamental categories of our knowledge or else with his idea that it is possible to establish one unique such list. And, as anyone at all familiar with the literature on Kant's work will recognize, there are a great many other features of his argument for the objective validity of the categories that are open to question. Nevertheless, even if all of these worries are taken with great seriousness, considerable interest would reside in any plausible argument that a categorial structure (whether or not a unique or an *a priori* such structure) must belong to the objects that we can know. Since Kant's proof of the objective validity of the categories, even when it is stripped of its doubtful or questionable aspects, may still perhaps offer such an argument, a great deal of interest for contemporary philosophy still belongs to that proof and to the Deduction as a whole.

In addition, a number of the individual topics that Kant considers in the course of his proof are of exceptional importance to contemporary philosophical concerns. Among these topics are his views on apperception and the *I think* and the relation of his account of thought and of what he calls synthesis of the manifold of intuition to modern views on the logical phenomenon of intensionality. We will return to these and other such topics later in this book.

2. THE KANTIAN PICTURE OF KNOWLEDGE (I): INTUITION, CONCEPT, SENSIBILITY, AND UNDERSTANDING

Kant solves the problem of the Transcendental Deduction, in the *Critique of Pure Reason*, by attempting to prove, deductively, that the categories apply necessarily to any object that we do or can know. I will develop his proof in detail in Chapters Six through Ten. Put in rough terms, this proof proceeds from the assumption that we do or can know a certain object, together with various premises about the nature and cognitive capacities that belong to us, the knower. Through the use of that assumption and of such premises, Kant attempts to establish the necessary subjection of the object to the categories. He then takes it in a certain way to follow that, because that object is an arbitrarily selected one, any object whatsoever that we do or can know falls necessarily under the categories. As we will see in more detail in Chapter Four, this argument amounts to a proof of the objective validity of the categories from the 'possibility of experience.' And, as we have just seen, it establishes the objective validity of the categories specifically (and only) with respect to the set of all objects of our possible experience.

As many commentators have noted – and as Kant himself notes somewhat cryptically at Axvi–xvii – alongside the above deductive argument for the objective validity of the categories he also offers a detailed explanation of how it comes about that the categories apply with necessity to all the objects that we can know. This explanation is couched in terms of his theory of the mental entities and operations that underlie our knowledge, and if his deductive proof of the objective validity of the categories amounts roughly to what he calls the 'objective deduction' of the categories, the explanation itself comes roughly to what he describes as the categories' 'subjective deduction' (Axvii; also Chapter Four below). Taken together, the proof and the explanation introduce central Kantian notions like unity of apperception; synthesis of the manifold of intuition and its necessity; the concept of an object in general; judgment, objectivity, and the logical functions and the logical forms of judgment; the categories; and imagination and its operations in synthesis. By so proceeding, the proof and explanation raise most of the philosophical and exegetical questions that concern Kant's theoretical philosophy as a whole.

To see the above points about the Deduction clearly and in detail, we need in this chapter and in the next several chapters to lay out various points about Kant's treatment of knowledge that either are presupposed in the Deduction itself or else deserve independent discussion. I begin now with a sketch of some main aspects of the picture of knowledge that Kant has adopted in the first *Critique* by the start of the Deduction. I should emphasize that, like my preceding remarks, the discussion here and in Section 3 *is* no more than a sketch. Only some of the central matters that bear on Kant's account of knowledge will be noted; these matters will not always receive a full discussion; and various philosophical or interpretive complications will be ignored. The further ramifications of Kant's account must wait their statement until later in our interpretation.

The picture of knowledge on which Kant focuses throughout the first *Critique* is that of the knowledge of a *being like us* – a being, whether or not specifically human, that possesses, and gains its knowledge by means of, our own sort of cognitive apparatus.[2] That cognitive apparatus he describes in broadly Cartesian terms. According to Kant, we know by means of representations (*Vorstellungen*) in our mind – which are mental entities, equivalent in general to Descartes' or Locke's ideas, that occur in our mind in such a way as to represent the objects to be known. In Kant's view, as in that of his predecessors, our knowledge via such representations occurs only insofar as the representations represent the objects to a certain sort of inner consciousness that is possessed by our mind. We can, however, ignore this point for the present. We need only note, now, that Kant takes there to be, in our human case, two fundamental and distinct kinds of such representations, both of which are involved in our knowledge: intuitions and concepts.

An intuition (or *Anschauung*) is what Kant calls a singular, immediate representation. Such a representation represents to the mind a single, individuated object (say a single, particular house) and stands in a direct relation to that object that is not mediated through the relation that the intuition bears to any further thing that is itself related in some way to the object.[3] There are both empirical and *a priori* intuitions. Empirical intuitions are those, like the ones that represent to us a house or a tree now standing before us, that involve sensations. Our *a priori* intuitions, as we will note in Section 3, are certain sensation-independent forms – in fact, space and time – that serve to structure, with necessity, all our empirical intuitions or the objects that they represent to us.

In contrast to an intuition, a concept (or *Begriff*) is a general, mediate representation. To speak a bit roughly, a concept presents to the mind a general property (for example, the general property of being a tree or of being triangular) that does or can belong to various individual objects (for example, to all the various individual trees or triangular things). Kant calls such a general property a 'mark' or 'characteristic' of those objects. He supposes that through its presentation of the general property to the mind, the concept represents to the mind, in a general fashion, those objects. The concept represents those objects simply as being those things, whichever they may be, that do or can possess the general property. Thus that concept is related mediately to those objects through the relation that it bears to that general property which they themselves possess.[4]

As all readers will remember, Kant takes concepts themselves to be either empirical or *a priori*. Empirical concepts, like those of being a house or of being a tree, are, in general, concepts that are abstracted by our understanding from the objects of our empirical intuitions. We will see the details of this abstractive process, which involves our understanding's attention to the properties that these objects have in common, in Chapter Three. *A priori* concepts include both the categories and various mathematical concepts, like those of being triangular or of being spherical. The latter concepts, although they do not apply to *all* objects with necessity, nevertheless characterize, in a sensation-independent way, features of what we will see to be the *a priori* intuitions, space and time. So Kant counts them as *a priori*. Both our empirical concepts and our *a priori*, mathematical concepts he supposes to specify general types or kinds – say the type or kind *house*, *tree*, or *triangular thing* – into which fall the objects that are represented to us by our (empirical) intuitions.

Kant connects the possession and use of intuitions and concepts in our human case with our possession and use of the two mental faculties of sensibility and understanding. Sensibility is our mind's passive capacity to receive representations through the 'affection' of our mind by objects (A19/B33). Understanding is our mind's active, spontaneous capacity to produce thoughts (A50/B74). These faculties, and their operations, are utterly distinct from each other. Moreover, in our human case all our intuitions are yielded in one way or another by our sensibility, which can do nothing but intuit; and all our concepts are yielded by the activities of our understanding, which can do nothing but generate and operate discursively with general concepts.[5]

These last points mean that insofar as we human beings, and beings like us, are to know single, individuated, particular objects as such – for example, the single house that is before us – we must do so always through intuitions that are yielded by our sensibility and thus our sense experience. Again, insofar as we are to know objects as being of general types or kinds – for example, as being of the general type *house* or *tree* or *triangular thing* – we must do so always through concepts that are generated by our understanding and thus through our thought. As we have just seen, after all, our understanding cannot intuit; and our sensibility cannot think. Thus through no conceptual presentation of any conjunction of general properties can our understanding, by itself, give us any representation that represents one single individuated object as such. And through no heaping up of the data that are given to us through the affection of our sensibility can our sensibility by itself yield any representation that presents a general type or kind of thing under which individual things do or can fall.

Although our sensibility and understanding are thus utterly distinct faculties of mind, Kant is of course famous for holding at A51/B75–76 and elsewhere the basic position that our human knowledge of objects requires the use of both these faculties and of the intuitive and conceptual representations proper to them. In non-Kantian terms, that is, he holds that our human knowledge of objects requires both sense experience of those objects and thought; and that knowledge is always about entities that it recognizes as being single, individuated objects that are of general types or kinds. In fact this position is slightly weakened by Kant on occasion, for he of course wishes to allow for examples in which we know objects that we are currently unable directly to sense. But because such examples pose no strikingly unusual problems for the Transcendental Deduction, I will largely ignore them now. Thus for the moment we may take Kant to hold simply the position that I have stated above about the required role, in all our knowledge of objects, of sensibility and understanding and hence of intuition and concept.

We will say more later about this position. It is fundamental to Kant's thought, and one can argue at length about whether he himself sees it as supported by, or rather as itself simply implying or being equivalent to, various others of his fundamental views. It is not necessary to embark on any such discussion here. But it is worth noting that a view in some ways resembling this position is found already in the 1770 Inaugural Dissertation, in which Kant in effect argues that our having knowledge requires at least the possibility of our – or of some being's – recognizing the object known as the particular, individuated thing that it is by intuiting that object.[6] And I myself think that this Inaugural Dissertation view, like the specific first-*Critique* position just mentioned, is independent of Kant's other fundamental views.

Here, however, we need not delve into questions about the status of that position for Kant, nor need we investigate the ultimate plausibility of such a position. Rather, in carrying out our task of producing a clear, comprehensive interpretation of the Deduction, we can simply restrict our attention, until later, to cases specifically of our knowledge, through intuition and concept, of single, individuated objects as being of general types or kinds. So proceeding, we can see whether Kant succeeds in demonstrating necessary category application to the objects of such knowledge. We can then determine how far, by appeal to the above sort of position, he can show necessary category application in those cases in which we know objects that we are currently unable to sense. And we can consider, as is necessary, any further questions that are raised by that position or by the existence of such knowledge.

Kant's own pattern of exposition in the first *Critique* in fact supports the preceding interpretive strategy. As noted above, he himself is willing to allow certain exceptions to the specific A51/B75–76, first-*Critique* position that our human knowledge of objects requires intuitions of those objects and thought. As we have already indicated, these exceptions seem not to pose any strikingly unusual problems for the Deduction itself; nor does Kant anywhere worry that they do. And they come in the first *Critique* (at A225–26/B273–74 in the Postulates of Empirical Thought in General) well after the Deduction itself and hence significantly later than his major statement of the A51/B75–76 position in the famous second paragraph of the Transcendental Logic. Kant is thus himself happy to carry out the argument of the Deduction before dealing with such exceptions to that position. I will therefore follow the above strategy and focus attention, for the present, on cases of knowledge, achieved via intuition and concept, of single, individuated objects as being of general types or kinds.

3. THE KANTIAN PICTURE OF KNOWLEDGE (II): SPACE, TIME, AND TRANSCENDENTAL IDEALISM

Kant's picture of our knowledge as achieved via intuition and concept of course becomes quite distinctively and even idiosyncratically Kantian in detail. But in many of its fundamental respects the overall picture, as I have so far sketched it, is quite close to the earlier, and related, pictures that were developed by other philosophers in the Cartesian tradition – philosophers like Descartes himself and Locke. This fact, which it is now important to examine more carefully, can be seen by noting a series of points about such pictures of knowledge that either were not emphasized or else were ignored in our Section 2 discussion.

In particular, if we abstract from fine points of difference concerning such matters as innate ideas, primary and secondary qualities, and the role of concepts and of sensation in our knowledge, then philosophers like Descartes, Locke, and Kant can all be interpreted as holding that we know objects via certain representations in our mind, representations that are *of* or *about* those objects and represent them to our mind *as being* things of one sort or another (say as being a single, individuated object that is a house, a tree, or a triangular thing). In addition, all three of these philosophers suppose that we thereby gain knowledge of these objects as being of the sorts that the representations represent them to our mind as being. These three philosophers suppose further that, in our human case, our mind possesses a certain inner consciousness that can grasp various of the mind's own properties and representations.[7] And they take it that our knowledge of the object in fact arises specifically in those cases in which one of our representations is so grasped by our mind's inner consciousness that that representation comes to represent the object to that inner consciousness as being of a certain sort.

Finally – and here we reach a matter that looms large for the interpretation of Kant – all three philosophers make the same supposition concerning the object that the representation is *of* or *about*, the object that is represented to the mind by the representation. All of these philosophers suppose that that object may be said to have an existence apart from or independent of the fact that that object is represented to – or is otherwise grasped or apprehended by – that or any other mind. Or again, and to put the matter in Descartes' terms, that object has a formal and not merely an intramental, 'objective,' reality or existence.[8] In Locke's terms it has a 'real existence' 'without us.'[9] And in Kant's terms it has an existence in itself.[10]

Because these last remarks raise complex questions and bear crucially on major differences between Kant and his Cartesian predecessors, it is worth illustrating them in further detail. To this end, consider some arbitrary being, *H*, whose cognitive capacities and operations are like ours. Imagine that *H* has knowledge of the single, individuated tree that is before *H*. Ignore the specific facts about intuition and concept application that, as we have just seen, Kant takes to underlie this knowledge of *H*'s. Ignore also the way in which, for Kant, *H*'s knowledge involves category application to the object known. Then, as has been suggested in the last two paragraphs, *H*'s knowledge occurs, according both to Kant and to philosophers like Descartes and Locke, when *H* acquires a representation *r* that is *of* or *about* a certain object *o* which *r* represents to *H* as being a single, individuated tree.

Moreover, by the preceding discussion, since *H* knows via *r*, *r* must be so grasped by *H*'s inner consciousness that *r* comes to represent *o* to that inner consciousness as being a single, individuated tree. When *r* is grasped by *H*'s inner consciousness in that way, *H* then knows *o* as being of the sort that *r* represents *o* as being – namely, as being such a tree. Finally – and here we return to what is now a crucial fact – both Kant and those of his predecessors like Descartes and Locke make a basic supposition about *o*, the object that *r* is *of* or *about*. All three of these philosophers suppose that that object *o* in fact has an existence apart from or independent of the fact that *o* is in any way represented to (or is otherwise grasped by) *H*'s mind or any other mind.[11] That is, and as we have suggested, it is supposed by such philosophers that *o* has an existence in itself.

If, however, the philosophers that we have mentioned in effect all take the object *o* to have an existence in itself, Kant's treatment of the knowledge that *H* achieves of *o* via *r* of course diverges radically from the treatments developed by Descartes and Locke. Although the details of their theories are rather different, both Descartes and Locke can be said to regard *r* as giving *H* knowledge of *o* in the form that *o* takes as *o* exists in itself. These philosophers suppose, that is, that as *o* so exists, *o* is a single, individuated object that is a tree. And they take *r*'s representation of *o* to *H* as being such a tree to give *H* knowledge of *o* as *o* so exists. But Kant is notorious for denying, in the Transcendental Aesthetic of the first *Critique*, that the knower *H* can ever gain, either via *r* or via any other sort of representation, any knowledge of *o* as *o* exists in itself.

As *o* exists in itself, Kant argues, *o* is not a spatiotemporal thing. But *o*, as *o* is known by *H*, *is* a spatiotemporal thing, a single, individuated tree. Hence in representing *o* to *H* as being such a tree, *r* cannot be giving *H* knowledge of *o* in the form that *o* takes as *o* exists in itself. Rather, *r* gives *H* knowledge of *o* simply in the form that *o* takes insofar as *o* is represented to *H* by *r* – namely, in the form of the single, individuated tree. Or, to express this last point in slightly more Kantian language, *r* gives *H* knowledge of *o* not as *o* exists in itself but only as *o appears* to *H* via *r* – or only as *o* is *represented* to *H as being* via *r*. And *o*, as *o* is thus known by *H* – that is, *o* as *o* appears to *H* or is represented to *H* as being – is itself a wholly intramental, mind-dependent entity.

The various arguments that Kant develops in the Transcendental Aesthetic for the above view are both complex and open to numerous exegetical and philosophical questions. We can abstract here from almost all of these questions. But because the view itself and portions of Kant's argument for it affect the underlying development of the Transcendental Deduction, we should note his most basic claims before proceeding. As every reader knows, these claims turn on Kant's treatment of space, time, and outer and inner sense. And they also bear directly on the sort of transcendental idealism that he takes to be supported by that treatment and by his above conclusions about the mind-dependent status of object *o* as *o* is known by *H*.

Kant's account of these matters is at its most plausible and its clearest in the Aesthetic's Transcendental Expositions of Space and Time.[12] Kant argues there that mathematics yields us *a priori* knowledge of substantive (and so synthetic) necessary truths that concern space and time and the objects in space and time. (Thus classical Euclidean geometry is supposed to yield such a knowledge of space.) Yet, he holds, it is impossible that, in the absence of any sense experience of a group of objects, we could nevertheless know substantive, and necessary, truths that concern those objects, at least insofar as those objects are taken to exist in themselves in complete independence of the fact that they are represented to (or are otherwise grasped by) any mind. Hence space and time, since they are the subject of such truths, cannot be entities as those entities exist in themselves. Instead, space and time must be entities that exist *solely* within our mind in the form that those entities take as they are represented to or are grasped by our mind. And, as they so exist, they must be subject to a sense-experience-independent, knowledge-yielding inspection by our mind.

As Kant further argues, however, all the single, individuated objects that we know are, as we know them, spatiotemporal things in all their features. Thus by knowing space and time, *a priori*, to be subject to substantive, necessary truths of mathematics, we evidently do or can know, *a priori*, that all such single, individuated objects are subject to those truths. We can indeed know, *a priori*, that all the various such objects that we have not yet sensed or even considered fall under those truths. But, Kant asks, how can we conceivably be held to know, *a priori*, that such truths apply to *all* those objects, including objects that we have not yet sensed or considered? And, he answers, there is only one way in which such knowledge is possible. That way is to suppose that space and time, as entities that exist solely within our mind, are *a priori*, sense-experience-independent *forms* that function to structure spatiotemporally all the single, individuated objects that we do or can know. And if we recall the above picture of knowledge, we can easily see how space and time can be such forms.

Kant takes it as obvious that all the single, individuated objects that we know are known through representations that are produced in our mind by the affection of our sensibility and involve sensations. That is, all such objects are known through empirical intuitions like *r* above. However, because space and time exist only in our mind, the single, individuated objects that we know cannot, as they exist in themselves, be spatiotemporal things. Yet all such objects, as we know them, are spatiotemporal. Hence the spatiotemporal features of the objects of our knowledge must be features that those objects acquire simply insofar as they are represented or appear to us via our empirical intuitions. And space and time themselves, as *a priori* forms of the above sort, must then be entities that function within our mind to structure all objects as those objects thus appear to us. Moreover, because single, individuated objects are, as we know them, spatiotemporal in all their features, these objects, as we know them, must be no more than objects as those objects appear to us via our empirical intuitions. And because such objects are spatiotemporal, unlike objects as they exist in themselves, such objects must be, in the forms in which we know them, wholly intramental and mind-dependent things – things that do not exist in independence of their being represented to (or otherwise grasped by) our mind.

The position that Kant here arrives at in the Transcendental Aesthetic he calls transcendental idealism. By this position he means the view that space and time, together with all the single, individuated objects of our knowledge (in the forms in which we know those objects), are nothing but wholly intramental and mind-dependent entities of the sort indicated above – entities that do not in any way belong to or occur among objects as objects exist in themselves. Kant couples this transcendental idealism with an empirical realism with respect to space, time, and the objects of our knowledge. By empirical realism with respect to space and time he means the position that space and time really belong to (and do not, for example, merely seem to belong to) the objects of our knowledge (in the forms in which we know those objects). And by empirical realism with respect to those objects he means the view that those objects, in the spatiotemporal forms which they are represented to us as having, are indeed genuine objects that we know with certainty, through our empirical intuitions, to exist. Kant takes it as clear that, on his own theory, such an empirical realism is true. And he regards his theory as being, in this respect, more satisfactory than either of two previous theories: Descartes' theory that (as Kant interprets it) we can only infer with probability the existence of the spatiotemporal objects of our knowledge; and Berkeley's theory, which Kant misinterprets as asserting that space, and hence each object in space, is a completely impossible thing, a nonexistent *Unding*.[13]

Kant of course considerably extends his preceding Transcendental Aesthetic line of argument (whose ultimate success we have yet to discuss), and in the process he renders more specific than above his account of space and time as forms in the mind. He does these things through appeal to his division of our human sensibility into its two components of outer and of inner sense. Outer sense is our mind's capacity to be affected by objects in such a way as to be yielded representations – in fact, intuitions – of those objects as being 'outside us and all without exception in space' (A22/B37). Inner sense is our mind's capacity to intuit itself in such a way as to gain intuitions of 'itself or its inner state' (A22/B37).

As the first of the phrases just quoted implies, our outer-sense intuitions invariably represent objects to us as being distinct from us and as in space. Consequently on the basis of the preceding line of argument Kant takes space itself specifically to be the form, within our mind, of outer sense. As he sees the matter, space is thus associated with the structure of our outer sense in such a way as to guarantee that all objects, insofar as those objects appear to us via our outer intuitions, are spatially structured.

Again, Kant supposes that our inner intuitions invariably represent to us properties and parts of our own mind as being in time – for example, as succeeding one another temporally. He in addition takes it that only our inner intuitions thus represent entities as being in time. Therefore, given the preceding line of argument, he regards time specifically as being the form, within our mind, of our inner sense. For Kant, time is thus associated with the structure of our inner sense in such a way as to guarantee that all entities, insofar as they appear to us via inner-sense intuitions, are temporally structured. Because time is simply the form of inner sense and inner sense is the means whereby we know our mind and its parts and properties, we can know our mind only as it appears to us in temporal form and not as it exists in itself. This result is of course explicitly affirmed by Kant in many places in the first *Critique*.

Kant develops his treatment of space, time, and outer and inner sense considerably further than I have just been noting. We can ignore the details, which are often both complex and obscure. I should observe, however, that Kant regards our representations as having the ontological status of 'inner determinations' or 'modifications' of our mind – that is, as being, roughly, properties of our mind that constitute cases of mental states, activities, or awareness.[14] In conjunction with his account of time and inner sense, this classification leads to an important consequence that needs immediate attention.

As we have seen above, Kant, like various of his predecessors, takes us to know via our representations only insofar as those representations are grasped by a certain inner consciousness belonging to our mind. In fact, he usually proceeds as though that inner consciousness, or at least an invariably requisite part of it, were our inner sense. (Another requisite part turns out to be our understanding's concept-utilizing acts of thought, but we are ignoring such acts of thought here.) We have just observed, however, that our inner sense operates by means of inner intuitions that are *of* various properties and parts of our mind and represent those properties and parts to our mind as being in time. Hence it follows that we have knowledge via our (non-inner-sense) representations only when those representations are themselves represented to us, by means of inner intuitions, as occurring in time.

This consequence affects various aspects of the Transcendental Deduction. Moreover, it is applied by Kant in particular to those of our representations that he counts as outer intuitions. Thus insofar as he proceeds in the above way, Kant takes outer intuitions, when they yield us knowledge, to appear to us in inner sense as standing in time relations. And he holds that we attain knowledge via our outer intuitions when we grasp them mentally as they appear to us in that way.[15] This conclusion involves complications both for his view of the interaction of outer and of inner sense and for his view of the role of inner intuitions in knowledge. We will note some of these complications in Chapter Two.

A final group of points about Kant's account of space, time, and outer and inner sense is worth noting in a preliminary fashion here. As I have observed several times above, space and time, as forms of outer and of inner sense, are entities that function within the mind to structure objects as objects are represented by the relevant empirical intuitions. This observation does not, however, indicate exactly how space and time function in that way, and Kant himself never makes absolutely clear the precise nature of that functioning. Furthermore, the question of the nature of such a functioning is complicated by the fact that he takes the generation of specific, definite intuitions in our mind to require not only the functioning of space and time as forms but also the category-governed synthesis of the manifold of intuition, an operation to which we will return later in this book.

If, however, we abstract from all these complications, then we can derive a plausible picture of how Kant supposes the above functioning of space and time to occur. In the case of space (on which I focus exclusively here), his basic idea is that space operates as a form in the mind with respect to the mental materials that are given to outer sense. Space operates in such a way as to guarantee the construction, from those materials (in conjunction with the operations of synthesis), of single, definite outer intuitions of the sort discussed above.[16]

This basic idea of Kant's carries puzzles with it, for (as my language has been meant to bring out) space or spatiality is supposed to belong to, and hence surely should be supposed to structure, *objects* as objects are represented by outer intuitions. Space (or spatiality) is surely not supposed directly to belong to or to structure the outer intuitions (or their mental materials) themselves. Moreover, and as we will note in Chapter Two, those outer intuitions presumably have existence in themselves as they occur in the mind; and thus space itself, as operating with respect to them, would surely have to have such an existence, completely contrary to Kant's own views. Yet not only does Kant not consider this point, but also he in fact regards space (and also time) as being itself an *a priori* intuition in the mind, as well as being an *a priori* form of intuition. He does so largely on the grounds that space is an essentially single, unique entity that we know *a priori*. But it would certainly seem that, as being such an intuition, space should – presumably like all intuitions in the mind – have an existence in itself.

We cannot resolve all these puzzles at this point. Nor can we comment on the exact relation that Kant takes to hold between space (or time) as a form of intuition and space (or time) as itself a single, individual intuition. As I will note again later, however, it seems clear that in his above views Kant is identifying various intuitions with the objects that appear via those intuitions. Or, in other words, he is identifying various intuitions with objects in the spatial (or temporal) forms that those objects take as they are represented to us by those very intuitions. Thus it seems obvious that Kant should take space itself to belong to, or somehow to structure, such *objects*. But what structures the *intuitions* that represent such objects, or what brings it about that those intuitions invariably represent objects as being spatial, should not be space itself. Rather, it should be some formal factor in the mind that operates directly on those intuitions (or on the mental materials out of which they are constructed) and determines that the intuitions should come to represent objects only as being spatial. Yet Kant takes space itself to structure outer intuitions and so to determine that this latter situation obtains. And hence he seems to identify the outer intuition itself with the object that is spatially structured.

Again, what exists in itself in the mind (if, as I argue later, intuitions indeed so exist) should not be space (or time) itself, held to be an *a priori* intuition. Instead, what so exists should be an *a priori* intuition that represents space. And space (or time) should then have no existence whatsoever except insofar as it is represented to us by that *a priori* intuition. This last result is, after all, what Kant's own theory has in effect been seen above to require. Yet, as we have just noted, Kant seems to take space to exist in itself in the mind in the form of an *a priori* intuition. And thus he seems to identify the *a priori* intuition of space with the object – namely, space itself – that is represented by that intuition.

Although Kant in places allows a distinction between the empirical intuition and the spatial (or temporal) object that appears via that intuition, he does not maintain such a distinction in any consistent, explicit way. And, as far as I can tell, he never even implicitly proceeds in a manner that allows one to distinguish space (or time) itself from the *a priori* intuition that should represent space (or time) to us. Kant's refusal to maintain these distinctions – or at least to maintain them explicitly and clearly – has deep connections to other central parts of his views (and indeed to the practices of some of his Cartesian predecessors). In interpreting and evaluating the Transcendental Deduction, we cannot afford to follow Kant in this practice blindly, however. Thus, and for the present, we will ourselves distinguish an intuition (and, indeed, a representation in general) from the object that is supposed to appear to us via that intuition. We will return to this distinction in Chapter Two and later.

With the exception of Kant's views about synthesis, which we will consider beginning in Chapter Three, and the various issues to be considered in Chapter Two, I have now completed the discussion of those points about space, time, outer and inner sense, and the objects of our knowledge that Kant develops in the Transcendental Aesthetic and that we need to note before coming to the Transcendental Deduction itself. I will not comment here on the philosophical success of every one of these points. For example, Kant's treatment of representations as properties of the mind, like many of the details of his account of outer and inner sense, becomes important only at certain points of the Deduction and may largely be ignored elsewhere. Moreover, what is important in such Kantian views (many of whose details can be seen to raise serious philosophical questions) can often be expressed in ways that do not require the acceptance of every single Kantian point about, say, the manner in which our outer intuitions are supposed to appear to us (and are supposed first to function in our knowledge) in inner sense. We will avail ourselves of such modes of expression whenever that seems desirable.

Again, Kant's view that the objects of our knowledge exist in themselves but are unknowable as they so exist is a source of familiar, and grave, philosophical difficulties. And, if anything, even more troubling difficulties are raised by his more particular view that we cannot know our mind as it exists in itself, when that view is taken in conjunction with his willingness to theorize about faculties and representations that it would seem can belong to our mind only as it so exists. We cannot avoid considering these difficulties and the extent to which Kant – or at least the overall argument that he offers in the Deduction – can escape them. But because he adopts such views in pre-Deduction parts of the first *Critique*, we do best to take these views as Kantian givens at least at the beginning of our interpretation of the Deduction. We can observe in later chapters the problems with – and possible Transcendental Deduction alternatives to – such views.

Finally, Kant's philosophical difficulties about objects as they exist in themselves of course flow directly from the conclusions of his preceding argument to establish transcendental idealism and hence the wholly intramental and mind-dependent status of space, time, and the individual spatiotemporal objects of our knowledge. We need not evaluate every detail of that argument here, but it is clear that it can be challenged at many points. Thus, for example, many philosophers will reject, in whole or part, the basic Kantian account of our mathematical knowledge of space and time, an account from which Kant's argument begins. Even if one accepts that entire account, it is not clear from the above argument why there should not be synthetic truths, known or knowable *a posteriori*, that characterize space or time. Yet it seems clear that such truths, were they to exist, might well describe, although perhaps only in a contingent fashion, a structure of (physical) space or time that belongs to objects as they exist in themselves.[17] Hence Kant's argument would have to allow that space, time, and the spatiotemporal objects that we know might have an existence in themselves insofar as we could know them *a posteriori*. The most that the argument can prove would therefore be that space, time, and such objects must have a mind-dependent status to the extent that they are the subjects of various synthetic necessary truths known *a priori*. And that conclusion is insufficient, in the light of the foregoing remarks, to establish any full-fledged transcendental idealism.[18]

Furthermore, and as my preceding comments suggest, Kant's argument involves an identification of physical space and time with certain sorts of mathematical (and perceptual) space and time. Whatever one thinks of his treatment of our mathematical knowledge of space and time, the rise since Kant's time of alternative geometries and of related matters surely undermines that identification in the precise form in which he accepts it.

Finally, on the face of it, it is not immediately clear why, even given Kant's own identifications of physical with mathematical (and perceptual) space and time, there might not be synthetic necessary truths that would characterize a structure of space and time that belongs to objects as objects exist in themselves.[19] Indeed, it seems conceivable that those synthetic necessary truths might be known – and known *as* necessary – in an *a priori* manner by minds that through a process of something like Darwinian evolution would have come to have built into themselves that knowledge. Through this evolutionary process, the relevant knowledge states in these minds might have been nonarbitrarily and accurately aligned with the (necessary) spatial and temporal structure belonging to the objects existing in themselves. So the necessity that these minds would attribute to those synthetic necessary truths would be, as Kant would require, objective and not merely subjective and 'felt.'[20]

One can of course develop Kantian responses to the above objections. A complete discussion of such responses would occupy many pages. It seems clear, however, that even in the absence of such a discussion the above objections (and there are others) suffice to render Kant's argument very doubtful. And because his other arguments (which we have not examined) for transcendental idealism are also open to question, we must consider at various places below how the Transcendental Deduction looks when we abstract from all the basic claims of that idealism (and not simply from his views about the unknowability of objects as objects exist in themselves). Because, however, Kant assumes the truth of transcendental idealism in developing the Deduction, our interpretation of the first-*Critique*'s presentation of the Deduction must follow him in that assumption. As I hope to show, this interpretive procedure is not without its own philosophical and exegetic interest. In any case nothing in our discussion has demonstrated that something akin to a Kantian form of idealism (as against Kant's own arguments for such an idealism) is positively to be rejected.

One last set of comments should be made before I sum up. The view of objects as existing in themselves and appearing to the mind in spatiotemporal forms that I have attributed to Kant needs to be distinguished from a recent, mistaken interpretation. According to my view, when we know an object, that object is one thing that plays two roles. First, the object has existence in itself, in the sense explained earlier. Second, the object exists in the form it takes as it appears to us through (or as it is represented to us by) the sensible intuition through which we know it.

The present idea of existence in itself is the idea of the ontological independence of the object from the cognitive or other mental states through which we apprehend it.[21] According to this idea, the object exists with various properties, say F and G, and its existence with F and G does not depend on the occurrence of any sensible intuition or other apprehension of the object and would have occurred even if that intuition had not occurred in us. As Kant says, speaking in an ontological vein, when it is considered as existing in itself, the object is taken 'apart' from its relation to our intuition.[22] The object also is regarded as having its 'own nature *[Beschaffenheit an sich selbst]*,'[23] 'reality,' and 'existence.'[24] This object appears to our mind, through the affection of our sensibility and the relevant sensible intuition, as having certain properties, say *M* and *N*. We then know this object in the form that it takes as it thus appears to us – that is, we know the object as having *M* and *N*.

On the present view there is nothing in the precise notion, *taken by itself*, of the object as existing in itself that implies that the object, as it so exists, is unknowable by us. In other words, the notion of the object as it exists in itself is not 'by definition' the notion of something that we cannot know. If the properties *M* and *N* that the object appears to us to have were among the properties that the object has as it exists in itself, then in knowing the object through the intuition as having *M* and *N* we would be knowing the object as it exists in itself. For example, if the object as it exists in itself were cylindrical and if that object appeared to us as being cylindrical, then in knowing that object as it appears, we would be knowing it as it is in itself.[25] And this sort of fact is accepted at various points in the first *Critique*. Thus Kant's objection to the Leibnizian view that through our senses we know, in a confused way, objects as they exist in themselves is not that the very notion, taken by itself, of an object as it exists in itself rules out such knowledge. Rather, such knowledge is ruled out by the nature of our sensibility and its utter difference from our understanding, that nature (and that difference) preventing us from grasping the nature of objects as they exist in themselves.[26] Moreover, as I have urged above, Kant's principal and best reason for holding that we cannot know objects as they exist in themselves is that we cannot account for our *a priori* knowledge of synthetic necessary truths about space and time if we suppose that our knowledge concerns objects as they so exist. It is not that an object as it exists in itself is by definition unknowable by us.[27]

The recent interpretation that I reject has some superficial similarity to my above view but is fundamentally very different.[28] According to this interpretation (and so far in agreement with my earlier comments), in making his theoretical distinction between things as they appear and things as they are in themselves, Kant is not distinguishing between two distinct groups of objects (say, appearances in the mind and things existing in themselves that produce those appearances).[29] Nor is he distinguishing (as my own view requires) between two distinct realms of objects, a realm of objects as objects appear to us and a realm of objects as objects exist in themselves, the objects as they appear to us having also, in general, an existence in themselves. Rather, he has in mind just one group of objects, the ordinary spatiotemporal objects of our knowledge, and two different ways of *considering* those objects.

The recent interpretation holds that when we consider the object of knowledge, say the tree before us, as appearing to the mind, we consider that object as conforming to the mind's *a priori* necessary conditions for our knowledge of objects (in particular, to the condition that the object is a spatiotemporal thing). When we consider the object as it exists in itself, we consider that same object as existing independently of the mind and of the fact that it satisfies such necessary conditions. That we consider the object as existing independently of the mind means that we consider it simply as a 'something,' a bare *x*, without our making any appeal or reference, within our consideration of it, to its relations to the mind or to its satisfaction of the above necessary conditions.

According to the recent interpretation, it follows from these points that a thing as it exists in itself is by definition unknowable by us. The reason is that a thing as it exists in itself simply is, by definition, a thing that is considered independently of the mind in the above sense. But such a thing is a thing that is considered without reference to the fact that it satisfies necessary conditions for our knowledge of objects. So the recent interpretation concludes that a thing as it exists in itself is by definition a thing that is unknowable by us.[30]

Defenders of the recent interpretation hold that its way of understanding Kant's distinction between objects as they appear and objects as they exist in themselves best fits the overall claims of the first *Critique*. I hold that this interpretation is textually implausible and logically flawed.

As I have emphasized above, Kant's descriptions of the object existing in itself speak, in what certainly sounds like an ontological vein, of an object with a real existence and nature independent of that object's representation to or processing by the mind.[31] This reading of these descriptions also is consistent with Kant's original as-it-appears/in-itself distinction in the Inaugural Dissertation. Yet there his distinction clearly is to be understood in an ontological rather than in a recent-interpretation way; and I know of no texts that acknowledge the great change from the Dissertation view – and from the related views on formal reality and existence-in-itself held by earlier philosophers like Descartes and Locke – that would have occurred had Kant adopted recent-interpretation ideas in the first *Critique*.[32]

Moreover, various of Kant's specific descriptions of objects existing in themselves do not fit the recent interpretation. A case in point is his treatment of God. Kant implies that God is an object, having existence in itself, that cannot appear to us in sensible intuition.[33] So God is an example of an object existing in itself that we do *not* arrive at simply by considering an object of knowledge while not referring to that object's relations to our mind and to its satisfaction of the relevant necessary conditions. Such examples are set aside by the recent interpretation as 'special cases.'[34] But they fit seamlessly into my sort of ontological-independence view, and it seems preferable by far to accept an account that matches Kant's own examples rather than to eliminate various of those examples in order to adjust his work to match one's own account.

Again, it is made clear already in the Transcendental Aesthetic that objects existing in themselves affect our mind (as it is in itself) and give rise to intuitions through which we know those objects as they appear to us.[35] The recent interpretation can argue that in speaking of an object existing in itself as doing this affecting, Kant is merely talking of an object of knowledge considered without reference to its relations to the mind; he is therefore not introducing any new sort of thing distinct from ordinary spatiotemporal things.[36] But to argue in such a way leaves unaccounted for the affection relation itself, which holds between entities as they exist in themselves. This affection relation is not in any obvious way plausibly described as being a relation – a relation which we know as holding between objects of our knowledge – that we consider without referring either to its holding between those objects or to its meeting the necessary conditions for our knowledge of it.[37] Nor is any other treatment of the affection relation directly apparent on the recent interpretation. But my view can immediately allow that such a relation of affection, distinct from any phenomenal relation, occurs in the realm of things existing in themselves and holds there between such things.

The recent interpretation also faces a severe logical difficulty. Suppose that, in considering a thing, I make no reference or appeal to the fact that that thing satisfies necessary conditions for our knowledge of objects. Then all that follows is that it is *not* the case that I *consider* the thing *to* satisfy those conditions. It does not follow that I positively *consider* the thing *not to* satisfy those conditions. Or, to speak in a slightly different way, from the fact that the object, as I consider it, *lacks* the property of satisfying those conditions (and so of being known by me) it does not follow that that object, as I consider it, *has* the property of *not* satisfying those conditions (and so of *not* being known by me). (Rather, neither the property of satisfying the conditions nor the property of not satisfying the conditions occurs on the list of properties that the object, as I consider it merely as a 'something,' has.) But in reaching its result that objects existing in themselves are by definition unknowable by us, the recent interpretation appears to argue precisely that such fallacious conclusions do follow.[38]

When it is thought through carefully, the recent interpretation thus faces numerous problems – not the familiar sort of problems about our apparent knowledge of the unknowable that Kant's own claims about things in themselves raise, but self-created problems of textual evidence and logical validity. One could raise other difficulties for this interpretation. But I have said enough to indicate why I prefer, and will continue to follow, the view of objects as they exist in themselves that I have developed above.

4. SUMMARY

Let me review the main points of the preceding discussion. As we have seen, Kant adopts, in those parts of the first *Critique* that precede the Transcendental Deduction, a fundamentally transformed version of a picture of knowledge that might without the transformation be accepted by various of his predecessors like Descartes and Locke. According to Kant's picture, we know single, individuated objects by means of those mental representations that Kant calls empirical intuitions. Such intuitions are *of* or *about* objects that they represent to our mind as being single, individuated things of certain sorts. In the case of those of our empirical intuitions that are yielded through our outer sense, the objects are represented as being single, individuated spatial things. In the case of those of our empirical intuitions that are yielded through our inner sense, the objects are various of our own properties and states of mind, including our own outer intuitions; and these objects are represented as being in time. We know such objects simply in the spatial or temporal forms that they are represented to us as having by our outer- or inner-sense intuitions. But because space and time themselves are mere forms in the mind that serve only to structure objects as those objects are represented as being by our intuitions, those objects themselves, in the forms in which we know them, are mere mind-dependent things that do not exist in themselves. We thus reach Kant's transcendental idealism and his view that we know objects only as they appear to us and not as they are in themselves. Given our comments in Section 2, it is to such objects, as they appear to us – that is, it is to such objects of our possible experience – that Kant will attempt to demonstrate the necessary applicability of the categories in the Transcendental Deduction. And Kant will attempt to show also that such category-subsumed objects constitute the whole of the objects of our possible experience and that we can have no grounds for taking ourselves to know any other objects by means of the categories.

CHAPTER TWO

INTUITIONS AND THEIR OBJECTS

1. THE TRANSCENDENTAL DEDUCTION AND CATEGORY APPLICATION

As we noted in Chapter One, in the Transcendental Deduction Kant attempts to prove that the categories apply to all the objects of our possible experience – to all objects that we do or can know. He attempts also to demonstrate that the categories cannot be shown or known to apply to any other objects.[1] The basic strand of his argument runs as follows. Kant assumes that we know some arbitrarily selected object – of course through some arbitrary given sensible intuition. From this assumption together with various points about us, the knower, he infers that this object is necessarily subject to the categories. As we see in detail in later chapters, he does so by appealing to the way in which we can think in the first-person of our experiences as ours. Because we can think of all our experiences in that way, the arbitrary given intuition in question must be so generated in our mind that its object is necessarily subject to the categories. Because this object is an arbitrarily selected one (as is the intuition in question), Kant generalizes and concludes that all the objects that we do or can know through sensible intuitions are necessarily subject to the categories. And because the objects that we can know are simply the objects that we can know through our sensible intuitions, he thus holds that the categories apply to all the objects of our possible experience. He goes on to infer that we cannot know or show the categories to apply to any other sort of object.

For the preceding line of argument to be successful, Kant obviously must not beg the question by building into the description of the object known, in the argument's starting assumption, any properties that are equivalent to category satisfaction. Instead, he must start with only what we might call the minimum Deduction assumption that, roughly, by means of an arbitrary given sensible intuition an object is known (but not an object into whose description in this assumption we build category satisfaction). In subsequent chapters (and especially in Chapter Eight, when the issues here become exceptionally important), we will examine this minimum assumption with care. And we will deal with a specific problem about showing, in the way just sketched, category application to the objects that we can know. But for the present it is sufficient, where required, simply to suppose that the Deduction starts with the above minimum assumption and to ignore the problem in question.

We can see from the above remarks that, in order to understand exactly how the Deduction proceeds, we must understand both Kant's views on first-person thought and his account of the generation of intuitions within our minds. Kant's views on first-person thought are best discussed within our exposition of the Deduction as a whole. But we cannot begin that exposition without having some further grasp of his treatment of the role of intuitions in our knowledge. In the present chapter we consider (i) various additional points about Kantian representations, including especially the status of our intuitions and concepts, of our knowledge, and of ourselves (as knowers) vis-à-vis the fundamental Kantian distinction between objects as they exist in themselves and objects as they appear via intuitions; (ii) Kant on appearing – and appearance – theories and the nature of the object that he takes an intuition to represent to us; (iii) a serious problem that besets his treatment of that nature; and (iv) various difficulties about the spatiotemporality of the objects that we know and the notion of a nonspatiotemporal, unknowable existence in itself. In Chapter Three (and also in parts of Chapter Four) we will then examine Kant's views on the generation of intuitions within our mind and in particular his treatment of the manifold of intuition and its synthesis.

2. KANTIAN REPRESENTATIONS IN OUR KNOWLEDGE: THINGS EXISTING IN THEMSELVES OR THINGS MERELY APPEARING TO US IN TIME OR BOTH?

As we saw in Chapter One and have emphasized above, Kant supposes that we know objects through mental representations of those objects. His treatment of representations – and, in particular, of intuitions and concepts – is, however, subject to a number of ambiguities. We should notice these ambiguities here if only not to be confused by them. With one important exception, to which I will come shortly, we can deal with the ambiguities simply by noting their existence and then specifying the way in which we will ourselves understand a Kantian representation.

Thus, and for the record, I will observe that Kant's notion of a representation admits of a wider and a narrower sense. His narrower and principal notion is the one that we have emphasized earlier. It is that of a representation as an inner determination of the mind that signifies an object and can itself become the object of further representations, including in particular further representations of inner sense.[2] His wider notion is that of a representation as *any* inner determination of the mind. This wider notion classifies as representations not only representations in the narrower sense but also mental properties such as sensations that are, as such, 'related solely to the subject as the modification of its state' (A320/B376) and so do *not* signify objects.[3] The wider notion becomes important when we consider, in the Deduction, Kant's theory of how our mind, through category-involving acts of thought, 'refers' or 'relates' various of its inner determinations – that is, various of its representations in the wider sense – to objects. But for the present we may continue to focus on Kantian representations in the narrower sense.

We also may note, and then simply indicate our resolution of, various further ambiguities that will be familiar both to readers of Kant and to students of earlier Cartesian discussions of ideas.[4] These further ambiguities include those between (a) a representation as a concrete mental entity and a representation as, in its role of object-signifier, something akin to a meaning or a meaning-like abstract entity; (b) a representation as a *mental object* that represents *to* the mind's inner consciousness its object and a representation as itself an *act* of inner consciousness that *directly grasps* its object (say directly grasps that object as occurring in time); (c) a representation as the *product* of certain mental processes or acts of representing and a representation as itself such a *process* or act of representing; and (d) a representation as a *token* of a given type of representation (say a token of the type: intuition, in the visual mode, that represents a conical, needle-bearing red spruce) and a representation as itself being such a *type*. For working purposes and except where noted below, we may take these last four ambiguities to be resolved properly if we regard a Kantian representation as (a) a concrete mental entity and (b) a mental object which through being grasped by the mind's inner consciousness comes to represent an object or a group of objects. In so proceeding, we will also regard a representation as (c) a product of certain mental processes or acts of representing and (d) a token of a given representational type.

Aside from these ambiguities, there is, however, one complication in Kant's notion of a representation that we cannot ignore here. That is the problem of how to understand the status of the intuitions and concepts through which we know, as well as of our own states of knowledge and of ourselves as knowers, in the light of Kant's distinction between objects as they exist in themselves and objects as they appear via intuitions. As I explained in Chapter One, an entity has existence in itself insofar as it has an existence apart from or in independence of its representation to or its grasp by any mind. And an entity has existence as it appears insofar as it has an existence that in one way or another depends on its representation to a mind. But then any existent entity ought to have either existence in itself or existence as it appears to some mind or else both an existence in itself and an existence as it appears. Thus if we now consider, in their roles in our knowledge, intuitions, concepts, our states of knowledge, and ourselves as knowers, the question arises of into which of these three groups these entities fall.

This is a deeply puzzling question for Kant.[5] Central parts of the theoretical position in the first *Critique* and related texts strongly support the view that the above items have, in their roles in knowledge, existence in themselves. Yet straightforward philosophical considerations, along with nontheoretical (and some theoretical) texts, support the opposing position that such items have, in their roles in our knowledge, the existence merely of entities appearing to the mind in inner sense. In addition, further texts and reasoning suggest that all such items, in those roles, should be regarded as having both existence in themselves and existence as they appear. Moreover, the argumentative situation here is made especially difficult by the fact that Kant nowhere in the first *Critique* considers the present question with anything like full explicitness, so that any answer to it must be stated carefully and must be to some extent inferential.

In favor of attributing an existence in themselves to the above items, in their roles in our knowledge, are several arguments based on Kant's theoretical picture of knowledge. In the first place, it is fundamental to that picture that we gain knowledge through intuitions (including inner-sense intuitions) and concepts. Suppose that those intuitions and concepts, as they yield us knowledge, had the status merely of entities appearing to us. Then, given Kant's theory of inner sense, those intuitions and concepts would evidently have the specific status of entities appearing to us through some group of inner intuitions. Since Kant does not in general count our representations as self-representational, this group of inner intuitions would have to be a *further* group of intuitions beyond those original intuitions (including the original inner-sense intuitions) that yield us knowledge. And he should describe the intuitions and concepts that yield us knowledge as *appearing to us* via this further group of inner intuitions.

However, nowhere in the first *Critique* does Kant describe any further group of intuitions via which the intuitions and concepts in question must appear to us. Indeed, the postulation of such a further group of inner intuitions would rapidly lead to an unacceptable regress, were those further intuitions themselves to be conceived of as yielding knowledge. Furthermore, at various places Kant explicitly denies that we have any *other* intuitions besides the usual outer and inner intuitions that yield us knowledge of spatial objects and of our own inner, mental states.[6] Hence it seems that intuitions and concepts, insofar as they yield us knowledge, do not have (or at least do not have merely) the status of entities appearing via intuitions. Rather, intuitions and concepts in their roles in our knowledge should have the status of entities existing in themselves. Specifically, they should at least be entities belonging, as inner determinations, to our mind as it exists in itself. And they should be used by our mind, as they so exist, to know objects as things that appear to our mind.

Moreover, Kant takes our states of knowledge to be, on the side of our mind, made up of intuitions and concepts, those intuitions and concepts occurring together in a judgmental relation. Therefore, given the above argument, our states of knowledge should evidently have (or should have at least) the same status of entities existing in themselves. And so we ourselves should also have that status, in our roles as possessors of those states of knowledge and utilizers of those intuitions and concepts.

In the second place, consider a sentence in the Kantian style like the following one, which expresses a major part of the first *Critique*'s theory of self-knowledge:

> I appear to myself in inner sense as being a temporal thing (or as being a temporal sequence of inner states), and *I* can know myself only as being such a thing (or only as being such a sequence) and not as I am in myself.

Given Kant's theory of inner sense, the reference of the first occurrence of 'I' in this sentence is clearly to myself (or to my mind) as I exist in myself in a nontemporal (and nonspatial) form. But then, by the usual referential practice in both English and German, the second, italicized occurrence of 'I' should refer to the *same* entity as does the first 'I.' Like the first occurrence of 'I,' this second occurrence of 'I' is, after all, an occurrence of 'I' simpliciter, not an occurrence of 'I' in a term like 'I, as I appear to myself in inner sense' (a term which would, of course, designate myself simply as I so appear). Moreover, this second occurrence of 'I' is obviously strictly comparable to the italicized occurrence of 'I' in the sentence

> I (in myself a thin person) appear to myself in the distorting mirror as being a fat person, and (given the way that I am looking at myself in the mirror) *I* now can describe myself – on the basis just of my visual experience – only as being such a fat person.

And in this latter sentence that italicized occurrence of 'I' evidently refers to the same entity as does the first occurrence of 'I,' namely to myself as I exist in independence of my representation by the distorting mirror and describe myself visually as I appear to myself in that mirror.[7]

If, however, the second, italicized, occurrence of 'I' in the above Kantian sentence refers to the same entity as does the first occurrence of 'I' in that sentence, then the second occurrence of 'I' refers to myself as I exist in myself. Thus it should be *that* Kantian I that knows itself as it appears to itself in inner sense. It should be *that* Kantian I that so knows itself just as much as it is the thin, mirror-image-independent I that describes itself as it appears to itself in the distorting mirror. Furthermore, *that* nontemporal (and nonspatial) Kantian I, the I as it exists in itself, is distinct in its properties from the I as the I appears to itself in a temporal form in inner sense. So there should be no temptation to suppose that the I that knows itself – the I as it exists in itself – *is just* the I in the form that the I takes *as* the I appears in inner sense. There should be no more temptation to draw that conclusion than there is to conclude that the thin, mirror-image-independent I that describes itself as it appears in the distorting mirror *just is* the I in the fat form that the I takes *as* it appears in that mirror.

We can conclude from the above reflections that the I that *knows* itself as it appears *to* itself in inner sense *is* (or *is at least*) the I as it exists in itself, there being no reason to suppose that the I that so knows itself is just the I as the I *appears to* itself in inner sense. Because Kant makes no distinction between the entity that knows itself in inner sense and the entity that knows spatial objects in outer sense, it must therefore be that same I that knows spatial objects. Thus, once again, we, in our roles as the possessors of our states of knowledge, should be taken to have (or to have at least) an existence in ourselves. And it will then also be reasonable to take our states of knowledge, as well as the intuitions and concepts through which we know, to have such an existence.

Besides the above two lines of argument, central Kantian texts also support the attribution of an existence in themselves to the above items, in their roles in our knowledge. In many places Kant discusses the factors that belong to a group of representations that constitutes a piece or state of knowledge in our mind.[8] In these discussions, he affirms that one factor is a concept – for example, the concept of a tree or of a triangular thing – that is generated and applied to the object of knowledge by the active faculty of understanding. And he takes that faculty itself to belong to the mind that possesses the state of knowledge. Again, he affirms repeatedly that another factor is a thought-act-constituted unity that exists among the representations in question. This unity he argues to derive from the subjection of those representations to unity of apperception. That subjection requires the operation of the active faculty of apperception, a faculty that he takes to belong to the mind that possesses the state of knowledge – and a faculty that he often identifies with the understanding. Moreover, he takes this unity to arise directly from the application, *within* that state of knowledge, of the categories – the pure concepts that belong to and are applied by the active understanding of the mind that possesses that knowledge.

For example, in the A-Transcendental Deduction Kant writes at A124–25 (with my italicization) that

> actual experience [which by B1, B147, B165–66, B218–19, A157/B196, the remainder of this A124–25 paragraph, and related texts, is here empirical knowledge of objects] ... *contains in* recognition, the last and highest of these merely empirical elements in experience, *certain concepts* [namely, the categories] which render possible the formal unity of experience ....[9]

And he says many similar or related things elsewhere in both the A- and the B-Deductions.[10]

Suppose, however, that our states of knowledge do require the presence, within themselves, of concepts that are generated and applied by the active faculty of understanding that belongs to the mind that possesses those states. Suppose also that those states of knowledge require the presence, within themselves, of the categories and of the category-derived unity of thought – a presence that is effected by the active operations of the faculties of apperception and understanding that belong to that same mind. Then those pieces of knowledge must clearly belong to the mind that possesses and utilizes those active faculties.

The mind in question cannot be, however, the empirical or phenomenal mind, the mind as it appears to itself in inner sense. That mind Kant describes as being no more than the passive, inert string of temporally ordered sensations and representations that appear in inner sense. As being such a string of elements, that mind possesses no active faculties at all (let alone even a passive faculty of sensibility).[11] Rather, the mind in question must be the mind as it exists in itself, the mind that possesses and utilizes the active faculties of understanding and apperception (and the passive faculty of sensibility). It is, after all, clearly to this mind that we must ascribe those faculties if we are to make sense of Kant's claims, in the introductory parts of the Transcendental Aesthetic and the Transcendental Analytic, that on the basis of the mind's *intuitive* and *conceptual operations* the mind knows objects (including the mind itself) that *appear to* the mind (and *not* objects that 'appear to the mind as the mind appears to itself'). And it seems also obviously to be inner sources in this mind that Kant means when he considers, 'in their transcendental constitution *[Beschaffenheit*], the subjective sources [which yield, among other things, the categories] which form the *a priori* foundation of the possibility of experience' (A97). If, however, our states of knowledge belong to this mind, those states will have, like this mind itself, an existence in themselves. The intuitions and concepts that make up those states will also have such an existence. And so will we, the knower, who will simply be this mind.

A final textual argument that points towards roughly these same conclusions should be mentioned briefly. As we have noted several times above and will see again in Chapter Ten, Kant holds that the categories – and thus, in particular, the category of existence or actuality (A80/B106, A145/B184, A218/B266) – can be taken by us to apply only to phenomenal objects, objects *as* those objects appear via sensible intuitions. Observing this fact, H. A. Pistorius in effect objected against Kant in 1784 that we can therefore attribute existence neither to the object (in itself) that *appears to* us nor to the self (in itself) *to* which that object appears. Hence Kant cannot really speak of our *knowledge* of *existent* appearances (*Erscheinungen*), for on his theory even the self, insofar as it can be said to *exist*, reduces to an appearance, and so 'there will be nothing but illusion [*Schein*], for nothing remains to which anything can appear.'[12]

We will consider Kant's answers to this acute criticism in Chapter

Seven. I can observe here, however, that in replying to Pistorius Kant undertakes to show that a certain existence is properly attributed to the self to which objects appear – a self that Kant in his reply does *not* identify with the self as the self appears to itself in inner sense. Because this self to which objects appear is surely the self that knows those objects as they appear *to* it, it follows that Kant does *not* identify the self that knows with the self as the self appears in inner sense. Without pursuing the details of his reply further here, we can see that this reply at least strongly suggests that we, as knowers, together with our states of knowledge and the intuitions and concepts via which we know, are *not* to be taken as having the existence of things as things so appear. And thus again it is plausible to take all these items to have, in their roles in our knowledge, an existence in themselves.

Let me now turn to the reasons for attributing *not* such an existence to the above items, but rather merely the existence that entities have as they appear in inner sense. These reasons are straightforward. First, we think of ourselves, as knowers, as existing in the same world as do the objects that we know. We also regard our states of empirical knowledge as being caused, in part, by our causal interactions with those objects. In addition, we conceive of ourselves as gaining or losing knowledge over time. And we think of ourselves as being sometimes able to acquire knowledge about our own states of knowledge, as well as about whatever mental representations may belong to those states. On Kant's theory, however, the world in which occur the objects of our knowledge is the spatiotemporal world of objects as they appear to us via intuitions. It also is that world, and only that world, within which we can take the categories of causality and of causal interaction to apply, as well as the whole notion of the gain or loss of a mental state over time. And it is only with respect to the entities in that world – including our own states of mind as they appear in inner sense – that we can acquire knowledge. Since for Kant we can know those states of mind only as they appear in inner sense, it thus seems that the specific existence merely of entities as they so appear must belong to ourselves, as knowers, to our states of knowledge, and to the intuitions and concepts which belong to those states of knowledge.

Second, the preceding reasoning is reinforced by the fact that in the first *Critique* Kant himself implies that we think of our knowledge in the above ways. There is no need to list all the texts, which come largely but not exclusively from passages where he is focused on some task other than that of directly developing his own theory. But note Bxii-xiii (the cases of Galileo, Torricelli, and Stahl), where he implicitly allows that human knowers gain knowledge over time and occur in the same world as do the objects that they know.[13] Note also his knowledge-claiming descriptions, throughout the first *Critique*, of our cognitive capacities and representations. (Here observe Axv on no hypotheses, but only certainty, as being permitted in his transcendental investigations of our *a priori* knowledge.) And notice finally texts like A28, A98, A115, B124 ff., B145, B208, A168/B210, A213/B260, and A491–92/B520, where he talks in terms of our causal interaction with the objects that we know, objects that affect our sense organs in such a way that we have sensations of those objects.[14]

As indicated above, a third group of arguments and texts suggests that the items that we have been discussing above should all have, in their roles in our knowledge, *both* an existence in themselves *and* an existence merely as they appear in inner sense. Or, to note a variant of this idea, our intuitions and concepts exist in themselves; but they constitute states of knowledge (and so belong to ourselves as knowers) *only* as they so appear. This idea or its variant can obviously be supported by combining (with the adjustments necessary to achieve consistency) our arguments for treating the above items as having an existence in themselves and our arguments for treating those items as having merely an existence as they appear. Thus it might be held that such an idea manages to reconcile as far as possible these in many respects opposing arguments. Moreover, both the idea and its variant are in harmony with various texts.

For example, although Kant does not particularly emphasize the matter, his official A22/B37 definition of inner sense (and related texts like A23/B38, A33/B49–50, and A34/B50) obviously allows thoughts and concepts, as well as outer-sense intuitions, sensations, and other 'inner states,' to appear in inner sense. He makes this point explicit at A342/B400, A357, A359, and A364, and in two Reflexionen written after the first *Critique*.[15] Such texts thus allow for our knowledge-yielding thoughts, along with the concepts which those thoughts apply, to occur in inner sense. There would consequently seem to be no difficulty, as far as these texts go, with the idea that the items we have been discussing exist both in themselves and as entities appearing in inner sense. Nor would these texts seem to pose difficulties for the variant of this idea noted above.

Which of the preceding treatments of the above items should we adopt? Kant's views are obviously in a fundamental and not wholly resolvable conflict here, for his basic theoretical considerations push him toward one answer to this question at the same time that the normal supposition that our knowledge is temporal and knowable pushes him toward a different answer. None of our above treatments is in fact perfectly satisfactory *both* exegetically *and* philosophically. But from the standpoint of interpreting the actual position of the *Critique of Pure Reason*, it seems clear that we must reject the claim that our states of knowledge, along with the other items above, have *merely* an existence as they appear in inner sense. And we must reject also the variant idea, noted a few paragraphs ago, that our intuitions and concepts exist in themselves but constitute states of knowledge only insofar as they so appear to us.

We must reject that claim and that variant for essentially the same reason. The reason is that neither the claim nor the variant can really deal with the strong arguments and textual grounds that we have seen above for taking all the items in question, including our states of knowledge, to have (or to have at least) an existence in themselves. As we have seen, after all, those arguments include the fact that the mind that knows must possess an active, concept-utilizing understanding (and faculty of apperception); and they include also the fact that Kant's reply to Pistorius does not at all identify the mind that knows with the mind as it appears in inner sense. Given these facts – and given the central position of all such facts and arguments in the actual first-*Critique* picture of knowledge – we cannot suppose that Kant holds that our knowledge states, and so on, have merely an existence as they appear in inner sense. Nor can we suppose that he accepts the variant idea noted above. Rather, we must hold that all such items have at least an existence in themselves. And we must regard the arguments (and texts) that suggest the *mere* existence of these items as they appear in inner sense as simply posing a great philosophical problem for Kant and not as delineating the true position of the first *Critique*.

Since the mind that knows is, for Kant, the mind that possesses an active, concept-utilizing understanding (and an active faculty of apperception), it would seem that the mind that knows simply *cannot* in the end be the mind merely as it appears to itself in inner sense.[16] And I think that that conclusion is finally correct. Despite the philosophical difficulties that ensue, the ultimately most coherent account of Kant's overall position is that the mind that knows, its states of knowledge, and the intuitions and concepts via which it knows, all have, in their roles in our knowledge, an existence in themselves.[17] But there is no need to be rigid here. Suppose that the knowing mind is the mind as it exists in itself, and suppose also that the preceding items exist in themselves, in their roles in our knowledge. Then, acknowledging the texts cited three paragraphs ago, we can certainly still allow that these items can appear to the mind in a temporal form in inner sense and can be known by the mind as they so appear. Moreover, we can also allow that such items belong, in a derivative sense, to the mind as the mind appears in inner sense. We can allow this last point simply because such items will or can be members of the passive string of temporally ordered representations and other inner states that constitutes the mind as the mind so appears.

To allow this point may seem exegetically odd, for that passive string is not at all the active, concept-using mind that Kant takes to have knowledge. But as long as we recognize that it is the latter mind that he primarily takes to know, there is no great difficulty in maintaining the point. And I will so proceed below as is required. We will thus take all the items that we have been discussing to have primarily an existence in themselves. But where it seems appropriate or required (for example, in the interpretation of synthesis as a time-governed operation), we will take those items also to exist in a temporal form in inner sense. In addition, we need not hesitate to focus on the philosophical bearings of Kant's treatment of knowledge while abstracting from the details of the distinction between objects as they exist in themselves and objects as they appear in inner sense. When we proceed in that way, we can then follow our normal belief that our states of knowledge are temporal, knowable entities, without puzzling over the issues that have concerned us here. In any case in Section 4 we will consider some of the difficulties that arise when we conjoin Kant's views on objects as they exist in themselves with our present view that our states of knowledge (and so on) have, for Kant, primarily such an existence.

3. THE OBJECT THAT KANT TAKES AN INTUITION TO REPRESENT TO US: THINGS AS THEY APPEAR AND APPEARANCES

We must now turn from considering issues about the ontological status of our states of knowledge to considering issues about the nature of the object that Kant takes an intuition to represent to us. Such issues concern, primarily, what we will see to be his simultaneous acceptance of two distinct theories of that nature – namely, an appear*ing* theory and an appear*ance* theory. Like the preceding issues, the issues about such theories are ultimately important for understanding the exact way in which the arguments of the Transcendental Deduction proceed.

Moreover, and as we will see in Section 4, these appearing- and appearance-theory issues lead to a grave problem for Kant when they are combined with the results of our Section 2 discussion and with his views on the nonspatiotemporality of objects existing in themselves. And this problem also must be discussed and as far as possible resolved before we can hope to evaluate the exact success of the Deduction.

Before we examine these issues, I should note that their discussion is complicated by what we saw in Chapter One to be Kant's position that our knowledge of spatiotemporal objects really involves our awareness of outer-sense intuitions as those outer-sense intuitions are represented, by means of our inner-sense intuitions, as occurring in time. However, this position and its complications do not really alter, but simply render more involved, the basic points about appear*ing* and appear*ance* theories that I have to make below, as well as the points that I will make about the problem for Kant that I have just mentioned. So for simplicity I will continue to speak of our knowledge of spatiotemporal objects as being achieved via our intuitions. And I will return to the matters about outer and inner sense only in Section 4.

To develop the issues about appearing and appearance theories I need recall only that the object that we know via some given intuition – the object that that intuition represents to us – we have so far characterized as being an object that *appears to us* via that representation. On this view, that object exists in itself, in an unknowable, nonspatiotemporal form. The intuition then becomes *of* that object, represents it to our mind as being some single, individuated spatiotemporal thing – for example, a single, individuated tree before us – and we then know the object simply as being such a thing. Moreover, for reasons that we noted in Chapter One, that single, individuated spatiotemporal thing should really be distinguished from the intuition via which we know it, although Kant often identifies the two.

Now the theory that Kant here presents of the object of knowledge is in fact one version of what has recently been called an *appearing* theory of that object (and of our knowledge).[18] Such theories characteristically take the object of knowledge to have (or at least to be able to be conceived to have) an existence in itself. They take that object then *itself* to be grasped by and known by the mind as *that object itself appears to* the mind via the operations of the mind's cognitive apparatus. These theories, as such, do not require that that object should appear in the form that it takes as it exists in itself in independence of its grasp by the mind. And, as I interpret them, these theories, as such, allow for the possibility that the object appears to the mind only through the mediation of some representation which the mind also grasps. But such theories do hold that, whatever form the object takes as it appears and however its appearing comes about, it is *that object itself* that *appears to* – and it is *that object itself* that *is* grasped and known by – the mind. It is not *simply* some idea, representation, or other mental simulacrum of the object which the mind grasps and knows.[19]

Evidently, however, besides the above appearing-theory treatment of the object known, there is another way of understanding that object. This other way focuses on Kant's frequent identification of the object known with the intuition via which we know that object. And it yields a distinct new theory of the object of knowledge, a theory that for simplicity I have not mentioned above at all.

This new theory agrees with Kant's appearing theory on one important point which we have not so far emphasized. That point is that objects existing in themselves affect the sensibility of the mind as it exists in itself and thereby yield the mind intuitions.[20] But Kant's new theory denies that those intuitions function as vehicles through which those objects are *themselves* presented to the mind, in spatiotemporal forms, and thus are *themselves* grasped and known by the mind. Rather, it is *those intuitions simpliciter* that are the precise entities that the mind grasps and knows. According to the new theory, those intuitions simpliciter are spatiotemporal entities that are completely distinct from the objects, having existence in themselves, which generate the intuitions in us. In particular, those intuitions are *not* to be regarded as *being* objects – objects having also an existence in themselves – in the forms that *those* objects take as *they* exist in the mind.[21] Rather, those intuitions themselves are the *only* objects that the mind grasps and knows.

This new, or second, Kantian theory of the objects of our knowledge is a form of what recent writers have called an *appearance* theory of such objects (and of our knowledge). Appearance theories, like appearing theories, characteristically take various objects to exist in themselves. Unlike appearing theories, however, appearance theories deny that *those objects themselves* appear to the mind (or are grasped or known by the mind) in any form. Rather, those objects generate *appearances* of themselves – mental entities like ideas, sensations, or sense data – in the mind. These appearances are totally distinct from the objects. And it is these appearances which the mind knows, and not the objects themselves.[22]

Appearing and appearance theories – and Kant's versions of such theories – clearly represent distinct and incompatible treatments of the objects of knowledge. But in Kant's hands the theories are not sharply distinguished. (Nor does Kant deal with the philosophical problem that we will see in Section 4 to afflict his acceptance of *both* such theories within his own account of knowledge.) Indeed, it often seems that Kant simultaneously accepts both theories, not realizing that they are quite different. Whatever is the exact explanation of these matters, his holding of both such theories, and his frequent identification of them, can be seen directly from the texts. These texts have already been considered in detail by earlier commentators, but it is nevertheless worth noting some representative samples briefly here.[23]

The following passages bear the imprint of Kant's appear*ing* theory. Kant writes, for example, of 'the distinction, which our *Critique* has shown to be necessary, between things as objects of experience and *those same things* as things in themselves'; he says that 'the things which we intuit are not *in themselves* what we intuit *them* as being'; he holds that 'our sensible representation is in no way a representation of things in themselves but only of the way in which they *appear to* us'; and he remarks that 'the mind intuits *itself* [in inner sense] ... as *it appears to* itself, not as it is.'[24]

There are also many passages that illustrate Kant's appear*ance* theory. Kant says that 'what is first given to us is *appearance*'; he notes that we ought to 'treat the empirical intuition as itself *mere appearance*, in which nothing that belongs to a thing in itself can be found'; he comments that 'since a mere modification of our sensibility can never be met with outside us, the objects, *as appearances*, constitute an object which is merely in us'; and he remarks on '*appearance*, that is, a mere representation of the mind to which an unknown object corresponds.'[25]

Besides such texts, there are also numerous passages in which Kant seems to move unwittingly from one theory to the other or else identifies the two theories. At A27/B43, for example, he begins a sentence by talking of the special conditions of the *appearances* of things but ends by remarking that 'space comprehends all things that *appear to us* as external' (my italics). And he sometimes uses the *term* 'appearance' *(Erscheinung)* where he seems to be expressing his appear*ing* theory. Thus he writes, for instance, of 'appear*ance*, which always has two sides, the one by which the object is viewed in itself (without regard to the mode of intuit*ing it* ...), the other by which the form of the intuition of this object is taken into account.'[26]

Texts of the above sorts – and they could be multiplied considerably – strongly support the presence in Kant's picture of knowledge of both appearing and appearance theories; and they also show his frequent identification of, or his refusal to distinguish between, these theories. The philosophical interest of these facts is, for our present purposes, twofold.

First, Kant's acceptance of both these theories yields him, whether or not he realizes it, two quite different characterizations of the ontological status of the object of knowledge: (i) as a thing that appears via, and in a certain way depends for its existence on, a representation – and a representation from which that object nevertheless is distinct; and (ii) as an inner determination, or a property, of the mind, an inner determination that the mind regards as an object (or a synthesized set of such inner determinations that the mind regards as an object). The difference between these characterizations has significant consequences for Kant's picture of knowledge which we will see as our discussion develops. Second, Kant's acceptance of both theories connects directly with a philosophical problem that seriously affects the understanding that we have so far achieved of the Kantian object of knowledge. It is to this problem – which we have mentioned above and which has to do with the nonspatiotemporality of objects existing in themselves – that we must now turn. In stating this problem I will continue to ignore Kant's special position about the relation of outer- and inner-sense intuitions, for, as I have indicated earlier, this position complicates but does not essentially alter the points that we must make about the problem. At the end of our discussion I will then note the implications of Kant's position for the present matters.

4. A PROBLEM FOR KANT

Consider again Kant's appear*ing* theory of our knowledge, on which I will focus for the present. Crucial to this theory is the idea that the objects that we know are entities that *both* exist in themselves *and also* appear to us via our intuitions. This idea Kant himself frequently emphasizes. Thus he writes at A42/B59, as we have noted in Section 3, that 'the things which we intuit are not *in themselves* what we intuit *them* as being' (my italics); and (as we have also noted) he speaks at Bxxvii of 'the distinction, which our Critique has shown to be necessary, between things as objects of experience and *those same things* as things in themselves' (my italics). There are many further passages to the same effect – for example,

Bxxvii again: 'the object is to be taken *in a twofold sense*, namely as appearance and as thing in itself'; and B306, where Kant speaks of distinguishing 'the mode in which we intuit [objects] from the nature that belongs to *them* in themselves' (my italics). And we can note that Kant often puts this idea to quite significant philosophical use – for instance, in his reasons for distinguishing his form of idealism from Berkeley's.[27]

If we thus suppose with Kant's appearing theory that any given object that we know both exists in itself and also appears via some given intuition, then on Kant's theory we are evidently accepting a certain *identity* – namely, the identity between an object that, as it exists in itself, we do not know, and an object that, as it appears to us in a spatiotemporal form, we do know.[28] In accepting this identity we are obviously not supposing, contrary both to Kant and to all logic, that within some single world or realm of objects there somehow obtains an identity between, say, (a) a thing that is completely nonspatiotemporal and is unknowable by any being like us and (b) a thing that is a conical tree and is known by us. Rather, it seems clear that we are accepting a version of what is often called a two-worlds or two-realms doctrine.[29] We are supposing that there is one world or realm, call it W, of objects as they exist in independence of their representation to or grasp by any mind. We are supposing also that there is another world or realm, call it W', of objects as they appear to us in spatiotemporal forms via our intuitions. And we are assuming that there is a strict identity between (c) an object o that in its occurrence in W is a nonspatiotemporal, unknowable thing and (d) an object o' that in its occurrence in W' is the conical tree that we know.

The world W is of course the world of objects as objects exist in themselves (the world that is often called the Kantian noumenal world). The world W' is the Kantian world of phenomenal objects, the 'field of appearances' of A40/B57. More could be said about the precise understanding of these worlds – and especially about the proper interpretation of the phenomenal world W' – in terms of the sorts of worlds or realms that one may appeal to in interpreting certain intensional notions that are centrally involved in Kant's theoretical views.[30] But my intention here is simply to use the worlds W and W', and the identity that we have just seen to hold between objects o and o', in order to develop, first, a severe problem for the appearing-theory version of Kant's views. (I will subsequently develop an equally severe problem for the appearance-theory version of those views. And, in conjunction with the problem for the appearing-theory version, this problem for the appearance-theory version will then lead directly to the final, overall problem for Kant that is mentioned in the heading for the present section of this chapter.) The specific problem for the appearing-theory version is this. Within his appearing-theory account of knowledge, Kant cannot consistently maintain such an *o-o′* identity in conjunction with his central claim that objects as they exist in themselves are nonspatiotemporal and unknown by us.

In order to argue for this result, I need to observe three points about our knowledge of a phenomenal object like the conical tree. First, such knowledge takes a propositional (or judgmental), that-clause form, according to Kant.[31] In particular, we can suppose, without any loss of generality, that our knowledge of the tree takes the specific form

(P) *H* knows that the tree is conical

where, as in Chapter One, *H* is some arbitrary being whose cognitive capacities and operations are like ours. We can note also that Kant clearly holds that our knowledge includes such facts about shapes as are asserted in claims like (P), for he says explicitly at B69–70 note, for example, that 'the predicates of space ... are rightly ascribed to the objects of the senses.' And his discussions of geometric knowledge also agree completely with this point.

Second, it is an evident and familiar principle about our knowledge that all claims of the form

(Q) If *H* knows that *p*, then *p*

or, formally:

$$H \text{ knows that } p \supset p$$

hold true. This principle is, indeed, so deeply embedded in our conception of knowledge that we would, it seems, simply – and rightly – refuse to count as knowledge any cognitive states for which it fails. Moreover, Kant himself seems clearly to accept the equivalent of this principle. He holds at A58/B83, for instance, that 'truth consists in the agreement of knowledge with its object, ... knowledge is false if it does not agree with the object to which it is related.' And elsewhere he makes many other, (Q)-accepting claims.[32]

Third, and as I have already argued in Section 2, in interpreting Kant's own account of knowledge we must attribute (at least) an existence in themselves, in their roles in our knowledge, to our states of knowledge, as well as to the intuitions and concepts via which we know and to ourselves, as knowers. Thus we must take the state of *H*'s knowledge that is expressed in (P) to have such an existence, and so we must regard that state of knowledge as itself existing within the world *W*.

Notice now, however, where our observation of the above three points has got us. By the conjunction of the first two points, any world or realm at which claim (P) holds true must also be a world at which

(R) the tree is conical

holds true. Yet, by the third point above, (P) expresses a state of knowledge that exists in the world *W* of objects as objects exist in themselves. Hence (P) must itself hold true at world *W*. And consequently (R) must hold true at *W*.

The consequences for Kant of the truth of (R) at the world *W* are, however, horrendous. In order for (R) (regarded as a consequence of (P) via our first two points) to hold true at *W*, it must be the case that (i) the singular term 'the tree' in (R) designates, at the world *W*, an object that can be truly said, at *W*, to be conical; and (ii) this object is in fact the object that, by (P), we are assuming *H* to know.

But then from points (i) and (ii) trouble follows immediately. The object that we are assuming *H* to know is of course the object o' that in its occurrence in the phenomenal world W' is a conical tree. Yet, as we have just been supposing, that object o' is *strictly identical to* the object o that in its occurrence in world *W* is a nonspatiotemporal, unknowable thing. (Here see the solid line in Figure 1, which represents this $o = o'$ object.) Hence, by the transitivity of identity and the intersubstitutivity of identicals, 'the tree' in (R) must designate, *at W*, the object o that, in its occurrence *in W*, is such a thing. But then by point (i) the object that in its occurrence in *W* is such a thing can be truly said, at *W*, to be conical (and a tree). It is clear, however, that if an entity that occurs *in* a world *U* can be truly said, *at U*, to have a certain property, then that entity has that property in its occurrence *in U*. Therefore o is, *in* its occurrence *in W*, conical (and a tree). (Indeed, given (P) and our above reasoning, o is, in its occurrence in *W*, an object that *H* knows as such a conical tree.) And that result is of course flatly inconsistent with Kant's central claim that objects, as they exist in themselves, are nonspatiotemporal (and unknown by us).

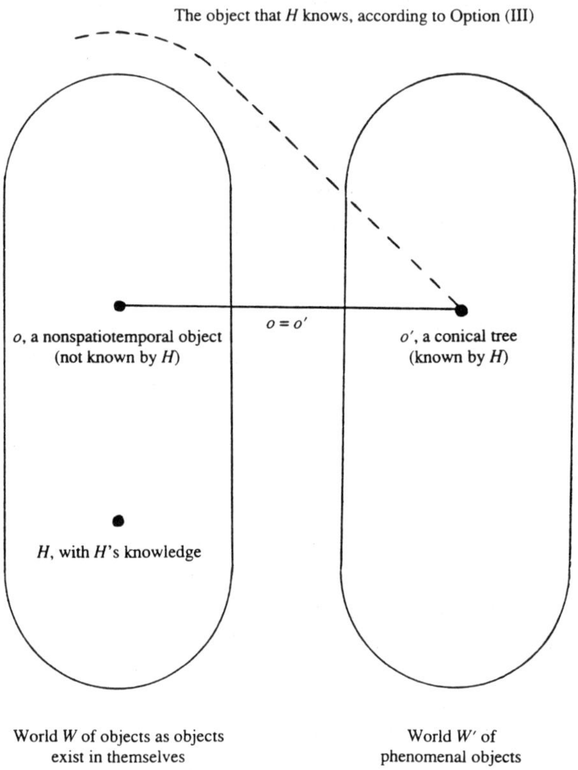

Figure 1.

We thus see that the three points that we have noted above lead to contradiction with that central Kantian claim when those points are taken together with the identity that Kant accepts, within his appearing-theory account of knowledge, between the object that, in its occurrence in a spatiotemporal form, is known, and an object that, in its occurrence in the world of objects existing in themselves, is unknown. But these three points can themselves be regarded simply as parts of – or as theoretical commitments made by – Kant's overall account of knowledge, on either its appearing- or its appearance-theory version. Hence we can detach our specific reference to these three points. We can talk simply of that overall account of knowledge on either of these versions. And so we now reach the problem that I have indicated earlier. The appearing-theory account of knowledge cannot consistently maintain the identity in question while also granting the basic Kantian position that objects, as they exist in themselves, are nonspatiotemporal and completely unknowable by us.

The above problem is of course a problem – and, I think, a quite severe problem – for Kant's appear*ing* theory. It might be thought, however, that Kant could completely avoid this problem simply by adhering throughout to his alternative, appear*ance*-theory account of knowledge. According to that account, after all, the object known is an appear*ance*, an intuition in the mind that is *not* to be taken as identical to the object, having existence in itself, that has produced this intuition in the mind. So within the appearance-theory account Kant could accept all three of the points that we have noted above and yet not be forced into the conclusion that the object of knowledge has an existence in itself in a spatiotemporal (and knowable) form.

This way of avoiding our above problem does indeed enable Kant to escape that precise problem. But to suppose that he adheres strictly to the appearance-theory version of his account of knowledge is to create an equally severe, and a closely related, problem. For the reasons that we have seen above in Section 2, our intuitions have an existence in themselves in their roles in our knowledge. Hence the appearances that we thereby know also have such an existence. But, as we have seen from Chapter One on – and as the first two points that we have noted above make vividly clear – Kant takes spatiotemporal properties, for example the property of being conical, to be possessed by the objects of our knowledge, and so by appearances. Thus if he adheres throughout to the appearance-theory version of his account of knowledge, he arrives at an inconsistency similar to the inconsistency that we have just derived from his appearing theory.

The actual position that Kant develops in the *Critique of Pure Reason* thus faces a dilemma. As he presents it, that position takes either – and only – an appearing-theory or an appearance-theory form. (Or so it certainly seems from the texts.) But either of these forms leads quickly to a flat contradiction with his basic first-*Critique* views about the unknowability and nonspatiotemporality of objects existing in themselves. And thus the actual position of the first *Critique* would seem to be untenable.

This dilemma represents *the* ultimate, and major, problem for Kant to which I wish to draw attention here.[33] It is, I think, a problem that is seldom, if ever, noted in its full extent. Moreover, the force and severity of this problem should not be underestimated. In order to indicate that force – and to show as clearly as possible the depth of the issues that this problem raises for Kant – it helps to note some possible responses to it and why these responses fail.

One obvious response would be to attack the assumption, on which both horns of the above dilemma depend, that objects, as they exist in themselves, are unknowable, nonspatiotemporal entities. Yet, just because this assumption is so basic to his entire Critical Philosophy, Kant cannot himself escape the dilemma in this way. Again, it seems impossible to find a satisfactory way between the horns of the dilemma. Within Kant's picture of knowledge, the only alternative to the appearing theory is the appearance theory; and just as moving from the appearing to the appearance theory does not eliminate the basic difficulty that the objects of knowledge turn out to exist in themselves as spatiotemporal, so, too, abandoning the appearance theory can only precipitate Kant back into the form of that difficulty that attends the appearing theory. Furthermore, one cannot escape the dilemma by undercutting one of the points used above in deriving the separate horns of the dilemma.

In particular, one cannot avoid the above problem in its appear*ance*-theory form. One might suggest, for example, that the intuitions via which we know, and thus spatiotemporal appearances, do not have an existence in themselves. But our Section 2 discussion eliminates both this suggestion and the related suggestion, for the appearing-theory version of the problem, that *H*'s states of knowledge lack such an existence. Hence the above problem is inescapable for the appearance-theory version, given that appearances are intuitions in the mind.

Our problem for Kant therefore can be avoided only by undermining the reasoning that we used to develop the difficulty for the appear*ing*-theory version of his account of knowledge. But here again the prospects look bleak, as long as we remain within his own version of that account. As we have seen, after all, in that reasoning we take (P) ('*H* knows that the tree is conical') to hold true at *W* and to express a state of knowledge that exists *in W*; then we apply (Q) ('if *H* knows that *p*, then *p*') and infer the truth of (R) ('the tree is conical') at *W*; and, finally, we argue that if (R) is true *at W*, then an object existing *in W* is, contrary to Kant, a spatiotemporal, knowable thing. Because of our Section 2 discussion, such reasoning is not plausibly attacked by denying that (P) holds true at

W or that (P), if true at *W*, expresses a state of knowledge that exists *in W*. Thus we can avoid the problem for the appearing theory only by (I) denying that (P) is really a correct formulation of the knowledge that *H* claims to have (and the knowledge that will exist *in W*). Or we can avoid that problem only by granting that (P) correctly formulates that knowledge. But then we must argue either that (II) (R) does *not* also hold true at *W* or else that (III) (R), although true *at W*, does *not* concern an object that exists *in W*. Yet none of the options (I) to (III) is a happy one.

Option (I) rejects the idea that *H*'s knowledge of the conical tree really is expressed in terms of the flat predication, concerning that tree, that it is conical. Instead, it might be urged, that knowledge is really only to the effect that *H* knows, say, that *it appears that* the tree is conical. Thus, given (Q), only the claim that it *appears* that the tree is conical will hold true at *W*. (R) – the claim that the tree *is* conical – will or need only hold true at the phenomenal world *W'*. So the problem for the appearing-theory version of Kant's account of knowledge is avoided. However, this option fails, for – as I have emphasized – Kant takes the flat claim (P) itself to give a proper expression of *H*'s knowledge. He does not think that a claim about anything's *appearing* to be the case is part of the content of ours, or of *H*'s, ordinary knowledge of the shapes of things.

Option (II) suggests that (P) ('*H* knows that the tree is conical') but not (R) ('the tree is conical') must be understood to hold true at *W*. Thus Option (II) denies the fundamental knowledge-principle (Q). In place of this principle it might be proposed, for example, that if *H* knows that the tree is conical, then it follows *not* (as (Q) would have it) that the tree *is* conical simpliciter but rather that *it appears that* the tree is conical. So if (P) holds true at *W*, only this 'it appears that' claim also holds at *W*; (R) will hold, or need be conceived to hold, only at *W'*. However, not only have we seen (Q) to be a fundamental principle that holds true concerning *any* mental or other state that constitutes knowledge. But also we have seen that Kant himself seems clearly to accept (Q). Thus Kant says that truth consists in the agreement of knowledge with its object. He does *not* say that truth consists of the agreement of knowledge with *its appearing to be the case* that the object is so and so. Hence Option (II) must be rejected.

Option (III) is defended most plausibly in terms of the following subtle and intriguing idea. While (Q) holds true and both claims (P) ('*H* knows that the tree is conical') and (R) ('the tree is conical') are true *at* the world *W* of objects as objects exist in themselves, (R) is true *at* that world about an object that does not exist *in* that world but does exist *in* the world *W′* of phenomenal objects. (Compare the idea, familiar both in ordinary terms and in various forms of tense or modal logic, that I can say truly *at* the time 1991 that Kant was a great philosopher even though Kant himself does not exist *in* the set of objects that are existent *in* 1991 but only in a set of objects that are existent *in* an earlier time.) This idea comes closest, I think, to defending the appearing theory against the reasoning above; and it captures a great deal of the actual content of Kant's account of knowledge. Indeed, I will suggest below that, with certain important reservations, some such idea is the best way to help a Kantian theory out of the present problem. But what I must now urge is that this idea, and Option (III) in general, is not anything that *Kant himself* can accept while maintaining central tenets of the exact appearing theory presented in the first *Critique*.

To see this point, note that the present idea is *an* appearing theory, for it takes the object of knowledge to be a thing that is *itself* presented to the mind's awareness, in the form of a conical tree, via an intuition from which the object is distinct. But this idea denies the view that the object that is thus presented to the mind occurs in any form *in* the world *W* of objects existing in themselves. (Compare the suggestion that when I, in the real world, see 'in' a picture some object that I know to be purely fictional, I can still properly describe that object as appearing to me via the picture, even though I know that the object has no real-world existence.)

In denying the view just mentioned, however, the present idea denies *Kant's* own appearing-theory doctrine that the objects of our knowledge *both* appear to us in the form of spatiotemporal entities *and* have an existence in themselves. Or, equivalently, the present idea denies that there is an identity between an object that in its occurrence in the phenomenal world *W′* is a conical tree and an object that in its occurrence *in W* is a nonspatiotemporal, unknowable thing. Yet, and as we have seen, that doctrine is an important part of Kant's own appearing theory, and a part which he often puts to significant philosophical use.[34] So the present idea abandons something of consequence to Kant's picture of knowledge. Moreover, this idea also puts the noumenal world, the world of objects existing in themselves, into a strange, convoluted relation with the objects of our knowledge, a relation for which there is no model within Kant's own texts and no apparent independent rationale.

In order to explain this last remark, I should note that, according to the present idea, the object that *H* knows *cannot* be identified with the object *o′* that in its occurrence in the phenomenal world *W′* is a conical tree. The object that *H* knows cannot be so identified simply because object *o′* is itself identical to the object *o* that occurs in *W*, and, as the present idea has it, no such identity obtains between the object of *H*'s knowledge and any such object *o*. Moreover, according to the present idea, *H* cannot know the object *o′* itself as *o′* appears through the intuition by means of which *H* knows the actual, genuine object of *H*'s knowledge. Nor can *H* know the object *o′* as *o′* appears through some other intuition of *H*'s. After all, in the light of the *o-o′* identity, either sort of knowledge of *o′* by *H* would create the inconsistency that we have developed above for Kant's appearing theory. Yet, given that we can avoid our overall dilemma for Kant only by adopting the appearing theory, it is clear that if *H* can know the object *o′* at all, *H* can know *o′* only as *o′* appears through one or the other of the sorts of intuition that I have just mentioned. Hence, according to the present idea, *H cannot in any way know the object o′ itself*, on pain of our facing the inconsistency in question. Rather, the object that *H* knows – and can know – simply is an object that *coincides* with object *o′* *in* the phenomenal world *W′* but *diverges from* object *o′* (= *o*) by not *itself* existing in, or being identical to any object existing in, the world *W* itself. (Here see the dashed line in Figure 1 above, which represents this object and its divergence from the *o-o′* object.)

Suppose now that we couple our present idea with Kant's considered, first-*Critique* views on nonspatiotemporal, unknowable objects existing in themselves. Then we must conclude that the object that we actually know – the dashed-line object in Figure 1 – has no existence in itself but coincides within the phenomenal world *W′* (the only world where that object does or need exist) with *another object* that we do *not* know and are *not* in any way acquainted with – namely, the *o-o′* object, the solid-line object in Figure 1. But, as I have suggested above, this result does indeed put the world of objects as they exist in themselves into a strange and convoluted relation with the object of our knowledge. Moreover, nothing in Kant's texts suggests that he himself ever entertains any such contorted view of our knowledge and its objects. In addition, it is hard to conceive what independent and plausible philosophical arguments there could be for such a view.

Option (III) must thus be rejected. And with it we have exhausted our suggestions for undermining the reasoning that we used to develop the difficulty for appear*ing*-theory versions of Kant's account of knowledge.

Above we saw no satisfactory way to avoid the problem for the appear*ance*-theory version of that account. Although one could suggest further, and even more recondite, ways of avoiding our overall problem for Kant, there is no reason to think that any of these ways is likely to succeed any better than have the ways discussed above. Thus the appearing- and appearance-versions of Kant's account do lead to a deep-seated and insoluble problem when they are taken in conjunction with his position concerning the nonspatiotemporality and unknowability of objects as they exist in themselves.

What then should we do? As pure Kant interpreters we must simply acknowledge this problem and its grave implications for his theory. But of course we must also ask whether there is any way of modifying that theory in order to avoid the problem while retaining as much as possible of his position. For reasons suggested above, the best hope here lies not with Kant's appearance theory but with some form of his appearing theory, in particular with the modified form of that theory developed in Option (III). Although it leads into strange complications and is in important ways non-Kantian, this option at least retains the idea that the object known appears via an intuition from which it is distinct. So it avoids not only our problem for the appearing theory but also our problem for the appearance theory. And it also clearly retains Kant's idealism about the object of knowledge and his view that that object does not exist in itself as a spatiotemporal thing.

In fact there is a version of Option (III) into which Kant himself can easily slip, unawares, from his own form of the appearing theory. On Kant's appearing theory, the object that we know, *in the spatiotemporal form that that object takes as it appears to us and as we know it*, exists *only* in the phenomenal world *W′*. This fact does not mean that the object of knowledge *itself* exists only in *W′*, for on Kant's appearing theory – and to ignore the above problem for Kant – the object of knowledge *itself* exists also, in a nonspatiotemporal and unknowable form, in *W*. But this fact creates a temptation to hypostatize the object of knowledge, *as* it appears to us and as we know it, and to create a *new object*, which does exist *only* in *W′* – namely, the object-of-knowledge-in-the-spatio-temporal-form-that-the-object-takes-as-it-appears-to-us-and-as-we-know-it.[35]

This new object, which will now be taken to be *the* proper and genuine object of knowledge, will coincide with the *o-o′* object in *W′* and will exist nowhere else; and indeed this new object will be most naturally *identified with* a certain particular *part*, or 'world slice,' of the *whole o-o′* object – specifically, with that *part* or *world-slice* of the whole *o-o′* object that occurs *exclusively* in the phenomenal world *W′*. It is clear that if one does introduce such a new object as the proper object of knowledge, one arrives at a version of Option (III). (Moreover, this new version of Option (III) differs from our previous version only in that the previous version, unlike the new version, leaves it open whether the object of knowledge, which coincides in *W′* with the *o-o′* object but does not exist in *W*, in fact exists in some other world or realm than *W′*.)

A hypostatized object of the above sort can be introduced deliberately and clear-headedly (although it remains to be seen what plausible philosophical rationale could be given for creating such a convoluted view of the object of knowledge). But such a hypostatized object also can be introduced in a disreputable way, through a subtle linguistic confusion that I suspect is made by Kant (and indeed by other philosophers writing on topics like the present one).

This linguistic confusion turns on expressions like 'object *o*, as object *o* occurs in (or exists in) the phenomenal world *W′*.' As that expression is originally used in Kant's appearing theory, it designates the *whole o-o′* object – the *whole* solid-line object of Figure 1 – considered in *its* (in *that whole object's*) occurrence in *W′*. Similarly 'I, as I appear in the distorting mirror' is most naturally taken to designate the *whole object* Robert Howell (who also exists outside his appearance 'in' that mirror), considered in *his* (in that *whole object*'s) appearance 'in' that mirror. Again, 'the plank, as the plank occurs in the bucket of blue paint' is most naturally taken to designate not simply the plank's blue-paint-covered *end* in the bucket but rather the *whole plank*, considered in that *whole plank's* occurrence, at *its* end, in the bucket. (Thus one can speak of '*the plank*, which now exists also *outside* the bucket, as *it* occurs *in* the bucket,' and here one clearly does not mean to be speaking just of the *end* of the plank that *is* in the bucket, for that end does not now exist also *outside* the bucket.)

There is, however, a hypostatizing use of 'object *o*, as object *o* occurs in (or exists in) the phenomenal world *W′*' that is easily confused with this first use. According to the hypostatizing use, this expression – which should now be read as 'object-*o*-as-object-*o*-occurs-(or-exists-)in-*W′*' – does *not* designate the whole *o-o′* object considered in *its* occurrence in *W′*. Rather, it designates just *that portion* of the whole *o-o′* object that *does* occur exclusively in the phenomenal world *W′* – namely, that portion of the whole *o-o′* object that is entirely and exclusively a certain conical tree. Similarly, a hypostatizing use of 'I, as I appear in the distorting mirror' would designate just *that portion* of the whole object Robert Howell that so-to-speak has its occurrence or existence *entirely* and *exclusively* 'in' the mirror (or 'in' the mirror image) – namely, that portion that is entirely and exclusively a certain fat thing displayed by the mirror. And in its hypostatizing use 'the plank, as the plank occurs in the blue paint bucket' would designate *just that end* of the plank that occurs entirely and exclusively, with a blue color, within the bucket.

Now this hypostatizing use of 'object *o*, as object *o* occurs in (or exists in) the phenomenal world *W′*' is easily run together with the normal, nonhypostatizing use; and I suspect that some such linguistic confusion occurs at places in Kant's work.[36] Whether or not this suspicion is justified, it is clear that through such a confusion we can easily come to think of the object of knowledge in the hypostatizing way while beginning from a point where it is thought of nonhypostatically. If, however, one does this, then one is, surely unawares, identifying the object of knowledge with the particular part, or 'world-slice,' of the whole *o-o′* object that occurs exclusively in *W′*. Hence there is indeed a version of Option (III) into which Kant can slip, unawares, from his own form of the appearing theory.

This last point of course does not mean that one cannot introduce the above, hypostatizing version of Option (III) deliberately and clear-headedly. But Kant does not introduce that version in such a way, for it conflicts with his own appearing-theory view that the object of knowledge exists both in the phenomenal world *W′* and in *W*. Furthermore, there is no textual evidence that he takes account of any such version except momentarily, on the basis of the above confusion, and unawares. So the hypostatizing version is not *Kant's own* conscious, developed view of the objects of our knowledge. Nevertheless, the pressures of our above problem for Kant make Option (III), and its hypostatizing version, a convenient momentary refuge, to the extent that that refuge is reached without any realization of its convolutions or its conflict with significant parts of Kant's position. The above linguistic confusion then provides a very natural way to enter that refuge. Hence if we take Option (III), with all its difficulties, to provide the best way out of our overall problem for Kant, we are not suggesting a way out that differs utterly from the spirit or the letter of his work or that is simply anachronistic.[37]

5. OUTER AND INNER SENSE AND THE PROBLEM FOR KANT

In concluding the present discussion, we need to observe the bearing of Kant's views about outer- and inner-sense intuitions on appearing and appearance theories and our problem for Kant. Briefly, and as seen in Chapter One, Kant's official view of our knowledge of an outer object (the case on which I concentrate here) is this. An object, existing in itself, affects our sensibility and generates an outer-sense intuition, existing in itself, in our mind. This outer intuition then is represented to our mind by means of an inner-sense intuition as occurring in time. By grasping this outer intuition as it is thus represented to us, our mind grasps and so comes to know the spatial object (now, for reasons we can ignore, itself represented as occurring in time) that the outer intuition represents.[38] Evidently this treatment of our knowledge admits of appearing- and appearance-theory versions both with respect to the idea that the outer intuition is represented by the inner intuition and with respect to the idea that the outer object is itself represented by the outer intuition. Moreover, these versions lead either to our above problem for Kant or else to very similar problems.

Thus suppose that the outer intuition is represented by the inner intuition as occurring in time. The situation here is hardly different from the appearing- or appearance-theory situation already discussed. On the one hand, and as on the appearing theory, the outer intuition may exist in itself in a nonspatiotemporal form in the mind in itself and then *appear* via the inner intuition as occurring in time. But the knower can know that that outer intuition (as it appears via the inner intuition) *is* in time.[39] And then an argument like that given above will show that, contrary to Kant's basic views, the outer intuition, as it exists in itself in *W*, is temporal. So we reach the first horn of our problem for Kant. On the other hand, and as on the appearance theory, the outer intuition, as it occurs in time, may be *identified* with the inner intuition, which is considered to be an *appearance* utterly distinct from what we were previously calling the outer intuition as that outer intuition exists in itself in *W*. But then of course the inner intuition exists in itself in *W* and is here identified with the temporally occurring outer intuition. Hence, contrary to Kant's basic views, the inner intuition, as it exists in itself in *W*, *itself* occurs in time. And consequently we reach the second horn of our problem for Kant.

As one would expect, a specific instance of that problem thus emerges directly from the idea of the outer intuition as being represented by the inner intuition as occurring in time. Although I will spare the reader the details, one can show that a very similar problem emerges from the idea that the spatial object of knowledge is represented to the mind by such an inner-intuition-represented outer intutition. We therefore see that, as I noted at the beginning of Section 3, Kant's position about the relation of outer- and inner-sense intuitions in our knowledge really does not alter, but only renders more involved, our previous conclusions about his appearing and appearance theories and the fundamental problem that such theories face.

The problems that we have noted for that position of Kant's are as serious as our original problem for Kant. Like that problem, these new problems have no really adequate solutions within his own framework. Perhaps the best partial solution would be to adapt Option (III) to the case of the outer intuition's representation by the inner intuition. We would say that the outer intuition, as it occurs in a temporal form and gives us knowledge of the spatial object, *appears* via the inner intuition. But (adopting either a nonhypostatic or a hypostatic form of Option (III)) we would also suppose that the outer intuition does not exist *in* the world *W*. We then would have a fairly large range of alternatives with respect to the case of the spatial object's representation by the outer intuition. For example, we could take the spatial object to *appear*, in an Option-(III) way, via this Option-(III)-treated outer intuition. Or (since the Option-(III)-treated outer intuition does not exist in itself in *W*) we could allow the spatial object simply *to be*, in an appearance-theory way, the Option-(III)-treated outer intuition. Or other possibilities might be considered.

We need not examine these possibilities here, for it is intolerably complicated to carry along at all stages in our discussion the fine points that we have just been noting. Thus I will continue to speak simply of our knowledge of spatiotemporal objects; of the appearing- and appearance-theory forms that Kant, without clearly distinguishing between them, gives to that knowledge; and of the problem for Kant that these forms pose. Just because Kant does not clearly distinguish between these forms, it is best to proceed in terms of such forms, recognizing but not attempting to eliminate the serious problem that so proceeding creates. This procedure is also justified by the fact that, as can be seen from Chapter Eight on, Kant's Transcendental Deduction argument is carried out in terms of a proof of category application to an object that the knower *thinks* to have the features that are presented by the elements of the manifold of intuition. And while it will be important to note how this object and those features are understood in appearing- or in appearance-theory terms, in order to interpret the Transcendental Deduction we will not need to reach any final decisions about which understanding to adopt.

There also is no need always to restrict ourselves rigidly to just the possible treatments of Kant's views that I have emphasized above. For reconstructive purposes it may be best, at points, simply to ignore Kant's views on the nonspatiotemporality and unknowability of objects existing in themselves, views which help to create our problem for Kant. Or again, and as noted in Section 2, we may on occasion wish to ignore the conclusions of that section and to consider our states of knowledge and intuitions to exist merely as they appear in inner sense. This treatment of those entities would also eliminate the problem for Kant.

As a pendant to this discussion, note finally that, as we will see in Chapter Eight, Kant's ultimate view of the representation of objects by intuitions is that our understanding, through its activity of thought, 'refers the intuition to the object.' This means that our understanding *thinks* there to be a single object to which belong the various features presented by the elements of the manifold of intuition; and so the intuition (or its manifold) comes to function for us as *an* intuition, a representation that represents that single object. This view adds further complications to Kant's appearing and appearance theories. But here we need note only that such a view does not really alter our above results about our problem for Kant.

After all, the object that our understanding thinks for the intuition is one of two sorts. That object, *in our thought* (or in the world as the world is represented to us by our thought), is an object that *appears* to us via that intuition. Or else that object, *in our thought* (or in the world as the world is represented to us by our thought), is an object that is, as on the *appearance* theory, identical to that intuition. But then this situation is roughly the same, in its basic logical structure, as is a situation that we have implicitly touched on above – that is, it is roughly the same, in its structure, as the situation in which we consider whether, in the world as it is represented to us by the inner intuition, the spatial object is to be taken to *appear* via the temporally occurring outer intuition or else is simply to be taken, as on the *appearance* theory, to be identical to that temporally occurring outer intuition. As I have already suggested (without giving details), one can show that this latter situation leads to our above problem for Kant (or to a very similar problem). And a roughly parallel line of reasoning demonstrates that so too does the former situation, in which our understanding thinks an object for the intuition.[40] Moreover, the basic sort of partial, Option-(III) solutions to the problem for the spatial-object-and-outer-and-inner-intuition situation may also be mobilized for the problem arising in the object-thought-for-the-intuition situation.

6. THINGS IN THEMSELVES: A PRELIMINARY COMMENT

The above account of Kant – the pre-Transcendental Deduction picture of knowledge presented in this and in the preceding chapter – depends heavily on claims about objects (and our minds) as they exist in themselves. We have seen, for example, that the objects of our knowledge are supposed to have an existence in themselves, are supposed to affect the sensibility of the mind as it exists in itself, and are supposed then, as they appear to the mind, to be known by the mind as it so exists. Moreover, the knowledge that the mind thus achieves is supposed to be arrived at via intuitions and mental operations that belong to the mind in itself. In such claims as these there is an obvious and a thoroughgoing reliance on views about what exists in itself. Nevertheless, and as everyone knows, this reliance yields Kant a source of unending difficulties.

For example, how can we know that objects exist in themselves and behave in the preceding ways when Kant holds that we cannot know objects as so existing? For the same reason, how can we know that the mind, as it exists in itself, possesses the various cognitive faculties – and operates in the various cognitive ways – that we have seen above? Again, Kant argues in the Deduction and later that the categories, and hence the categories of existence and causality, apply only to phenomenal objects. How then can we even think of, let alone claim to know, objects as *existing* in themselves and *affecting* the mind, as the mind itself so exists, in what seem to be quasi-causal ways? Yet again, objects as they exist in themselves are for Kant nonspatiotemporal. Thus the cognitive operations of the mind, insofar as they have such existence, must be 'operations' that take place outside time as well as outside space. But what sense can be made of atemporal 'operations' or 'acts' of mind?

Finally, and again given that objects as they exist in themselves are unknowable, other familiar questions arise. For instance, how can we be so sure that there is a one-to-one correlation between objects appearing to the mind via intuitions (or between the appearances of objects in the mind) and the members of some group of objects existing in themselves? Instead, why should not a merging be possible – several distinct objects, existing in themselves, appearing to the mind as one spatiotemporal object (or several distinct such objects producing one spatiotemporal appearance of themselves in the mind)? Or why should not a branching be possible – one object, existing in itself, appearing to the mind as several distinct spatiotemporal objects (or one such object producing several distinct spatiotemporal appearances of itself)?[41] Or why indeed suppose that we can talk in any justified way of distinct, individuated objects existing in themselves?[42]

The above are questions that any interpretation of Kant must confront, except perhaps for some purely reconstructive efforts that wholly disregard Kant's views on objects existing in themselves. These are also questions that are especially pertinent to an account of Kant that, like the present one, seeks to understand his own presentation of the Transcendental Deduction and its philosophical interest. Nevertheless it would be premature to develop Kant's answers to such questions, or to evaluate their adequacy, until we have seen the Deduction's account of the way that we use the categories to think objects corresponding to intuitions. When we have considered that account, and when we have a better idea than we do now of Kant's overall picture of knowledge, we can then return in Chapter Ten to Kantian claims about objects in themselves. In the meantime, I will note only that, despite the interest of these claims, I do not think that in the end they can be defended in the form in which Kant makes them. Thus any discussion of the philosophical interest of the Transcendental Deduction must consider how the Deduction fares when such claims are abandoned or are modified in some way. But for the present we must consider the Deduction and Kant's pre-Deduction picture of knowledge as Kant actually presents them in the first *Critique*. So we will continue to utilize the above sorts of claims about objects existing in themselves.

7. SUMMARY

In the present chapter we have discussed numerous issues preliminary to the Transcendental Deduction itself. We saw that the Deduction will endeavor to show, in a non-question-begging fashion, that *all* the objects that we do or can know are necessarily subject to the categories. In addition, we argued for regarding our own states of knowledge, ourselves as knowers, and the intuitions and concepts via which we know as all having, in their roles in our knowledge, at least the status of entities existing in themselves. We noted further that Kant's picture of knowledge takes both an appear*ing*- and an appear*ance*-theory form. And we showed that each of these forms is inconsistent with his firm insistence on the nonspatiotemporality and unknowability of objects existing in themselves. As an attempt to resolve (although only imperfectly) this problem, we suggested the idea of treating the object of knowledge as an appearing-theory entity about which true knowledge claims can be made *at* the world of objects as objects exist in themselves, even though this entity itself does not exist *in* that world but, rather, exists *in* the world of Kantian phenomenal objects.

Finally, we have just discussed the serious difficulties that face Kant's views about objects as they exist in themselves. We will return to such difficulties later, once we are clearer than we are at present about his overall picture of knowledge and his arguments in the Deduction. But now we must turn to the final aspects of his picture that we need to consider before we embark on a discussion of the Deduction itself.

CHAPTER THREE

INTUITION, THE MANIFOLD OF INTUITION, AND ITS SYNTHESIS

1. INTRODUCTION

If we ignore the difficulties for Kant that we have so far noted, then the picture of the Transcendental Deduction that we suggested at the beginning of the last chapter runs roughly as follows. Kant seeks to show that the categories apply, with necessity, to all the objects that we do or can know. He seeks to show this conclusion in a non-question-begging manner by beginning with the minimum assumption that, by means of an arbitrary given sensible intuition, an object is known (but not an object that we assume to be category-subsumed). He argues that, because of the way in which we can think in the first-person of all our experiences as ours, it follows that that intuition must be so generated in our mind that the object that it represents to us (and the object that we know via it) necessarily falls under the categories. And he then infers that this same result holds for all the objects that can be represented by our sensible intuitions and so for all the objects that we can know.

We have not yet set out any of the details of, or the evidence for, this account of the Deduction. But it is clear from Chapter Two that this account must (with the qualifications there noted) see the sensible intuitions in question, and the knowledge that we attain via them, as having an existence in themselves. It is clear also that this account must enable both appearing-theory and appearance-theory versions of the Deduction to be stated. And it is clear, finally, that this account must come to grips with the fundamental problem for both these theories that we have seen in Chapter Two, as well as with the basic difficulties, noted there, about Kant's position concerning objects existing in themselves.

What we have not yet considered at all, however, are Kant's views on first-person thought and his treatment of the generation of intuitions in our minds. Kant's views on such thought will be indicated only briefly in Section 2 below, for these views are best examined within our overall exposition of the Deduction. But we cannot postpone further comments on the Kantian treatment of the generation of intuitions in our mind. I have in mind, here, Kant's account of the manifold of intuition and its synthesis. Moreover, I mean that account only insofar as it concerns intuitions that represent to us spatiotemporal things like trees. And I take that account in independence of the restrictions that we must impose on our description of the manifold of intuition at the beginning of the Deduction, before Kant has established the full details of his picture of knowledge. In particular, until the end of Section 5 I ignore the treatment that Kant should give of the manifold when it is considered in connection with the minimum Deduction assumption noted above – that is, when it is considered simply as the manifold of an intuition that represents an object (but not an object that we can assume to be subject to the categories).[1]

Kant's account of the manifold and its synthesis amounts roughly to the view that any intuition via which we know is given to us in the form of an unconnected manifold or variety (*Mannigfaltige* or, in Latin, *varia*) of elements. The view holds also that this manifold must be synthesized, or held together in the mind in a certain way, in order that we can know the single, individuated object that the intuition in question, when it is thus synthesized, represents to us. This view raises thorny exegetic and philosophical problems, and its details are often ignored by commentators impatient with Kant's own obscurities about synthesis or else with what they view as the philosophical implausibilities of his position concerning that notion. But many of the details are very important to any understanding of the actual argument that Kant presents in the Transcendental Deduction. And for that reason alone – and leaving aside the frequent philosophical interest of these details – we cannot ourselves ignore them here.

In order to interpret Kant's account of the manifold and its synthesis, we need ultimately to understand the nature of the elements of the manifold, the manner in which they are supposed to be held together in the mind, and Kant's reasons for adopting the position that he does on these matters. Because such topics connect to Kant's remarks, in the Transcendental Deduction, about the categories, we cannot at this point consider every aspect of these topics. In particular, we must wait to consider many of his views on the processes that take place in synthesis and his reasons for holding that such processes are required in our knowledge. But it is important to discuss immediately (i) Kant's distinction of our apperceptive, discursive thought-consciousness from our inner sense, and his account of concepts and their use by such a thought-consciousness; (ii) and (iii) the nature of the elements of the manifold of intuition and their relation, first, to concepts or properties in intuition-displayed objects and, second, to the spatial parts of such objects; and (iv) various problems that afflict Kant's present views, including a well-known regress yielded by his account of synthesis and a difficulty about the sense in which our discursive thought-consciousness can grasp individual spatiotemporal objects and their spatial parts. In Chapter Four we then turn to the further details of the Transcendental Deduction itself.

2. OUR DISCURSIVE THOUGHT-CONSCIOUSNESS AND THE NATURE OF A KANTIAN CONCEPT

As we have seen in Chapter One, Kant follows the overall Cartesian model by holding that we gain knowledge via representations only insofar as those representations present their objects to a certain inner consciousness possessed by our mind. But Kant distinguishes, as earlier Cartesian philosophers do not, between two sorts of inner consciousness: first, the inner consciousness that amounts to our inner sense and belongs to our faculty of sensibility; and, second, the inner consciousness that amounts to the acts of thought that are spontaneously produced by our understanding, or faculty of thought. Moreover, just as Kant takes our knowledge to require the joint operations of sensibility and understanding, so too he takes our knowledge to involve both forms of inner consciousness. On the one hand, this knowledge involves our inner sense, which enables us to grasp those inner, mental properties, or sensible intuitions, via which we know (including our own outer-sense intuitions). On the other hand, it involves our thought-consciousness, which applies concepts to the single, individuated objects that are represented by those intuitions and so enables us to know those objects as being of various general types or kinds.

According to Kant, our human thought-consciousness is both an apperceptive consciousness and a discursive consciousness (a consciousness that grasps and operates *only* with general concepts). The first of these points, which we will discuss in great detail later, means that various of the acts of thought that are produced by our understanding either contain the first-person representation of apperception *I* or *I think* or else are able to have that representation attached to them.[2] Or, in non-Kantian terms, and in regard to various of our acts of thought, this point means that one of two things is true. Either such an act of thought is one which we explicitly ascribe to ourself through our use within that act of some thought-equivalent of the first-person pronoun or a similar device (*'I think* that it is sunny out'; *'I think* that this is a table'); or else such an act of thought is one which we are at least *able* to ascribe to ourself in such a way. (I actually think simply that 'it is sunny out,' but I am able to ascribe this thought to myself in the explicit first-person form *'I think* that it is sunny out.') This fact holds for all human knowers, as Kant sees it; and he also takes it as definitive of any other being that is like us in possessing our sort of cognitive capacities.

The second of the above points – that our thought-consciousness is discursive and operates only with general concepts – is our main concern here. We saw in Chapter One that a concept is itself a general, mediate representation. It presents to the mind a general property that does or can belong to various individual objects, and through that presentation of the general property it represents those objects. The concept represents those objects in roughly the sense that it represents them, through the mediation of that general property, simply as being those things, whichever they may be, that do or can possess that general property. We saw also that Kant takes our empirical concepts to be abstracted by our understanding from the objects of our empirical intuitions. We have not, however, further developed any of these points. Because of their importance to Kant's treatment of the manifold of intuition we must now carry out that development, indicating as we do so Kant's views on our thought-consciousness and its discursive nature.

I restrict attention here to ordinary empirical concepts like those of a tree or of a house, for Kant's views on those *a priori* concepts that are the categories require separate discussion, and our further comments on *a priori*, mathematical concepts (like that of a triangle), as far as they are required, are best made in connection with his account of synthesis.[3] If we ignore issues about appearing and appearance theories that will become relevant later, then we can see that Kant's view of empirical concepts falls into a conceptualistic tradition that is also exemplified by philosophers like Locke.

Kant's basic view of such concepts begins with the claim that objects, as they are given to us through our empirical intuitions, do not possess properties (say the property of being a tree) that are inherently and actually general.[4] Rather, and at best, common to a group of such objects, and present *in* each of these objects, is a sort of conceptual matter, or feature, that is potentially general (or is potentially able to be treated as being general and as belonging to all these objects). As being present *in* the objects, alongside all the other such features that are present in those objects, this conceptual feature is what Kant calls a *Teilbegriff* or *partial concept* of the object. As he sees it, through an abstractive process that we will mention below our understanding focuses an act of thought on this feature and thereby assigns it a form, or generality. The generality of this feature, or in more modern terms its being a general property, thus conceptualistically depends on its being *thought* or *treated by* our understanding in a certain way.

The way in which our understanding's act of thought treats this feature in assigning it a generality is this. That act of thought makes the feature general by regarding it as a *Merkmal* – as a *mark* or *characteristic* – of the objects *in* which, as a partial concept, this feature itself occurs. When our understanding's act of thought so regards the feature, the feature is now, according to Kant, an *Erkenntnisgrund* or *ground of knowledge* of the objects; and whereas as partial concept the feature is contained *in* the objects, as ground of knowledge it contains the objects *under* itself. As such a ground of knowledge, the feature can now be used by our understanding to represent all of the objects that do or can possess it. And so our mind can make knowledge-claims about those objects.

Kant does not describe with absolute clarity the exact way in which our understanding's act of thought must regard such a feature in order to treat it as a mark and ground of knowledge of the objects in question. But various of his comments, including those cited below, suggest that the feature is regarded by that act of thought in the way that I have indicated earlier: namely, as being a feature via which we can represent to ourself, in thought, those objects – and via which we can represent those objects more specifically as being those things, whichever in particular they may be, that possess that feature (or to which that feature belongs as partial concept).

As I have noted earlier, Kant takes our understanding to arrive at empirical concepts themselves via a process of abstraction. The precise details of this process are not important here. But, for the record, we should note that Kant regards it (and thus he regards our understanding's assigning of generality to an empirical feature in an object) as a three-stage logical process. First our understanding compares different objects and notes the respects in which they differ; then it reflects on what they have in common; and finally it abstracts from all of the respects in which they differ and focuses attention exclusively on that common feature.[5]

Because of the importance of the above points to our interpretation of the Transcendental Deduction – and in particular because of the importance of Kant's view that a concept both is present *in* the objects of knowledge, as *partial concept*, and also contains those objects *under* itself, as *ground of knowledge* – it is well to cite textual support for these points at once. Evidence that what is given in intuition is always particular (or is always only potentially general, actual generality then being assigned to a feature only via an act of understanding) is present in texts like *Fortschritte der Metaphysik*, first section, and "Logik Philippi."[6] And this position can also be seen, for example, in *Logik*, § 5, Note 1. Kant there says that the form, or generality, of a concept concerns 'only how it can be referred [via acts of thought] to several objects'; and so he implies that this form, or generality, does not belong to the concept or feature as given but is first assigned to it in thought.[7]

Evidence that concepts function both as partial concepts occurring *in* objects and as general marks, or grounds of knowledge, containing objects *under* them, is prominent in *Logik*, Introduction, § VIII.C.[8] The relevant passage is worth quoting in full:

> Human knowledge [*Erkenntnis*] on the side of the understanding is *discursive*, that is, it takes place through representations that make what is common to several things the ground of knowledge, thus through *marks* [or *characteristics*: *Merkmale*] as such. We thus know things only *through marks*; and this means precisely *knowing* [or *recognizing*], which *comes* from *acquaintance* [*Erkennen, welches von Kennen herkommt*].
>
> *A mark is that in a thing which makes up part of its knowledge* or – which is the same – a *partial representation* so far as it is *considered as ground of knowledge of the whole representation*. All our *concepts* therefore are marks and all *thinking* is nothing other than a representing through marks.
>
> Every mark may be viewed from two sides:
>
> *First*, as a representation in itself;
>
> *Second*, as belonging as a partial concept to the whole representation of a thing and thereby as ground of knowledge of this thing itself.

The same position can be seen also, in whole or part, in *Logik*, § 1, § 2, § 4, § 5 and its Note, and § 8, Note. Various of these latter texts also refer to Kant's views on the generality that, through an act of understanding, a concept possesses as a ground of knowledge. Such views are especially clear in § 7 and its Note ('Every concept, as *partial concept*, is contained *in* the representation of things; as *ground of knowledge*, i.e. as *mark*, these things are contained *under* it.... The generality or general validity of the concept does not rest on the concept's being a *partial concept* but on its being a *ground of knowledge*.'[9]) They are found also in numerous Reflexionen.[10] And, it is important to note, the same views are present in the first *Critique*.

Thus we have B12:

> ... though I do not include in the concept of a body in general the predicate of weight, none the less this concept indicates [*bezeichnet*] an object of experience through one of its parts [i.e., through one of the parts of the experience of the object]...

(Compare A8 and also A105 on thinking an object through the predicates of a triangle.) Observe also B39–40:

> ... every concept must be thought as a representation which is contained in an infinite number of different possible representations (as their common mark), and which therefore contains these *under itself*...

and B133 note:

> If ... I think red in general, I thereby represent to myself a property [*Beschaffenheit*] which (as a mark) can be found in something, or can be combined with other representations...

And, like various of the *Logik* texts and Reflexionen noted above, the first *Critique* records the view that our understanding is discursive and so knows only via (and grasps only) concepts. Thus, see, for instance, A656 = B684: 'the understanding can have knowledge only through concepts: therefore... never through mere intuition.'[11]

The preceding texts show that Kant accepts the view of concepts and of our understanding's acts of thought that I have outlined above. There are a great many complications about this view that we can ignore here, including the question of how exactly to understand the notion of a property *as taken as general* by an act of thought or the related notion of the actualization of a property's merely potential generality by such an act. I should note at once, however, that in my initial, Chapter One exposition of Kant's treatment of concepts, and again above, I have spoken of a concept as a representation that *itself* in some way *presents* a general property (and a general property from which the concept itself evidently should be taken to be distinct). And I have said that through this presentation of the general property the concept represents to the mind the objects that possess that property.

This way of speaking is useful for expository purposes, and it corresponds to various of Kant's own descriptions of concepts.[12] For example, such a way of speaking can perhaps be seen in the first paragraph of the above quotation from *Logik*, Introduction § VIII.C (where Kant in effect talks of concepts as '*representations* that *make* what is common to several things the ground of knowledge,' my italics). It can be argued perhaps to be present in the B12 quotation (where what 'indicates an object of experience' – namely, the concept of a body in general – is one of the *parts* of the *experience* of the object). And in other texts Kant says things that seem even closer to this way of speaking, for instance in Reflexion 2278, where he writes that 'each concept always represents a general mark of certain things.'[13] The actual situation with regard to his treatment of concepts is, however, more complex than such texts suggest. As is shown both by my exposition of Kant on concepts and by a careful study of texts cited above, there is also a strong tendency in Kant *not* to take the concept to *present* to the mind a general property (a general property from which the concept is distinct) but, rather, simply *to be* the general property. Or, more accurately, there is a strong tendency in Kant to identify the *concept*, or *mark* or *representation* of the objects, with the *general property* that is present *in* all of the objects that fall under that concept, insofar as that property is taken as general by the understanding.[14]

This Kantian identification of concept with property – or this tendency to such an identification – can be seen in the above *Logik*, Introduction, § VIII.C text taken as a whole. In the first paragraph of that text Kant in effect identifies a mark with 'what is common to several things' and thus with a general property; and in the second paragraph he then identifies a concept itself, a representation of objects, with a mark as so understood. The same identification is also clear in B133 note, where he describes what he calls 'red in general' as a *property* that is a *mark* and so is (by B39–40 or the second paragraph of the above *Logik*, Introduction, § VIII.C quotation) a *concept*. One can also see this same identification in *Logik*, § 11, Note (compare also § 8, Note), where he speaks of iron, metal, body, etc. – those general properties – as being themselves *concepts* (and concepts which are, by texts like B39–40 or the second paragraph of the *Logik*, Introduction, § VIII.C quote, *marks*). And various other texts suggest the identification.[15]

Moreover, since concepts are, on any reading of Kant's theory, representational entities that occur *in* the mind (and are operated on *by* the mind), the effect of this identification is as follows. Concepts become general properties that indeed *occur in* (and are operated on *by*) the mind. We will return to this sort of point below, and again in Chapter Ten.

There are, I believe, various origins for these conflicting tendencies to treat a concept, on the one hand, as being a representation presenting a general property from which the concept is distinct and, on the other hand, as being a general property itself (when that property is taken by the understanding as general). These tendencies presumably arise, in part, from Kant's overall tendencies both to distinguish and also to identify a representation (and, in particular, an intuition) and an object. They also arise, I think, out of a not completely worked-out tension, within his philosophy, between a fairly pure representationalist and a more traditionally Aristotelian (or quasi-Aristotelian) view of concepts and properties, the latter view taking such entities to be 'forms' or general properties that are present, in knowledge, *in* the mind (and are there operated on *by* the mind). (Here see, also, Chapter Ten.) Moreover, such tendencies obviously interact with Kant's appearing- and appearance-theory pictures of knowledge in what are evidently quite complicated ways. We need not here work out these ways in detail. But because it is germane to our later discussion, I should note that each of these Kantian treatments of concepts can be combined *both* with Kant's appearing theory *and* with his appearance theory.

The idea that a concept somehow *presents* a general property to the mind while *representing* to the mind not that property but, rather, the objects that do (or can) possess it is of course not made absolutely clear by the above texts. But even without developing this idea further, it should be obvious that it can be combined with Kant's appear*ing* theory. It can be combined with that theory simply because, according to that idea, the concept is distinct from the property that it presents, just as on the appearing theory an intuition is distinct from the object that appears via that intuition. So by means of the property the concept can be taken to represent objects in the same sort of way that we have suggested above and will discuss again below. Again, this same idea evidently can be combined with Kant's appear*ance* theory. The combination can be made simply by identifying, as does that theory, intuition and object known, and then taking the concept to present to the mind some property that is possessed by intuition-representable objects – some property through which the concept represents to the mind various ones of those objects.

Suppose, now, that a concept is simply identified with the general property in question. Then Kant's appear*ing* theory can still be adopted, with its distinction between intuition and object known; and the general property (here identified with the concept) can then be taken to represent, in the way that we have suggested, those objects, appearing or able to appear via intuitions, that do or can possess that general property. Again, Kant's appear*ance* theory also can obviously be adopted. If the concept is identified with the general property and the intuition is identified with the object that it represents, then via that property the concept can be taken to represent objects (here identified with intuitions) in the way that we have noted.

Of course some of these combinations of one or the other of Kant's views on concepts with his appearing or appearance theory affect our understanding of his treatment of the roles of intuitions and concepts in knowledge. In particular, some of the above combinations destroy what might be called the presumed *parity* of intuitions and concepts as *representations* – namely, the presumed fact that intuitions and concepts, although respectively singular and general representations, nevertheless function representationally in the same basic ways. (Thus, given this parity, an intuition will be distinguished from the object that it represents just in case a concept is likewise distinguished from the property that it presents, and so on.) But a study of texts like those cited above strongly suggests that Kant does not always accept any such parity of intuition and concept. So the possibility of combining either of Kant's views on concepts with either of his main pictures of knowledge cannot be rejected on this ground.

The upshot of the above discussion is that Kant accepts two closely related, although different, treatments of concepts, each of these treatments being incorporable into his appearing- or his appearance-theory picture. We therefore cannot appeal to one or the other of those versions of Kant's picture in order to determine which, if either, of these two treatments we should favor. Moreover, a little thought shows that these treatments are, philosophically, roughly on a par. In considering this point, we may ignore (as Kant's subsequent use of the notion of a concept in the Transcendental Deduction allows) the basically abstractionist aspects of his approach to empirical concepts. We may also ignore the difficulties into which such approaches have been argued to fall. But then if we ignore such matters, each of the above treatments becomes an account of how, through its acquaintance with (and its use of) general properties, our mind is able to think of objects that have those properties. And, especially when Kant's treatments are taken as embedded in his historical view of such thinking, with that view's simultaneous representationalist and to my mind rather Aristotelian view of concepts and properties, it is hard to see that the one of these treatments is (at least when it is taken as so embedded) markedly superior to the other.

To modern readers it of course may sound odd to follow what I believe is Kant's Aristotelian precedent and to speak of concepts as properties or forms – and as properties or forms that are present *in* the mind (and are there operated on *by* the mind). And we will attend, below, both to the Kantian view of concepts as presenting properties to the mind and to the alternative Kantian view of concepts as being general properties themselves. However, at later points it will be necessary to focus specifically on just this alternative view. Indeed, the concepts-as-properties view becomes extremely important in our Chapter Ten discussion of the categories and our investigation of the final stages of one main argument in the B-edition Transcendental Deduction.

There is one other issue about Kant's account of concepts that we should discuss briefly here. It may seem quite puzzling, particularly from a modern standpoint on general properties, how a concept, whether as *presenting* a general property or as *being* a general property, really can *represent* to the mind the objects that fall under it. There is, however, a fairly simply answer to this puzzle. As one might expect from Kant's conceptualism, the answer is that no such representation of objects is really performed by what one might call the mere inert mental presentation of the general property or by the mere inert general property itself as it occurs in the mind. Rather, the understanding *uses* the general property, as it is presented mentally – or the understanding *uses* the general property itself, as it occurs *in* the mind – in order to form the *thought* of those objects, whichever they may be, that do or can possess that general property. (Indeed, given Kant's conceptualism, it seems that that general property is in fact presented to – or occurs in – the mind *just insofar as* the understanding forms such a thought concerning a feature that occurs in all of those objects.) And thus, via that thought, the understanding comes actively to think, and in that sense actively to represent to the mind, those objects.

Of course such a thought-effected representing of objects is unlike the picture, which may have been suggested by our earlier remarks on intuitions, of a representational entity as *by itself*, and as a result of its intrinsic nature, somehow setting before an act of inner consciousness the object or objects that it represents. But, as we will see in more detail when we return to such matters in Chapter Eight, that picture is not, in the end, an accurate picture even of the representational function of intuitions, according to Kant.[16] It is also worth noting that Kant's overall picture of general properties as being used by the mind to represent objects is, in one form or another, a doctrine that is common to a number of other Cartesian philosophers.[17] Thus it is not a doctrine that he himself would regard as novel or as otherwise likely to puzzle his readers.

3. THE ELEMENTS OF THE MANIFOLD OF INTUITION (I): MATTERS FOR CONCEPTS

The preceding discussion shows in some detail the discursive nature that Kant assigns to the operations of our understanding in knowledge: how our understanding utilizes (and must invariably utilize) general properties in order to think about, and thus in order to make knowledge-claims concerning, objects – and how (although we did not recall this point in the preceding section) our understanding, by noting the presence of such a general property in some intuition-displayed object, comes to recognize that single, individuated object as being of some general type or kind. We observed that these points bear numerous connections, many of which we will bring out later, to central concerns of the Transcendental Deduction. Of these connections, the one that we now need to consider is the relation of the preceding discussion to Kant's views about the synthesis of the manifold of intuition and in particular to the nature that he assigns to the elements of that manifold.[18]

In order to examine that relation, consider the elements of the manifold of an intuition as they are generated in the mind in independence of the activities of thought. Then just as an intuition represents an object to the mind, so too – and as on Kant's appearing theory – those elements may be said to *put before the mind*, for operations by those activities, features and aspects (distinct from those elements) that belong to the object that is represented by the intuition. (Or else, as on Kant's appearance theory, those elements may be said simply *to be* the relevant features and aspects.) Since, as we note below and in Chapter Eight, an intuition's representation of an object involves activities of thought, this putting-before-the-mind should not be regarded as a presenting, in the Kantian conceptual way, of the features and aspects in question. (The individual elements of the manifold, as they are generated in the mind in independence of the activities of thought, are obviously not Kantian concepts that, through activities of thought, present general properties to the mind.)

Rather, this putting-before-the-mind should be taken as a distinctive function of the elements of the manifold, a distinctive function which itself underlies, in part, the intuitive representation of objects and the conceptual presentation of properties.[19]

Given these last comments, we can now turn to the relation of our preceding discussion to Kant's views concerning the synthesis of the manifold of intuition. As we have suggested, Kant takes each intuition to be given in the form of an unconnected manifold of elements that must be synthesized in order that we can know the single, individuated object that the intuition, when it is synthesized, represents to us. This view of Kant's involves two basic claims. First, our knowledge of objects – and, more specifically, our intuition-mediated recognition of objects as being the single, individuated things that they are – always takes a sequential form. We know objects only by attending one by one to their various features and aspects, by perceiving them from successively different points of view and under potentially different conditions of observation, and so on. In the case of our knowledge of a given outer object – which is the case on which I focus for the present – Kant puts this point by supposing that the intuition through which we know this object is presented to us in the form of a disconnected manifold of representations that sequentially put before our mind (or sequentially occur before our mind and simply are) all of these features and aspects of the object. And then we have to synthesize this manifold in order to arrive at a single, unitary intuition that represents to us the single outer object in question as possessing all these features and aspects.[20]

Second, by the above view Kant is making another and a less obvious claim, a claim that concerns not only the sequential process of our recognition of the object but also, and more importantly, the nature of the object itself, in the form that that object takes as we know it. He is noting his Transcendental Aesthetic conclusion that the object, in that form, is a mere mind-dependent, phenomenal thing. He is then claiming that *this object itself* first occurs before our mind in the form of a disconnected manifold of data. If we accept the appearing-theory view, then this object first occurs in the form of a disconnected set of features and aspects that are sequentially put before our mind by various sequentially presented parts of an overall intuition that we have. Or else, if we accept the appearance-theory view, then this object first occurs before our mind in the form of a disconnected set of features and aspects that are identical to various sequentially occurring parts of such an intuition. Furthermore, Kant is holding that we have to synthesize this manifold of data in order that there should actually exist the single, individuated, and unitary phenomenal thing that *is* the object of our knowledge.

Both of these claims are important to the argument of the Deduction. The first, versions of which were accepted also by earlier Cartesian philosophers like Arnauld and Leibniz, is perhaps the less controversial, although even it runs contrary to the fact that we can take in at once a limited number of features and aspects of any perceived object.[21] The second is of course intimately connected with Kant's idealism and raises many questions. However, before discussing either of these claims, we must understand them more clearly than we now do. And, to reach that understanding, we need to consider exactly *what* Kant takes to be the elements of the manifold of intuition that are first given to us in a completely disconnected form.

Because Kant is not especially explicit about the nature of these elements, we can most easily grasp their nature by focusing first on some further details of his account of concepts. In particular, we need to anticipate our discussion in Chapter Ten and to note one overall intention of the Transcendental Deduction. That intention is to show that the categories of the understanding apply to the objects of empirical knowledge through a demonstration that the categories play a determining role in the synthesis of the manifold of intuition. To Kant's mind, this synthesis occurs correlatively with the synthesis of concepts in a judgment about the empirical object of intuition. By considering that synthesis of concepts and the related synthesis of the manifold of intuition, we can discover something about the nature of the elements of the manifold that is synthesized.

About these syntheses we can be reasonably brief. As we have seen, our understanding, and thus our apperceptive thought-consciousness, is discursive. So the knowledge-states that are yielded by that thought-consciousness must take the form of acts of apperceptive consciousness directed on concepts in our mind. But, for Kant, the knowledge that is thus yielded invariably takes a judgmental, that-clause form. It is a knowledge *that*, say, the tree is conical, and not simply a knowledge got by idly bringing the concept of a tree before thought-consciousness. Kant takes this last point to imply that concepts, as they occur before our apperceptive thought-consciousness in knowledge, must be related together in judgmental ways. More specifically, he takes there to be a group of logical forms of judgment that determine that a given set of concepts constitutes a judgment with a specified logical quantity, quality, relation, and modality.[22] And he holds that the concepts that we have been discussing must be related together, as they occur before our apperceptive thought-consciousness, according to various of these logical forms. For example, in the judgment expressed by the claim 'the tree is conical,' the concepts of being a tree and of being conical occur related in such a way that the concept of being a tree functions as the subject term of a judgment whose predicate term is the concept of being conical. And that judgment also has a singular quantity, affirmative quality, and assertoric modality.[23]

Kant describes this relating together of various concepts in a judgmental form before an act of apperceptive thought-consciousness as a *synthesis* or combination of those concepts in judgment.[24] This synthesis he regards as objective. It yields a judgment about an object or a group of objects. Moreover, the concepts that make up that judgment are related together not by subjective relations of mental association, which vary from mind to mind, but rather by logical forms that function in the same ways in all minds like ours and are in a certain sense necessary.

Now – and as we see in more detail in Chapter Ten – Kant connects this objective synthesis of concepts in judgment with the synthesis of the manifold of intuition. He makes this connection, in effect, by means of the fact that concepts, as they occur in judgments, *are* – or *present to the mind* – general properties, and general properties that themselves occur (in the sense explained in Section 2) *in* the objects of those judgments. More specifically, and as we have seen, Kant takes such general properties to occur, in only potentially general forms, *in* the objects that fall under the relevant concepts. One might call these properties, as they so occur, matters for concepts. Kant supposes that the concepts that occur in the judgments are, or present, general properties that differ from such matters for concepts only in that those general properties in the judgments simply are the result of our understanding's assigning a form, or generality, to the matters for concepts in the objects.[25]

As the form-assigned general properties occur as concepts in judgments (or are presented by concepts that occur in judgments), they (or those presenting concepts) then function as grounds of knowledge of the objects that fall under those concepts. In conjunction with the preceding discussion, this latter fact now indicates the sense in which Kant takes a synthesis of the manifold of intuition to occur correlatively with the synthesis of concepts in judgments.

In order to understand this sense, suppose that we, the knower, confront a tree in our perception; and suppose that we make the true judgment that the tree is conical. Then, given the preceding discussion, the concepts of being a tree and of being conical are held together, in a judgment occurring before our thought-consciousness, in a subject-term/predicate-term logical relationship. And Kant's view is that since this judgment is a true one, not only are we conscious simultaneously of the *matters* for these concepts as being contained *in* the tree that we intuit. But, also, we must be conscious that *in* that tree itself one of *its* elements, the *matter* for the concept of a tree, is functioning as *subject* in relation to another of *its* elements, namely the *matter* for the concept of being conical, which is functioning as *predicate*. Similarly we must be conscious that elements of the tree are functioning in ways that correspond to the presence, in our judgment, of the logical quantity of singularity and of the logical quality of affirmativeness.[26] And thus we must be conscious of a synthetic combination of elements in the tree that parallels, and is correlative to, the synthetic combination of concepts that occurs in the judgment that we make about that tree.

Again, suppose that we judge truly that all bodies are divisible. Then not only does our judgment require that the set of objects that we think as containing the feature of being a body should be contained within the set of objects that we think as containing the feature of being divisible (here compare Section 2 on thinking or representing objects through properties). But, also, our judgment, since it is true, requires that *in* each of the relevant *objects* the *element* or *feature* of being a body should itself function as *subject* to the element or feature of being divisible as *predicate*, just as in *the judgment itself* the concept of being a body functions as *subject term* to the concept of being divisible as *predicate term*. And thus again there must occur, correlatively to the synthesis of concepts in a judgment, a synthesis of elements that are present in the intuition-represented object or objects that the judgment concerns. (Here note Kant's well-known A79/B104 claim that 'the same function which gives unity to the various representations *in a judgment* also gives unity to the mere synthesis of various representations in an intuition'; and see Chapter Ten.)

Given the above anticipations of Kant's views on synthesis in judgment and in the manifold of intuition, we can now see that Kant must take our understanding, as it functions by itself discursively through its use of general concepts, always to know objects *only* through the properties or predicates that belong to those objects. We can also see that he should identify the elements of the manifold of intuition, or at least one group of such elements, with those features, or matters for concepts, that occur with potential generality *in* the objects and are then combined synthetically in parallel with the combination of concepts in judgment.

Or, to put this last point with more accuracy, given his appear*ing* theory, Kant should see each intuition as breaking down into (at least) a set of representations each of which puts before the mind one such potentially general feature, or matter for a concept (a potentially general feature that is itself distinct from the representation that puts it before the mind). And the object itself that is known through that intuition should be given in the form of (at least) such a set of potentially general features each of which is put before the mind by one of those representations. Again, given his appear*ance* theory, Kant should see each intuition as breaking down into (at least) the same sort of set of representations. Each of these representations will now itself be identified with one such potentially general feature. And the object that is known through the intuition, now identified with that (synthesized) intuition itself, will then be given to the mind in (at least) the form of such a set of potentially general features.

Once the potentially general features stand before the mind in either of the above ways, the understanding will then assign a form or generality to these features and so will yield (empirical) concepts to the mind. As indicated in Section 2, these concepts will themselves present to the mind (or else will simply be) those form-assigned, general features; and by means of them the concepts will represent to the mind the objects that possess those features. It should thus be clear that, on my present interpretation, the relevant elements of the manifold, which as they are given put before the mind (or are) the *matters* for concepts, are *not* themselves *the concepts* which subsequently, in the order of logic, present (or are) the general features that are the results of assigning forms to those matters. This result of course leaves open the question of what the exact relation is of the elements of the manifold and the concepts. Part of the answer to this question (concerning the case in which a concept is regarded as a general property) is implied by our above discussion. But since the overall issue of the relation of intuition-element and concept first becomes important in Chapter Ten, I will ignore that issue until then.

The preceding points about the elements of the manifold can be verified, at least in general, from the texts. In the first place, take the claim that our understanding, as it functions discursively by itself, knows objects only through the properties or predicates that it takes objects to have and of which it is aware in thought-consciousness. This claim is an obvious consequence of the discursive nature of our understanding, as we explained that nature in Section 2. That consequence is made explicit in first-*Critique* texts like A656 = B684, already quoted in Section 2 ('the understanding can have knowledge only through concepts: therefore... never through mere intuition'). It also can be seen at A105 (on thinking an object 'through the predicates of a triangle'), B12–14 and A8 (on 'the object which I think through the concept A'), and in the rather explicit A399–400 (where Kant asserts, with my italicization, that 'if I am to declare a thing to be substance in the appearance, *predicates* of its intuition must first be *given* me, and I must be able to distinguish in these the permanent from the transitory and the substratum (the thing itself) from what is merely inherent in it').

In the second place, consider the claim that Kant identifies the elements of the manifold of intuition, or at least one set of such elements, with matters for concepts. Other first-*Critique* texts make this claim evident, if it is not already so from the texts cited above. (And these other texts also themselves support the preceding claim about our understanding's knowing objects only through their properties or predicates.) See, for instance, B131 (on the 'unity of *given concepts*,' my italics); B140 (in the heading) and ff.; B143 (on 'the *manifold* of *given* representations (be they intuitions *or concepts*),' my italics); and *Prolegomena*, § 13, Note 2 (on, among other things, '*the qualities [Eigenschaften] that make up the intuition of a body*').[27]

Evidence for both of our above claims is also displayed in fragments from the 1770s and in a student's lecture notes said to date from 1784–85. Thus we read in Reflexion 4634 that

> We know each object only through predicates, which we say or think of it. Before [this knowledge occurs] that which is to be met with in us by means of representations is only to be accounted matter [*Materialien*] but not knowledge. Therefore an object is only a Something in general which we think to ourselves through certain predicates, which make up its concept. In each judgment, accordingly, there are two predicates, which we compare with one another. Of these, the one which makes up the given knowledge of the object is called the logical subject; the other, which is compared with it, the logical predicate. If I say: a body is divisible, then this means as much as: Something x, which I know [*kenne*] under the predicates which together make up a concept of body, I also think through the predicate of divisibility.[28]

Reflexionen 2281, 4638, 4645, 4674, 4676, and 5923 (dated by Adickes as after 1781) among other texts seem to express parts of roughly the same position.[29] This position is present also, in a much more sophisticated form making explicit reference to the role in judgment of the logical forms of judgment, in the lecture notes called "Metaphysik Volckmann," dated as from Kant's 1784–85 lectures. Kant is there reported as having said:

> All our knowledge is composed of judgments and these must have an object, the mere intuition is no knowledge. If I say that this or that belongs to some thing, so I know it, it thus is related to an object, and then it is a judgment, this however arises out of concepts. Now in considering the concepts which we receive from the senses it is arbitrary which form we want to use to judge, e.g., I can make the representation of body for myself as one or many or as all bodies. I can say this thing is a body also say it is not a body, the body is extended, also, the extended thing is a body. If, however, my sensible representations are to be referred to an object and my judgments about an object of the senses are to be referred to an object, then the form of the judgment can no longer be indifferent. For in our considering an object all our representations are to be regarded merely as predicates for possible judgments, which [object] is regarded merely as a something in general, and that [object] must be determined in all judgments which are passed about this something. In considering the object it must also be determined according to which form I thereby should judge, e.g., the representation of a body contains a variety [*enthält mancherley*], but it is determined only through predicates. Along with a solid house I must think a wall which encloses an empty space, etc. The representations are referred to something in general as predicates and [it is] determined, in considering the object, in which ways these representations can be predicated of it. Here I cannot consider the subject also as predicate but only as a subject.... Our experience is an entirely new product of our power of knowledge from sensible sensations, and representations of those rules according to which an object is determined in regard to its predicates.... Understanding alone thinks the object, and this can never be given except only as determinations, predicates of the same.[30]

We can thus safely conclude that (subject to certain qualifications introduced below) the elements of the manifold of intuition are – or are at least – matters for concepts, all objects being known by our understanding through such matters (or through such potentially general properties when they are taken as occurring in the objects). It is important to note that this result harmonizes with another strand in Kant's account of synthesis – namely, with his frequent descriptions of synthesis as a bringing together of perceptions or sensations in the mind.[31] Indeed, these descriptions make concrete – and in part confirm – our preceding conclusions. Thus Kant describes perceptions as representations with consciousness (A320/B376).[32] A sensation is any perception that is not referred by its possessor, in synthesis, to an object – that is, any perception that within its possessor's mind functions simply as an inner, mental property or what Kant calls a modification of the possessor's inner state (ibid.). As one can see from what we say in Chapter Eight, Kant regards the synthesis of a manifold of sensations as referring the sensations to an object in such a way that they come jointly to constitute an intuition that represents the object. So there should be an initimate relation between the manifold of sensations and the manifold of matters for concepts. In particular, one would expect Kant to take sensations themselves (or at least some sensations) simply to be the representations that put before the mind (or are) those matters for concepts.

Although Kant is not immensely clear about the relation between sensations and what we have been calling matters for concepts, this expectation is at least in general satisfied. Thus A86/B118–19 discusses 'the first strivings of our faculty of knowledge, whereby it advances from particular perceptions to universal concepts' and A374 asserts (with my italics) that 'perception is that whereby the *material* [*Stoff*] required to enable us to *think* objects of sensible intuition must first be *given*.' Again, Reflexion 3930 speaks of 'abstracted,' or empirical, concepts as abstracted from sensations. In addition, in the lecture notes "Logik Blomberg" Kant is reported to have said that 'the matter itself [for the concept] thus lies... in the experience [whose matter evidently is given via perception], the form of the universality however lies in the abstraction.'[33] (And here compare also the first-*Critique* texts cited earlier, as well as the earlier-quoted Reflexion 4634 and "Metaphysik Volckmann.") None of these texts definitively verifies the expectation above, but they are all the sorts of thing that Kant would say did he, in fact, regard at least some sensations as themselves being representations that put before the mind (or are) matters for concepts. Thus Kant's description of synthesis as a bringing together of sensations in the mind does help to confirm our above view of the manifold of intuition.[34]

Before we continue, a bit more should perhaps be said about the idea, important to that view, that elements of the manifold 'put before the mind' (or simply are) potentially general features, or matters for concepts. We have seen good reasons for attributing such an idea to Kant. Nevertheless, it raises questions. The most pressing turn on the givenness of the elements of the manifold. In particular, Kant does not in the end suppose that those elements, as they are given to the mind, actually function as a single intuition that represents a single object. Rather, Kant takes such elements to function in that way only through the mind's act of 'referring the intuition to an object.' But then how satisfactory can it be, on the one hand, to deny that such elements, as they are given, function together to represent an object while, on the other hand, affirming that various such elements, as they are given, actually manage to put before the mind (or to be) single, well-defined, potentially general features, or matters for concepts?[35]

This last question can be seen to turn, in part, on form-matter points that I consider in Section 5. And it raises many dark issues in Kant interpretation. Putting aside such issues, we can see from the texts that there is, in fact, a plausible answer to this question. Kant's reason for holding that intuition-elements, as they are given, cannot function to represent a single object is that the combination of such sequentially presented (and atomic) elements – and so their functioning for the mind as a single intuition – cannot be given. (Here see Chapters Four, Six, and Eight.) However, this reason does not apply to individual, given intuition-elements. Or at least it does not apply to such elements insofar as they are regarded as simple, noncomposite entities that put before the mind (or are) potentially general features that are themselves, as they initially occur before the mind, noncomposite. And A99 ('each representation, *insofar as it is contained in a single moment*, can never be anything but absolute unity') and A167/B209 ('sensation is that element in the appearance the apprehension of which does not involve a successive synthesis proceeding from parts to the whole representation'; see also A166/B208) certainly suggest that Kant regards the given elements in this way.[36] So the above question points to no real difficulty within his picture of knowledge.

As we will note in Sections 4 and 5, our present Kantian view of the manifold of intuition as consisting of, roughly, representations that put before the mind (or are) potentially general features, or matters for concepts, nevertheless leads to further questions. So I should emphasize finally that something like this view seems to have been a philosophical commonplace in Kant's time. Similar positions – or positions relating closely to one or another aspect of Kant's own views – are to be found in the work of philosophers like Aristotle, Locke, and G. F. Meier (whose logic text Kant used in his logic lectures), among others. In each of these philosophers there is the suggestion that we know objects only through properties – properties that are to be found *in* the objects and that (at least for Locke and Meier) themselves can be taken by the mind as *representational marks of* those objects.[37]

4. THE ELEMENTS OF THE MANIFOLD OF INTUITION (II): MATTERS FOR SPATIAL PARTS

The upshot of the preceding discussion is that in considering the manifold of intuition, we are considering at least two things: first, a group of representations that put before the mind (or are) matters for concepts, those representations being presented to us sequentially in sense perception; and, second, the group of those matters for concepts themselves, as those matters for concepts are put before our mind by the representations in the first group. (Hereafter I often drop the '(or are)' phrase.) This result is not, however, by any means a complete account of the elements of the manifold.

That more must be said can be seen quite simply. Suppose that we were to know objects only via representations that put before the mind matters for concepts. Then our knowledge would be perfectly general and could not concern single, individuated objects as such. We would know, for example, that *some* object or *other*, which has the property of being a tree, also has the property of being a conical thing. We might even know that *some unique object or other*, which has the former property, also has the latter property. But through such a general knowledge of an object as possessing various properties we would not know *of* or *about* any *particular*, individuated object *as such*, that *that* object has the properties of being a tree and of being conical. Furthermore, through such purely general knowledge we would not get the sort of direct-object confrontation with the object in sense perception that Kant, like other philosophers, will take to characterize our perceptual knowledge.[38] We might come via our understanding to *think*, for example, and even thereby to know, *that* some conical tree is directly before us. But such thought-effected knowledge-that would not amount to our *seeing*, direct-object fashion, *a conical tree occupying a place in space directly before us*, a conical tree various of whose spatial parts we would also, in the usual case, be seeing in such a fashion. So we cannot take the elements of the manifold simply to be representations that put before the mind matters for concepts. Or we cannot do so unless we agree that Kant is in serious difficulties here.

As we will see in Section 5, Kant's sharp separation of our understanding from our sensibility does create serious problems of the above sorts. But the problems are probably hidden from his eyes at least in part because he also has another view of the elements of the manifold. This other view is of various of those elements as putting before the mind the spatial parts that go to make up the object that appears via the synthesized intuition in question.

To state more accurately this other view, recall that space is the form of outer intuition and thus of outer objects. Distinguish, also, with Chapter One (and as Kant does not), between (i) the formal factor in the mind that guarantees that outer intuitions should represent, exclusively, objects in space and (ii) space itself as the form or structure that belongs to those objects. (At least we should make this distinction as long as we adhere, as I will for the present, to Kant's appearing rather than to his appearance theory.) Then Kant's other view is best represented as follows. An object existing in itself affects our sensibility in such a way as to yield us a set of sensations. These sensations, which are in themselves nonspatiotemporal, are then (in the order of logic) operated on by the above formal factor. The result is that they come to put before the mind potentially definite spatial object-parts potentially occurring at definite places in space. But these sensations do not put before the mind any *actually* definite spatial parts. Nor does space itself, as roughly the set of actual relations defining the locations of such spatial parts, exist as a single, unified thing in the mind.[39] In order for the intuitive representation of actually definite spatial parts, occurring at definite places in space, to come about – and in order for space itself so to exist – synthesis of the manifold is then required.

Various elements of the manifold thus are mental entities – in fact, sensations or groups of sensations – that put before the mind such potentially definite spatial parts, or what one might call matters for spatial parts, potentially occurring at definite places in space. Or, to state this result more carefully, on the appearing theory each (outer) intuition breaks down into (at least) a set of elements each of which puts before the mind one such potentially definite spatial part, or matter for a spatial part. And the object known via that intuition is given in the form of (at least) such a set of potentially definite spatial parts, each of which is put before the mind by one of those representational elements. Again, on the appearance theory each (outer) intuition breaks down into (at least) the same sort of set of elements. Each of these elements will itself be identified with one of the matters for spatial parts in question. The object that is known via the intuition – and that is now itself identified with that (synthesized) intuition – will then be given to the mind in (at least) the form of such a set of matters for spatial parts.[40]

Since, as we see below, Kant tends at least frequently to run together the elements of the manifold that put before the mind matters for spatial parts and the elements that put before the mind matters for concepts, a good deal of the textual evidence for the above results is contained in his discussions of the latter elements. I will note this evidence below when we see the evidence for the running-together that I have just mentioned. Other comments also will support our above results. In the meantime, one can observe, as upholding the parts of those results that concern, specifically, sensations and space as the form of outer intuition, Transcendental Aesthetic texts like A23/B38 and our overall account of the Aesthetic in Chapter One.[41]

One may also observe that our comments on sensations and elements that put before the mind matters for spatial parts are not in any real conflict with our earlier, Section 3 remarks on sensations and the elements that put before the mind matters for concepts. For one thing, the existence of the very running-together that is in question means that Kant tends to run together sensations considered as yielding elements that relate to matters for spatial parts and sensations considered as yielding elements that relate to matters for concepts. For another thing, Kant's discussions of the interaction of sensations and space in synthesis are by no means clear or detailed enough to let us see those discussions as really *contradicting* his views on the interaction, in synthesis, of sensations and the elements that relate to matters for concepts. On both of these counts (and there are others), our above comments, although not presenting a complete account of Kant on sensations, can thus stand, for present purposes.

Combining our present results with those of Section 3, we thus see that the elements of the Kantian manifold of intuition ought to amount *both* to representations that put before the mind (or are) matters for concepts *and* to representations that put before the mind (or are) matters for spatial parts. Indeed, comments in Chapter Ten will show that when the nature of the manifold is linked to Kant's discussions of the categories, we should see a third aspect of the Kantian manifold as emerging, in connection with the intensive qualities of objects.[42] Even restricting ourselves, as we will here, to the former two sorts of elements, we can see that, as was already intimated at the end of the last section, numerous questions can be raised about such elements.

The chief questions can be noted by remarking that (i), for Kant, no particular matter for a spatial part can be plausibly identified with any particular matter for a concept. Kant holds that through a concept one represents the whole object and not just part of that object; but, as it was explicated above, a (matter for a) spatial part is a (potentially definite) part of an object.[43] Moreover, (ii) from our above descriptions of the representations that put before the mind matters for spatial parts and matters for concepts, as well as from texts like A99ff., A166–67/B208–209, and A120 (where Kant says that the perceptions in the manifold of appearance 'occur in the mind separately [*zerstreut*] and singly'), it is plausible to suppose that Kant takes each element of the manifold, as it is generated in the mind in independence of the activities of thought, to put before the mind just one matter for a spatial part or just one matter for a concept (and not, say, both a matter for a spatial part and a matter for a concept).[44] Given points (i) and (ii) and others, however, questions about elements of the manifold now arise immediately.

Thus, (a) even given point (i), it seems impossible that there could be spatial parts that lack size, shape, and so on. It also seems impossible that there could be any matter for a concept (which, after all, is supposed to occur *in* an object) without there being some spatial part or object in which that matter for a concept occurs. So the occurrence of matters for spatial parts seems in general inseparable from the occurrence of matters for concepts. But how can this fact be reconciled with the idea in (ii) that each of the individual elements of the manifold puts before the mind just one single, separate matter for a spatial part or matter for a concept?

Again, (b) the idea that we can take in individual properties and spatial parts of an object only separately runs contrary to our ability to grasp at once a limited number of features of an object. Furthermore, such an idea obviously leads to an extraordinary view of our ordinary perception of objects. And, on a natural construal, this idea also seems to run contrary to important first-*Critique* texts in which Kant insists that objects are given as wholes in intuition and we arrive at intuitions of their parts only by mentally dividing the original given intuitions.[45]

Moreover, (c) Kant's view in the Transcendental Deduction (and Analytic of Principles) is that the things that intuition-elements put before the mind function as determinate spatial parts and determinate general properties only through the categorial structuring of the intuition-elements and what the intuition-elements put before the mind. But then consider the fact that spatial *parts* or general *properties* (even if only in potential forms) are put before the mind by intuition-elements. Why should that fact not be a consequence of the categorial structuring rather than (as suggested above) a consequence merely of what is given in intuition?

Furthermore, (d) Kant claims that spatial parts always contain further, smaller spatial parts.[46] But then it is not clear how an intuition-element, as it is given, can put before the mind a single, potentially definite spatial part. In the light of this Kantian claim, such an intuition-element, as it is given, would apparently put before the mind a manifold of (potentially definite) spatial parts as constituting one overall (potentially definite) spatial part. Yet such a result seems to conflict with at least the spirit of Kant's view, noted in Section 3, that intuition-elements, as they are given, do not function together to represent single objects. And, since all spatial parts contain subparts, this conflict cannot be resolved by supposing that given intuition-elements put before the mind simple, noncomposite spatial parts. Finally (e) questions like (d) and (c) (and aspects of (a) and (b)) are related to issues about infinite regresses and form and matter.

Up to a point some of the above questions can, I think, be resolved. Thus (c) raises well-known Kantian issues. However, for our present purposes we can bypass such issues and observe only that, for Kant, the specific spatial and qualitative character of the object known derives from the nature of what is given via intuition.[47] And the texts make it plausible that that nature should involve the above sort of potentiality: namely, the potentiality of what is put before the mind by intuition-elements to function as determinate spatial parts or determinate general properties of the object.

Again, consider (a). Suppose that matters for spatial parts and matters for concepts, considered as entities, are indeed inseparable in the way sketched above. Then, given various logical and exegetical points, it certainly seems to follow that matters for spatial parts and matters for concepts in their occurrence in the mind do stand in relations (relations of course involving potentialities) to other matters for concepts or matters for spatial parts. Nevertheless, this fact might be held to pose no fundamental problem.[48] Kant might claim that, even though these relations hold among the individual matters for spatial parts and matters for concepts, nevertheless in putting these matters before the mind for the operations of the mind's activities of thought, the intuition-elements do not thereby put before the mind, in such a way that it is also made available for those activities, the fact that these matters stand in such relations.

However, despite the existence of these sorts of answers to (c) and (a), many of the preceding questions seem unanswerable in any way that manages to acknowledge all of Kant's views about intuition and the manifold. Thus – and to postpone the issues under (e) until Section 5 – consider (b). All of the problems there observed certainly exist about the idea that we take in individual properties and spatial parts of an object separately (even if only in potential forms). Yet it seems impossible to interpret Kant in such a way as to eliminate that idea while also respecting basic Kantian claims that are important for the Transcendental Deduction.[49] Thus we have already observed that the idea in question is supported by a natural reading of A-Deduction A99ff. and by A166–67/B208–209 (and also by the A120 'separately and singly' point cited above). Here we may observe, as well, the Transcendental Deduction emphasis on the claims that (knowable) combination cannot be given (B130) and that the unity of the manifold of space and of individual objects in space is due to synthesis (B160–62).[50] Given Kant's picture of knowledge, these claims seem to lead to the result that individual spatial objects and individual parts of space are yielded only by the syntheses of subparts that those objects and parts of space contain (here note also his views as observed in (d) above). Yet, for Kant, such subparts (at least in potential forms) will themselves be put before the mind by (or will themselves be) representations. After all, it is with regard to such representations (or with regard to such representation-presented things) that he takes synthesis to operate. Hence when we know individual spatial objects or parts of space as such, the individual spatial parts of those objects are put before the mind separately by individual elements of the manifold (and similarly for the individual properties of the objects). This same conclusion is suggested also by various other texts.[51] So it seems that, to the extent one tries to respect the claims of such texts, one cannot interpret Kant's views about the manifold while also eliminating that conclusion.[52]

Since the conclusion in question creates the problems noted under (b), I think it likely that those problems really cannot be answered in a way that respects all of Kant's views about the manifold. Similar reasoning suggests that the questions raised under (d) also cannot be answered in such a way. It is worth noticing that the fact that Kant tends to run together elements that put before the mind matters for spatial parts and elements that put before the mind matters for concepts may at least partly explain why he fails to consider explicitly questions like those raised under (b) and (d). After all, the question under (d) – and the last question under (b) – concerns matters for spatial parts rather than matters for concepts. And analogues of these question for the case of intuition-elements that put before the mind matters for concepts either do not arise or else seem to have Kantian answers. (Thus the analogue, for the matters-for-concepts case, of the last question discussed under (b) does not arise. Kant never implies that each general property – or matter for such – is invariably given as a single property, in such a way that our awareness of subproperties contained as parts of the original given property is then reached only by our mentally analyzing the original property. And the analogue, for the matters-for-concepts case, of the question discussed under (d) seems answerable in a Kantian way, as our discussion at the end of Section 3 shows.)

Numerous texts show that Kant tends to run together the two sorts of elements of the manifold that I have noted. As suggested above, various of these texts also support parts of our account of elements that put before the mind matters for spatial parts. It is not, of course, that Kant is unaware of the differences, as such, between general properties (or concepts) and spatial parts. In the well-known Metaphysical-Exposition-of-Space arguments at A24–25/B39 and B39–40, he points out differences between space (and parts of space) and general concepts. And in his later remarks in B201–202 note on the different types of combination (or synthesis), he clearly distinguishes between the 'composition' of the manifold, where the relevant constituents are spatial, homogeneous parts, and the 'connection' of the manifold, where the relevant constituents are such conceptual or concept-like elements as substance and accident. But despite these facts, usually when he discusses the manifold and its synthesis he lumps together spatial-parts and conceptual elements, and he does nothing to resolve the sorts of problems that we have just noted.

Thus Kant's important A77/B102ff. discussion in the Metaphysical Deduction begins with the synthesis of the pure manifold of space and time (and so presumably with a synthesis of spatial – and of temporal – parts). But, without any indication that we are now dealing with a new sort of element of the manifold, the discussion then moves at A78/B103 to noting that the synthesis in question must be brought 'to concepts.' Again, A102 on drawing a line in thought obviously concerns my holding together various imagined (or thought) parts of the line. And neither here nor in his later discussions of related matters does Kant make it immensely clear how concepts (or *which specific concepts*) enter into this synthetic process – and indeed enter in such a way that they are also held together, judgmentally, within my mind so that I can know the line.[53]

Similarly, when Kant discusses at A105 my coming to know a triangle, he in effect notes that I proceed by conceptually *thinking* a combination or synthesis of *three straight lines*; yet these lines, which are spatial parts of the triangle, he later describes as 'the *predicates*... of a triangle' (ibid., my italics). Moreover, in the "Metaphysik Volckmann" notes quoted in Section 3 he is reported as having proceeded in a like manner, for he there introduces a *solid house*, *a wall enclosing an empty space*, and so on, as *predicates* that I must take to belong to the object known. And in Reflexion 2282 he says that a mark of a thing may be a *concept* of a *part* of a thing as well as a concept of a whole thing; but he goes on to say that *the hand itself* is a mark of the man, thus effectively confusing the issue of whether marks (which his theory requires to be conceptual) may not also simply *be* spatial parts.[54]

The above texts thus support our account of the spatial-parts manifold. But such texts also show that Kant fails to make any sharp, clear distinction, at least in his usual discussions of synthesis, between the elements that put before the mind (or are) matters for concepts and elements that put before the mind (or are) matters for spatial parts. We cannot follow him in this practice, however, just because these elements require distinction. Hence, below, I will continue to distinguish such elements. Because it is accepted by Kant in the Deduction, I also will proceed in terms of the idea that such elements separately put before the mind individual matters for spatial parts or individual matters for concepts. However, in doing so I will not try to resolve (more than I have above) the questions that this idea raises. As we have noted, most of those questions seem insoluble to the extent that one tries to acknowledge all that Kant says about the manifold and its elements. And while one could abandon this idea, when one pursues Kant's own reasoning in the Deduction it is simplest to follow his Deduction view of the manifold. Moreover, it is especially desirable to proceed in this way simply because the fundamental argument of the Deduction is the inference from the subjection of the manifold to unity of apperception to the categorial structuring of the elements of the manifold. And, as our later discussion in effect shows, the main reasoning in this argument does not really turn on the details of the above idea.

Once Kant has available a notion of the manifold as including elements that put before the mind (or are) matters for spatial parts, he can answer the problems that we developed at the beginning of our present discussion. Thus he in effect takes us to know something *of* or *about* some particular, individuated (outer) object as such, just in case, roughly, besides the other factors that are involved in this knowledge, the object in question is displayed in perception – and therefore in intuition – as occupying a definite, particular location in space. Such a display involves elements of the manifold that put before the mind spatial parts of the object occupying various sublocations of the overall location in space. So the introduction of the above sort of elements allows Kant to claim that our knowledge of (outer) objects is *not* a purely general knowledge but concerns particular, individuated objects as such.

Suppose, again, that, as we have suggested, elements of the manifold put before the mind spatial parts of objects occupying definite locations in space. Then Kant can also explain how our perceptual knowledge takes the direct-object form that it does. As is shown by numerous examples (for instance, those of drawing a line in thought, of perceiving a house, or of delineating the figure of a four-footed animal), he takes the above sort of elements to put before the mind, in a *direct-object fashion*, spatial parts of objects occurring at locations in space.[55] The result of the synthesis of these elements will therefore be a single, unified intuition that represents, in a direct-object fashion, its object as occurring at a spatial location. And thus via the synthesis of such elements we will get, in sense perception, the required sort of direct-object confrontation with the object.

In closing, I should note that our results about Kant's treatment of the manifold clarify the two Kantian claims with which we opened Section 3 of this chapter. The first claim was that our knowledge of objects always takes a certain sequential, synthesis-requiring form. The second was that the object that we know is first given to the mind as a sequentially appearing, synthesis-requiring manifold of data. Both of these claims can now be made somewhat more precise.

As can be seen by combining the Section 3 statement of these claims with our above results, the first claim amounts to the assertion that, in general, we can know an object only through a manifold-of-intuition-mediated survey, one by one, of the various properties and spatial parts that belong to that object. The second claim amounts to the position that the object of knowledge is *itself* given to us in the form of a sequentially appearing collection of general properties and spatial parts, a collection that our mind has to hold together before our consciousness in order that the object should exist as the single, individuated thing that we know. We will return to the details of these claims in Chapter Eight in their proper Transcendental Deduction context.

5. PROBLEMS AND LOOSE ENDS

Although the above comments complete our basic discussion of the Kantian manifold of intuition, in order to avoid confusion – and for future reference – it is worth noting six general questions that arise about that topic, beyond the questions already discussed in Sections 3 and 4.

First, one may wonder exactly what is meant by a spatial part. This question raises various complications. However, Kant's position seems to be that the spatial parts of an object are sections or regions of the object that are bounded by surfaces, lines, or points (that is, are bounded through what he calls at A25/B39 and elsewhere the introduction of 'limitations' into the one underlying space in which the object occurs). And, ignoring further issues, I will follow Kant in this view. This decision is made the easier by the fact that the fundamental line of argument in the Deduction does not turn on the fine details of his notion of a spatial part.

Second, one may ask about the nature of the manifold of inner sense, as against that of the outer-sense manifold that we have been attending to throughout this chapter. To consider this nature, recall that Kant generally proceeds as though we have knowledge via *outer* representations only insofar as those outer representations are represented, by means of *inner* intuitions, as occurring in time. Notice also that this same idea is the one he must appeal to, given his overall account of the mind's mechanisms, in order to explain how elements of the (outer-sense) manifold succeed in sequentially putting before our mind parts and properties of the object known. Hence what we have been discussing above have not been elements of the outer-sense manifold taken absolutely neat. Rather, they have been, officially, the elements of the *outer-sense manifold as those elements appear to us in inner sense*. And so if we now ask what are the proper elements of the *inner-sense* manifold, then one answer will be: *inner-sense* representations via which such outer-sense elements *appear to* us as in time, or else *inner-sense* representations that are, as *appearances*, identical to such time-ordered outer-sense elements. And, in addition, the inner-sense manifold will also include inner-sense representations of such (for Kant) noncognitive (and themselves nonrepresentational) properties of our mind as our feelings and emotions.

To express these last points more exactly than I just have, note that, in parallel with our account of outer sense, one would expect that elements of the inner-sense manifold would arise in the mind and would by themselves (and prior to the operations of synthesis) put before the mind only potentially time-ordered outer-sense elements (and emotions, feelings, and so on). Or else (and in accordance with Kant's appearance theory) elements of the inner-sense manifold would arise in the mind; and as they occur in the mind, those elements would be identical to such outer-sense elements occurring in only a potentially definite time order. Despite various complications and obscurities, this view of inner sense is supported by the texts. Moreover the texts indicate that the actual, definite time order of outer-sense elements arises only through synthetic operations of the mind. In particular, Kant holds that the potentially time-ordered outer-sense elements (and emotions, feelings, and so on) are acted on by the mind's thought-related faculty of imagination (and by other synthetic operations) in such a way as to be reproduced before the mind's acts of thought in an actual, definite time order. And that actual, definite time order is the time order of representations that we are aware of when we say, for example, that first we see the top of a tree (first we have the outer-sense element that puts that top before our mind), then we experience a feeling of delight (then we have that feeling), and finally we see the trunk of the tree (finally we have the outer-sense element that puts the trunk before our mind).[56]

The preceding interpretation of Kant's view of inner sense enters into obscure and controversial areas, for some writers see various texts – and the demands of an intelligible account of synthesis – as implying that inner sense *by itself* presents entities in an actual, definite time order. As I have just suggested, my own view is that the texts rule out such an interpretation (as indeed does Kant's basic position that a sense like inner sense cannot by itself present entities as standing related together in definite, determinate ways).[57] But for our purposes in examining the Transcendental Deduction, the disagreement between this latter interpretation and my own view is not of great importance. As we have noted, the Deduction starts with the minimum assumption that an object (but not an object that we take to be category-subsumed) is known through an arbitrary given sensible intuition. The Deduction then argues that a special, category-involving synthesis by the understanding is required because the elements of that intuition are subject to unity of apperception. This argument will proceed in the same basic way whether we take the elements initially to occur before the mind in a potentially definite time order (a time order that then requires imagination and other synthetic operations in order to become actual) or we take those elements initially to occur before the mind in an actually definite time order.[58]

Since the disagreement between my interpretation of inner sense and the alternative interpretation that I have just noted is, for present purposes, not of great significance – and since I think the texts support my interpretation – in the remainder of this book I ignore the alternative interpretation. Thus I suppose simply that inner-sense elements, by themselves, initially put before the mind outer-sense elements as occurring only in a potential time order. I also ignore further details of Kant's account of inner sense, for there is no need to burden our discussion with such details unless they are directly relevant to the argument of the Deduction.

Third, a thorny group of questions arises from Kant's basic idea that the elements of the manifold put before the mind, or simply are, *matters* for concepts or spatial parts – matters to which the mind, through synthesis, assigns a *form*. The nature of these questions can be indicated by noting that Kant intends his matter-form terminology very seriously, and in at least roughly the Aristotelian sense. Thus it ought to be impossible to speak of such *matters* as existing and characterizable in independence of the *forms* that belong to them. Form and matter as the Aristotelian conceives of them are, after all, inseparable. Thus while one can by an act of abstraction speak of, say, the prime matter (in one Aristotelian sense) that is contained in a gold coin as being that that underlies and persists through substantial change, one cannot regard this matter as able to exist in independence of any form, and one cannot characterize it except as formed in some way (for example, as being the stuff in the gold coin). But then similarly it would seem that we should not be able to regard the elements of the manifold – or what they put before the mind – as able to exist in independence of the informing activities of synthesis. And we should not be able to characterize those elements, *as so existing*, as putting before the mind or being matters for concepts or matters for spatial parts.

A problem therefore arises for our preceding discussion of Kant. In fact such a problem arises on several levels. (Thus for Kant the categories themselves are roughly forms that are imposed, in synthesis, on the already space- and time-formed manifold.) To a great extent most such problems will have to be ignored in this book. I also will have to ignore almost all of the obvious connections between such problems and the questions raised above in Section 4. It should be emphasized here, however, that these problems are largely a reflection of tensions in Kant's own handling of the form-matter contrast. For example, on the one hand Kant speaks, throughout his career, of the process of assigning a form to a matter for concepts through abstraction as simply being a process of focusing attention on a general property that does *not* exist in independence of the objects that possess it. And one should note that our above reflections (under question (a) of Section 4) on the relations of matters for concepts and matters for spatial parts clearly show that it would be very difficult at least for each of those types of matter really to exist in independence of the other. On the other hand, Kant insists in both the A- and B- texts of the first *Critique* (A90/B122, A90–91/B123, B145) that intuitions are given to us *independently* of the functions of thought, and he is of course famous for his general, and sharp, distinction between our sensibility and our understanding.[59]

The first of these sets of Kantian points would seem to favor, at least indirectly, an Aristotelian view of the manifold of intuition as inseparable from the form that synthesis gives to that manifold through the activities of understanding and thus of thought. But the second of these points clearly suggests that the elements of the manifold exist, and are characterizable, in independence of those activities. Problems like the one that we have noted above for our own interpretation hence arise directly within Kant's own work. And the difficulty that our interpretation faces seems simply to reflect this tension in his work between these two views of the manifold. After all, that difficulty arises precisely because we have, following Kant's views, described the elements of the manifold as putting before the mind (or being) *matters* that require a *forming* in synthesis. Yet at the same time we have accepted his sharp separation of intuition from thought. Following Kant, we have supposed that those elements can be given to us – and so can exist – independently of the synthetic (or other) functions of thought. And that supposition has led us to think that it must be possible to describe the nature of the elements of the manifold – or of what they put before the mind – insofar as they are given independently of such functions.

I therefore do not think that, from a purely exegetic standpoint, our above characterization of the elements of the manifold is to be rejected, although of course in giving it I have ignored a great many complications that attend Kant's views on form and matter. In any case, in Chapter Ten we will return to the parts of those views that are most relevant to the Transcendental Deduction – in particular to the idea that in synthesis the categories function as forms that are imposed on the manifold. Moreover, it should be noted that the notion of the manifold of intuition and its elements is meant to do work within Kant's system. And so even on the Aristotelian form-matter approach to that manifold one ought to be able to say at least that these elements put before the mind (or are) matter disposed to take on – and matter that actually has taken on – the form of being a determinate general property or spatial part. One should be able to speak in this way just as much as one can speak, within an Aristotelian approach, of the stuff that is disposed to take on – and that actually has taken on – the forms of being gold and a coin. But such a way of speaking is in many respects very close to the way of speaking that we have adopted above. It only drops a literal reading of Kant's central idea that intuitions are given to us in independence of the synthesizing functions of thought (and so are given to us as literally existing in independence of – and thus presumably as characterizable in independence of – those functions). I myself think it clear that Kant intends, and that we must therefore accept, a literal reading of that idea. But, as thought will show, much of the interpretation of the Deduction that I present below can be reformulated in terms of this alternative way of speaking. So I will not pursue the form-matter issues further here.[60]

The next – and fourth – question that we need to examine bears some relation to the above form-matter issues but I will not consider it in those terms now. It is a question about a familiar regress that seemingly afflicts Kant's treatment of the spatial manifold. This regress arises as follows. As we have seen, each intuition involves a set of elements each of which puts before the mind (or is) a (potentially) determinate spatial part of the object that the entire intuition represents. It would seem that, as putting before the mind a single, individuated part of the original object, each of these elements is itself an intuition and so contains a manifold of similar elements, to each of which the same sort of reasoning applies again. We thus reach an infinite regress of spatial manifolds of intuition. And from this regress difficulties arise at once.

Given Kant's theory of synthesis, after all, any intuition via which we know must be synthesized by our mind. But consider now the original intuition that we have been discussing above. It looks as though that intuition cannot be synthesized without the simultaneous or prior synthesis (in temporal order or in the order of logic) of the manifolds that belong to the various elements of the manifold of *that* intuition – and so on through each of the manifolds that occurs at some point in the above regress of manifolds. Yet suppose, on the one hand, that the original intuition must be synthesized with the simultaneous synthesis of all these other manifolds. Then in knowing the object of the original intuition we must simultaneously know, through the synthesis of the manifolds of all the subintuitions that are here involved, *all* of the object's infinite number of spatial parts in *all* of their spatial details. And such a strange result is contrary both to plain fact and to Kant's views. Yet suppose, on the other hand, that the original intuition's synthesis requires the prior synthesis of all the other manifolds. Then the above regress becomes vicious, for in order to know via the original intuition we must *already* have synthesized the manifolds of each of the elements of *its* manifold. And this situation goes on to infinity in such a way that we never arrive at the point of actually having synthesized, and so of actually having knowledge via, the original intuition.

Because concepts do not contain an infinite number of representations within themselves (B40), the above regress does not arise for Kant's treatment of the matter-for-concepts elements of the manifold. Together with Kant's general refusal, in discussing synthesis, to distinguish between the matters-for-spatial-parts and the matters-for-concepts elements, this fact may explain why Kant himself never considers these problems. The problems are nevertheless real, are obvious in one form or another even to beginning students of Kant, and cannot be shunted aside. Their full discussion would be extremely complex, for they touch on difficult issues, which are not here our concern, about the overall understanding of Kant's picture of knowledge.

Without trying to evaluate the plausibility of every step of preceding regress, I will simply note that the immediate difficulties seem to be removed if we adapt to our own purposes a reconstructive idea of Parsons' – namely, the idea of distinguishing, as Kant does not, between explicit and implicit intuitions.[61] Very roughly, an explicit intuition would be any intuition that is given as having a manifold of elements that put before the mind (or are) various spatial parts of the object of that intuition. An implicit intuition would be an intuition that is given as representing a spatial object (or a spatial part of an object) without putting before our mind any further, specific spatial subparts of that object (or of that spatial part). (Thus the usual explicit intuition would represent an object that together with various of its parts is at the center of our attention. The usual implicit intuition would represent a part of *that* object – or another object or object-part that falls within our field of perception – to whose details we are not attending.) Given this explicit-implicit distinction, the above problems disappear, for there is now no reason to suppose that every element of an explicit intuition must itself be explicit. The regress is stopped before it starts.

While the explicit-implicit distinction eliminates the immediate difficulties that we have just observed, the use of that distinction creates complications of its own. Thus as I am here understanding it, an implicit intuition, as it is given, contains no manifold of elements, contrary to Kant's A99 claim that each intuition contains a manifold.[62] Moreover, while containing no manifold, each implicit intuition nevertheless represents a whole spatial object or part of an object. And that fact infringes on the A162/B203ff. discussion of appearances as extensive magnitudes, 'the representation of [whose] parts makes possible, and therefore necessarily precedes, the representation of the whole.' Furthermore, the explicit-implicit distinction makes better sense within Kant's appearing than within his appearance theory. (On the appearance theory the spatial object that is known cannot straightforwardly be identified with an implicit intuition without being taken, insofar as it is so identified, to lack spatial parts. Yet Kant takes no spatial object to lack spatial parts.) Since Kant makes the claims just noted at A99 and A162/B203ff. – and since it is desirable to preserve the possibility, throughout our discussion of the Deduction, of treating him both as an appearing and as an appearance theorist – I will not adopt this explicit-implicit approach to the above regress. But just because that approach allows one to eliminate the regress, if one is willing to modify Kant's views, it is worth bearing in mind below. It also is worth bearing in mind simply because main parts of the Deduction turn out to be independent of the issues that give rise to the regress.[63] So (although I will not attempt this here) one could modify Kant as necessary in order to adopt that approach and still accept the general interpretation of the Deduction that I offer in this book.

A fifth question about the manifold of intuition arises as follows. As we saw at the beginning of Section 4, Kant's idea that we know objects through a manifold of matters for concepts raises problems about how our knowledge can concern particular individuated objects. It also raises the problem of how our knowledge can get a direct-object-style confrontation with those objects in sense perception. These problems were supposed to be solved, in the way sketched later in Section 4, through the idea that the spatial manifold of intuition represents to us, in a direct-object fashion, spatial parts of objects as occupying definite, particular locations in space. But even if that idea is granted, a further problem now arises, just because of the fact that our knowledge – and thus our knowledge through such a manifold – always requires the operation of our understanding as well as of our sensibility.

This fact creates a problem for the following obvious reason. Our understanding is a purely discursive faculty entirely separate from our sensibility. As such, it is a faculty that operates simply with general concepts. However, if our understanding operates with, and so presumably mentally grasps, *simply* such concepts, then how can our understanding conceivably grasp, as such, *single*, *individuated objects* or the spatial parts of such objects as those objects are represented by our sensible intuitions? Yet if our understanding cannot grasp such objects (and so cannot grasp them in a direct-object fashion), then despite Kant's introduction of the spatial manifold of intuition, our knowledge really *cannot* concern such objects and really *cannot* relate to them in any direct-object manner. Hence Kant's theory faces serious problems of just the sorts that we raised at the start of Section 4. Indeed, the problems go further than we have so far indicated. If our understanding grasps simply general concepts, how can our understanding ever attend to (actually not-yet-general) matters for concepts in such a way that it can assign them their form of generality? And so how can our understanding ever arrive at empirical concepts that it can use to acquire any empirical knowledge at all?[64]

Problems of the above sort pose a severe difficulty for Kant's picture of knowledge on his own understanding of that picture. Nevertheless one might suppose that Kant could escape such problems simply by weakening the understanding-sensibility contrast. Specifically, this contrast could be weakened to the extent of allowing our understanding to grasp intuited spatial locations and the intuited single, individuated objects and object-parts that occupy those locations. (Thus, despite his official theory of our understanding, Kant proceeds throughout the first *Critique* as though our understanding can do such things.) I think that no harm results if, in studying the Deduction, one allows the understanding to proceed in such a way. So we may ignore these further, and deeper, difficulties and assume that, by some such means, Kant manages to escape them.

The final, sixth, question about Kant's treatment of the manifold of intuition concerns the question how he should characterize that manifold from the point of view of the minimum assumption noted above in Section 1 and in Chapter Two. As we saw in Chapter Two, in order to avoid question-begging, the Deduction must start with only the assumption that via an arbitrary given sensible intuition an object is known, but not an object that is assumed to be subject to the categories. Yet of course this arbitrary given intuition must be given in the form of a manifold. So we need to know how the elements of this manifold should be described.

In answering this question, we should for the present follow Kant's basic position that the elements of any manifold through which we know are presented as occurring in a time order. And, accepting Kantian theoretical views in a way that is not question-begging given the goals of the Deduction, we should regard as implicitly attached to this position the claim that this presentation happens through inner sense. But, for reasons indicated in our discussion of inner sense, we will suppose that the time order in which the elements of the manifold are initially presented is, as far as it is due to inner sense operating alone, only a potential one.

Moreover, in making this last supposition, we should note a point that concerns both Kant's theoretical account of our knowledge of a genuine spatiotemporal object like a tree and the minimum assumption. That point is that neither on that theoretical account nor on the minimum assumption need one assume that the elements of the manifold, insofar as they yield knowledge of an object, have to be those elements regarded *merely* as they are initially presented through inner sense in a potential time order. In fact, on the theoretical account our knowledge of a spatiotemporal object involves the functioning of those elements together, before thought-consciousness, as elements of a single, unified intuition. That functioning itself – and, further, the subjection of those elements to unity of apperception – involves an operation of imagination (and of other factors in synthesis). And that operation makes those elements occur before our thought-consciousness in the sort of actual time order (of, say, our perceptions of various properties and spatial parts of an object) that was noted above. Again, on the minimum assumption there is nothing to stop us, if necessary, from building into the claim that an object (although not necessarily a category-subsumed object) is known, the further view that the elements of the manifold, insofar as they yield that knowledge, occur before the mind in an actual time order (of course through imagination and other synthetic operations). Indeed, as we will see in Chapter Five, there are strong reasons to build this idea into that claim.

Given these points, we can now turn to the specific question of how to characterize the manifold of intuition from the point of view of the minimum assumption. This question is complicated by the fact that, as we see in Chapter Eight, Kant's argument in the Deduction ultimately takes different forms depending on whether the minimum Deduction assumption is given a strong reading, as supposing that (i) a single object is known through (and is itself distinct from) the elements of the manifold, or the minimum Deduction assumption is given a weak reading, as supposing merely that (ii) the object that is known through the elements of the manifold may amount (as this assumption by itself goes) to no more than those elements themselves, as those elements are presented to the mind.[65] This question is also complicated by a restriction that, as we see in Chapter Four, Kant imposes in the B-Deduction on the arbitrary given intuition. We may, however, for the present ignore this restriction. And, if we do so, then we can see, in a preliminary way, that, depending on whether the minimum assumption is given the strong or the weak reading, two different characterizations of the elements of the manifold should be offered.

In particular, if Kant gives the minimum assumption the strong reading, then at the start of the Deduction he should assume no more about the relevant elements than that, as those elements are presented to the mind, they individually put before the mind (or are identical to) – in what is not assumed to be a category-involving way – separate, unspecified 'features and aspects' (or separate, unspecified general properties) of the object known. Any more specific characterization of those elements – at least any that is of real interest – will, it seems, bring in category-involving characterizations of the object. (Thus those elements must not be taken to put before the mind any features that make the object known automatically a substance or an extended magnitude.)

Again, if Kant gives the minimum assumption the weak reading, then he will ultimately have to show in the Deduction that there is a single object that is known through (and that is itself distinct from) the elements of the manifold as they are presented to the mind. So, in order to avoid question-begging, he must not assume at the start of the Deduction that those elements put before the mind (or are identical to) any features and aspects of such an object. And indeed it is simplest (and the Deduction argument based on the weak reading becomes most general) if, at the start of the Deduction, Kant assumes about those elements only that they are presented to the mind; and hence he makes no claim that those elements, as they are so presented, put any features, aspects, or general properties (or anything else at all) before the mind.

From the above discussion one can see that, as far as the minimum Deduction assumption itself goes, Kant should ignore the results of our previous discussion of the manifold of intuition as putting before the mind general properties and spatial parts of the object known. (After all, we arrived at those results by considering Kant's views about intuitions that represent spatiotemporal – and category-subsumed – objects like trees.) But this fact about what Kant should *assume* at the start of the Deduction is of course compatible with the Deduction's showing more specific points, later, about the elements of the arbitrary sensible intuition that the minimum assumption introduces. In particular, this fact is compatible with the Deduction's showing that those elements put before the mind general properties and spatial parts of a single, category-subsumed object. And, if it succeeds, the argument of the Deduction will allow such a point to be shown.[66]

In our own discussion, we can of course appeal, as is necessary, to our results about the manifold of intuition whose elements put before the mind general properties and spatial parts of the object. And, where it is required, we will proceed in terms of the manifold considered from the standpoint of the minimum Deduction asssumption. However, except for passing comments in Chapter Six, until Chapter Eight we will not need to discuss again either the minimum assumption or the manifold of intuition considered from its standpoint. So in intervening chapters I will often speak simply of the minimum assumption as claiming that via an arbitrary given sensible intuition an object is known, but not an object that is assumed to be subject to the categories.

6. CONCLUSIONS

In this chapter we considered a large body of material that concerns Kant's views on the manifold of intuition and hence, ultimately, the Transcendental Deduction and its theory of synthesis. This required the study both of Kant's account of concepts and of his views about the spatial parts of outer objects. We saw that empirical concepts, the ones that primarily concern us here, have a two-part nature. As partial concepts, they occur *in* objects, in the form of potentially general properties of those objects or of what I called matters for concepts. Through an act of thought directed to such a potentially general property, our understanding assigns it a form of generality and so takes it to be able to belong to many objects. Proceeding in this way, our understanding makes the property a mark or representation, and hence a ground of knowledge, of those objects. Kant's account allows us to take the general concept, as so arrived at, either to be a representation that presents the general property or else to be that general property itself. And both these views of concepts are compatible with both Kant's appearing and his appearance theory.

Kant's view of the spatial parts of objects was roughly that each object is represented as having spatial parts that occur at definite spatial locations that make up the overall spatial location of the object. This view of spatial parts can itself be rendered compatible both with his appearing and with his appearance theory.

Coming now to Kant's specific claims concerning the manifold of intuition – and focusing just on the case of intuitions that represent genuine spatiotemporal objects like trees – we noted that these claims amount to a two-part thesis. First, our knowledge of any single, individuated object always occurs through our attention, one by one, to various of the features and aspects that belong to this object. And this process of attention requires our mind to hold together, or to synthesize, the intuition-elements of the manifold that yields us this awareness of the object. Second, the object itself first occurs before our mind, through the manifold of such elements, in the form of a disconnected set of features and aspects. And our mind must synthesize this set in order that there should actually exist the single, individuated phenomenal object that we know. Through attention both to Kant's account of the synthesis of concepts in judgments and to his views about the spatial parts of objects, we concluded that the intuition-elements in question must be of two sorts: first, elements that put before the mind properties (or matters for concepts) that belong to, and occur in, the objects known; and, second, elements that put before the mind spatial parts (or matters for spatial parts) that belong to those objects. The resulting conception of the manifold of intuition we saw to be expressible in terms both of Kant's appearing theory and of his appearance theory. And we saw that Kant himself, though clearly accepting both sorts of elements, does not clearly distinguish them or indicate their exact relations to each other.

Finally, we noted a variety of further questions that affect our understanding of the manifold. These questions concerned the nature of a spatial part; the nature of the manifold of inner intuitions; Kant's form-matter distinction; a familiar regress that arises in connection with the spatial manifold; problems about how our knowledge can involve a direct-object grasp of single indivduated objects; and, lastly, the question of how Kant should characterize the manifold of intuition from the standpoint of the minimum Deduction assumption that, through an arbitrary given sensible intuition, an object is known – but not an object that we can immediately take to be subject to the categories.

A great deal more could be said about all of these issues concerning the manifold of intuition. But we have assembled enough information about that difficult notion. We are at last ready to turn directly to the study of the Transcendental Deduction itself.

CHAPTER FOUR

THE TRANSCENDENTAL DEDUCTION: ITS STRUCTURE, GOALS, AND OPENING CLAIMS

1. INTRODUCTION

We have now seen Kant's basic picture of knowledge, with its idea that we know single, individuated objects via intuitions and concepts and with its transcendental idealism – its claim that the objects that we know are mere mind-dependent, spatiotemporal things, either objects as those objects appear via our intuitions or else appearances identical to those (synthesized) intuitions themselves. We have also seen, and have just been summing up, Kant's position about the manifold of intuition. In the case of an intuition that represents a spatiotemporal object like a tree, Kant holds that that intuition is given to us in the form of a manifold that puts before our mind (or is identical to) properties and spatial parts of the object; and that object first occurs before our mind in the form of a manifold of such properties and spatial parts. However, we have observed that, to avoid question-begging, the Transcendental Deduction should at its start make only the minimum assumption that via an arbitrary sensible intuition an object is known, but not an object that is assumed to be subject to the categories.

As we noted earlier, in the Transcendental Deduction Kant wishes to demonstrate that the categories of the understanding apply with necessity and strict universality to all the objects that we can know, those objects that are the above sort of mind-dependent, spatiotemporal things.[1] Demonstrating this fact will demonstrate what he calls the objective validity of the categories with respect to that group of objects. Moreover, the nature of the demonstration will also establish that the objective validity of the categories, as far as it is our cognitive concern, is restricted to such objects. And demonstrating this fact will in addition establish or imply numerous other points of philosophical interest, for example, the anti-Humean point that certain necessary connections hold among the distinct existences that constitute the elements of the manifold of intuition.

Kant's arguments for the objective validity of the categories, and for these further claims, involve a mass of details about concepts, judgments, the manifold of intuition, apperception, and synthesis. We have already considered some of these details in earlier chapters, and now we will begin to show their relevance, and the relevance of still further details, to the arguments in question. We should recall immediately, however, that, as we will see in detail below, Kant considers the main argument in the Transcendental Deduction – the proof of the objective validity of the categories – to have the overall structure of a 'proof from the possibility of experience.'

That is, this argument is a deductive argument that starts from the assumption – which should be made in a minimum form – that via a given, arbitrary sensible intuition a being like us has empirical knowledge ('experience,' in one Kantian sense of that term). The argument notes various points about the cognitive capacities and operations of such a being, including the necessary subjection, to what Kant calls unity of apperception, of the sensible intuition. Given the assumption in question and those points, the argument infers the judgmental and ultimately the categorial structuring of the object that is known via that intuition. However, the intuition and hence the knowledge in question are arbitrarily selected. So Kant takes it to follow, as we see in detail in Chapter Six, that any object that a being like us does or can know through a sensible intuition is necessarily subject to the categories. He then applies this result in such a way as to infer, specifically, category application to the spatiotemporal objects that we human beings can know. And he thus takes the main argument of the Deduction to demonstrate the objective validity of the categories with respect to just the class of objects for which that objective validity is supposed to hold.

For reasons that I have already emphasized, this proof from the possibility of experience must of course employ some version of the above minimum assumption. We will consider this fact (and the exact forms that this proof takes depending on what version Kant uses) below and in later chapters, along with other details of the proof. It is worth noting at once, however, that the overall structure of Kant's proof suggests a method of establishing category application that is independent of various Kantian details and likely to be of philosophical interest in its own right. So it will be possible later to discuss the interest and success of Kantian-style arguments for category application without reference to all of Kant's own doctrines or to their difficulties.

In the present chapter, we first consider the Transcendental Deduction thesis that combination cannot be given. We will relate Kant's reasons for that thesis both to our Chapter Three remarks about the different sorts of manifold that – at least in a preliminary way – should be taken to go along with the minimum assumption and to our Chapter Three comments on the manifold of an intuition that represents a spatiotemporal object like a tree. After evaluating these reasons, we will examine Kant's own understanding of the problem of the Transcendental Deduction. We will consider his initial statement of that problem in the opening § 13 and § 14 of the Deduction (A84–94/B115–29) and the interpretation of notions like objective validity, the idea of experience as empirical knowledge, and the possibility of experience and a proof from the possibility of experience. We will also indicate the structure of the B-edition (1787) Deduction as a whole. Finally, we will note Kant's preliminary, § 14 argument (at A92–93/B125–26) for the Deduction. And thus we will be ready in Chapter Five to focus directly on the official Transcendental Deduction argument, in the B-text, for the objective validity of the categories.

2. 'COMBINATION ... CANNOT BE GIVEN' (B130)

Kant's B-Deduction § 15 thesis, quoted above, is important to the B-Deduction and in different words is featured also in the A-Deduction.[2] This thesis is important to the Deduction because, by appeal to it, Kant argues that the combination that belongs to the elements of the manifold in our knowledge is not present in those elements as they are given but is due to a synthesis-producing (and ultimately category-utilizing) act of mind. This argument applies both to the sorts of manifold of intuition taken to go along with the minimum Transcendental Deduction assumption and to the manifold of intuition taken to put before the mind matters for properties or matters for spatial parts. In order to prepare for the main argument of the Deduction, it is useful to discuss Kant's thesis and his argument for it here. To do so, we will consider, in order, his view of the given, his notion of combination, and his reasons for the combination-cannot-be-given thesis.

Philosophical theories of the given have taken two basic forms: first, psychological accounts that identify certain entities as being initially provided to the mind in a mentally unprocessed form so that the mind, by operating on those entities, may subsequently gain knowledge; and, second, epistemological accounts that identify certain entities as being entities our knowledge of which in one way or another evidentially grounds all of our further knowledge (or all of our further knowledge of some specified sort). Kant's notion of the given is a theory of the former, psychological (or, in his case, transcendental-psychological) sort, although his notion bears some very indirect relations to accounts of the latter, epistemological sort. This Kantian notion of the given we have in effect observed in earlier chapters, in our discussions of Kant's views on outer and inner sense.

Kant's basic view is that the given entities in our knowledge consist of the elements of the manifold as those elements are produced by means of our sensibility in independence of the processing operations of our understanding. In the case of our knowledge of a spatiotemporal object like a tree, we have seen that objects existing in themselves affect our outer sense, yielding elements of the outer-sense manifold that are then presented through inner sense as occurring in a potential time order. It is such outer-sense elements, as they are produced through the affection of outer sense and then as they are presented through inner sense, that count as the given entities, for Kant. Again, from the standpoint of the minimum Transcendental Deduction assumption the given entities evidently amount to (or at least include) the elements of the manifold that is introduced by this assumption, as those elements are presented through inner sense as occurring in a (potential) time order. (Once any of the preceding sorts of intuition-elements are operated on by imagination and other factors in synthesis so as to occur in an actual temporal order, those elements no longer count as given entities, for Kant.)[3]

As various texts show, including especially § 15 and later parts of the B-Deduction, the notion of combination applies to a set of entities just when the entities in that set are related together in such a way that they make up *one* thing.[4] About combination, as so understood, Kant's main point is that it cannot be given in the ways that we have explained above. That is, in the case of our knowledge of a spatiotemporal object like a tree, combination cannot belong either to given outer-intuition elements as they are produced via the affection of outer sense or to such elements as they are then presented through inner sense as occurring in a potential time order. (Nor can combination belong to object-features and aspects as they are put before the mind by such outer-intuition elements.) Again, in the case of the given regarded from the standpoint of the minimum Deduction assumption, combination cannot belong to a group of mental entities, initially presented before the mind via inner sense as occurring in a potential time order, through which an object (but not an object assumed to fall under the categories) is known. (Once intuition-elements are operated on by imagination and other factors in synthesis so as to occur before the mind in an actual temporal order, those elements then make up one temporal sequence and hence should be taken by Kant to form a combination.)[5]

We can best see Kant's reasons for these points in the case of the given manifold that puts before the mind properties and parts of a spatiotemporal object. As we proceed, we can note the related reasons that support such points in the case of the given regarded in terms of the minimum Deduction assumption. We also can note that Kant's reasons in the first case relate to the two basic Chapter Three claims that he makes in connection with the property-and-spatial-parts manifold. (Those claims were that, first, we human beings can know an object only through a sequential, manifold-of-intuition-mediated survey of properties and spatial parts that belong to that object, and, second, the object that we know is itself presented to us in the form of a sequentially appearing manifold of properties and spatial parts.)

In the case of the property-and-spatial-parts manifold, Kant's reasons for holding that combination cannot be given can be interpreted as follows. First, Kant will hold that (a) necessarily, combination does not belong to outer-sense elements as they are initially produced through the affection of outer sense. Second, he will offer a variety of grounds for supposing that (b) necessarily, combination does not belong to such outer-sense elements as they are then put before our mind through inner-sense elements. (And he will make a similar point about object-features and aspects as they are put before our mind by such outer-sense elements.) Through such grounds, Kant will have shown (b); and having established (a) and (b), he will have demonstrated that, in the case of the property-and-spatial-parts manifold, combination cannot be given. Furthermore, grounds resembling those for (b) also can be used by Kant to show that combination cannot be given in the case of the given regarded in terms of the minimum Deduction assumption.

Kant's reasons for (a) seem essentially to be that outer-intuition elements, as they are initially produced through the affection of outer sense, occur in the mind as unrelated entities in the same sort of way that (as we see in connection with (b)) such elements occur when they are put before the mind through inner-sense elements. (And, because he is interested precisely in forms of combination knowable or recognizable by us, he could in any case argue that outer-intuition elements, as they are initially produced through outer sense, have an existence in themselves and so cannot form a knowable or recognizable combination.) Hence in the case of the property-and-spatial-parts manifold, everything in Kant's reasons for the combination-cannot-be-given thesis turns on what grounds he can offer for (b). And, in the case of the given regarded in terms of the minimum Deduction assumption, everything evidently will turn on what grounds he can offer that resemble his grounds for (b). There are, I think, three main grounds that Kant has for (b) (and then three main resembling grounds that he has in the case of the given regarded in terms of the minimum Deduction assumption.)[6] Moreover, these grounds all relate to the two Chapter Three claims made in connection with the property-and-spatial-parts manifold.

Kant's three grounds for (b) are the following. (I) As explained in Chapter Three, Kant holds that we human beings can know an object only in the sequential way that is there indicated. (And he holds that the object can be given to us only in the way that is there discussed.) Moreover, he also holds that, necessarily, the individual stages of our sequential survey of the object, and the intuition-elements that mediate these stages, are atomic, isolated entities. That is, and necessarily, these individual stages and intuition-elements never by themselves recognize or put before the mind any relations that object-properties or spatial parts may bear to one another. And the individual elements themselves, as they occur in a potential time order through inner sense prior to the conceptual operations of the mind – and prior to the operations of imagination and other factors in synthesis that make that potential order an actual one – never stand in relations to one another. From these various observations, however, (b) follows. (And the similar point that concerns object-features and aspects also follows, given by Kant's idealism that the object known, with its properties and spatial parts, *is* the object that appears through, or that is identical to, the – synthesized – manifold of intuition.)

Furthermore, in the light of these observations, it is clear that Kant should argue in a resembling fashion in connection with the mental entities that constitute the given regarded in terms of the minimum Deduction assumption. He should hold that these mental entities are atomic, isolated entities in a sense analogous to the sense of 'atomic, isolated entities' just explained. So he should conclude that, necessarily, combination does not belong to these entities. In addition, and to return momentarily to the preceding observations in support of (b) itself, these observations clearly relate to the two Chapter Three claims, for they simply spell out the relevant parts of those claims in more detail than we have seen so far.

(II) Kant holds that, necessarily, the outer intuitions through which we human beings know exist in the mind in the form of sets of sensations, as those outer intuitions initially are put before our mind through inner sense as occurring in a (potential) time order. (Moreover, he holds also that, necessarily, the object, as we know it, is itself given to us, in the way noted in Chapter Three, as a manifold of sensation-presented properties and spatial parts.) But, necessarily, sensations are absolute, atomic entities that, as they occur in the mind prior to the operations of thought, bear no relations to one another.[7] And from these facts (b) follows. (And so, also, does the similar point that concerns properties and spatial parts.)

Furthermore, it is clear that Kant will identify the given, regarded in terms of the minimum Deduction assumption, with sets of sensations occurring in a potential time order in the mind. Hence reasoning like that above shows that he should infer that, necessarily, combination does not belong to the given so regarded. Moreover, and to note again the reasoning for (b) in the previous paragraph, this reasoning obviously relates to the two Chapter Three claims in the simple sense that it can be used to support those claims.

(III) In his B-Deduction § 15 discussion of combination and givenness, Kant asserts that combination 'is an act of self-activity of the subject' (B130; see also B134–35). This assertion suggests, along with the tenor of his conceptualism, that Kant accepts a version of the Leibnizian position that relations (or cases of the holding of relations) are 'works of the mind' (and, for Kant himself, are works or results specifically of – and dependent for their existence on – the activities of our thought and understanding).[8] But then, in the case of our human knowledge, (b) follows immediately, for no combination-relations can hold among outer-sense elements (or among properties and spatial parts) as those elements are initially put before our mind prior to the activities of our thought (and imagination).

Furthermore, given the above Leibnizian position, it also follows immediately that no combination-relations can hold among the mental entities that constitute the given regarded in terms of the minimum Deduction assumption. And, to return to the reasoning for (b) itself in the preceding paragraph, this reasoning of course does not demonstrate the two Chapter Three claims. But it at least establishes a view of combination and the given that harmonizes with the thoroughly sequential way in which those claims take us to acquire our knowledge (and take the object itself to be presented to us).

The above statement of Kant's reasons for the thesis that combination cannot be given could be sharpened. But we need not seek to improve the statement here, for even as it stands it is certain that it does not adequately support that thesis. The problem is not in the overall reasoning from (a) and (b) to the thesis. The problem is that even if (at least for the sake of argument) we accept (a), grounds (I) to (III) fail to demonstrate (b). (Similarly – though we do not need to consider this case separately – the use of grounds resembling (I) to (III) to demonstrate the minimum-Deduction-assumption version of (b) and Kant's thesis also fails.)

Thus consider (I). It is a psychological fact that, in coming to know an object, we can take in at once a number of the object's features and aspects, as well as various of their interrelations. So a sequential attention to initially isolated and unrelated properties and spatial parts is not required for our knowledge. Nor does it seem that, in order for us to know, the mental states or entities through which we know must first occur, in a (potential) time order but unrelated to one another, before our mind. So (I) does not demonstrate (b).

Consider next (II). (II) turns on an atomic view of sensations as given that is present in the work of pre-Kantian philosophers like Hume and Locke (on some interpretations). But for both psychological and philosophical reasons such a view of sensations (or of our mind as having atomic sensations given to it, in a psychological or quasi-psychological sense, for its further operations) is now to be rejected. So (II) fails to demonstrate (b).

Consider finally (III). As is well known, nineteenth- and twentieth-century discussions of relations have shown that there are no good reasons to suppose that entities, whether mind-independent or not, cannot stand in relations in ways that are independent of the activities of the mind. (Moreover, this fact is not undercut by, but itself tends to undercut, Kant's conceptualism, which in any case has independent problems and is officially applied by Kant merely to monadic properties.) So (III) does not demonstrate (b).

Just because (I) to (III) fail to demonstrate (b), the preceding reasoning from (a) and (b) does not establish the combination-cannot-be-given thesis for the case of the property- and spatial-parts manifold. (And, for similar reasons, we cannot rely on the reasoning, sketched above and resembling that for (b), that tries to establish the minimum-Deduction-assumption version of that thesis.) Nor do I know of other, better reasoning for Kant's thesis. Moreover, suppose that we remain within his overall framework of intuition and concept in knowledge. Then, even within that framework, it is not hard to imagine how intuition-elements or properties and spatial parts – or minimum-assumption mental entities – *could* be given in combination-implying relations. Again, suppose that we abandon our acceptance (which has been for purposes of argument) of (a). Then it is also easy to see how combination relations could belong to entities existing in themselves (for example, to outer-intuition elements initially existing in the mind through the affection of outer sense or to other sorts of entities existing in themselves). Furthermore, the problems about Kant's views on combination and givenness evidently go even further than the present ones. Recent philosophers' attacks on the notion of the 'given' would strongly challenge his idea – accepted also by many other writers – that there are thought- (or 'interpretation'-) unprocessed elements that are yielded in such a way that by operating on these elements, we can come to know. And whether or not these attacks ultimately undermine all versions of that idea, they raise serious questions about Kant's basic view that intuition-elements are provided to our mind (and are somehow identifiable by us in connection with our philosophical theorizing) in a thought-unprocessed form in such a way that we can come to know via them.

These problems for Kant's views on combination and givenness, and possible Kantian responses to them, could be examined for some time. However, I think that the problems are ultimately unanswerable, at least as they affect Kant's specific views. And an extended discussion of such problems is impossible here for reasons of space. Moreover, in any case we will observe later that (as various recent commentators have held) one can state central claims of the Transcendental Deduction in ways that are independent of his views. Since Kant appeals to those views throughout his presentation of the Deduction, it therefore seems best, in our own exposition, simply to note, without further comment, his specific uses of them.

3. 'HOW *SUBJECTIVE CONDITIONS OF THOUGHT* CAN HAVE *OBJECTIVE VALIDITY*' (A89/B122): THE TRANSCENDENTAL DEDUCTION IN KANT'S CONCEPTION OF IT

As we have said above and will examine in detail below, the Deduction takes the form of a proof from the possibility of experience. This fact emerges in the course of Kant's own introduction to the main themes of the Deduction, an introduction which relies heavily on the understanding-sensibility distinction and his treatment of the manifold of intuition. We need now to consider that introduction before we turn to the details of this fact itself.

Kant begins the discussion in § 13 of the first *Critique* with his well-known distinction between questions of right and questions of fact in the employment of concepts, and particularly in the case of the employment of *a priori* concepts (A84/B116 ff.). A concept may *in fact* be employed so as to have or to claim objective reality or validity – that is, so as to apply (or at least to be taken to apply) to certain objects. But there may still be no *right* to employ that concept in that way, for that employment may not be justified. And this point is especially important in the case of the *a priori* concepts of the understanding, the categories. Various philosophers may regard the categories as in fact applying to objects, but it is not clear how that employment of the categories is to be justified, for they will apply to objects (if at all) in complete independence of sense experience, and thus we cannot appeal to our experience of objects as falling under them in order to show the correctness of their employment.

We are therefore 'faced with the problem' of how such *a priori* concepts, including the categories, 'can relate to objects which they yet do not obtain from any experience' (A85/B117). And this problem will be solved, at least for the case of the categories, through a transcendental deduction. As Kant writes, adopting for his own purposes the legal term 'deduction,'

> The explanation [*Erklärung*] of the manner in which concepts can thus relate *a priori* to objects I entitle their *transcendental* deduction; and from it I distinguish empirical deduction, which shows the manner in which a concept is acquired through experience and through reflection upon experience, and which therefore concerns, not its legitimacy, but only its *de facto* mode of origination. (A85/B117)

As can be seen from the context (where, for example, he speaks at A84/B116 of the 'proof' of the objective reality of concepts), as well as from his subsequent procedure in the Deduction and his remarks elsewhere, Kant here means that a transcendental deduction as applied specifically to the categories will do two main things: first, it will offer a proof that the categories apply *a priori*, and so with necessity (A89/B121), to the relevant objects; and, second, it will offer an explanation, couched in terms of his picture of knowledge, of how such an application comes about.

Kant says that such a deduction of the categories is especially important just because the categories are unlike other *a priori* 'concepts' or representations that we have so far encountered, namely the pure forms of intuition such as space. The application of the latter representations to objects can be justified simply by inspecting those objects as they are exhibited, in pure intuition, as structured by those representations (A87-88/B120). But, unlike the forms of intuition, the categories are generated wholly by our understanding and attribute to objects properties (or 'predicates,' A88/B120) that are grasped through *a priori* thought. As Kant sees it, the categories therefore relate – or purport to relate – to *all* objects ('they relate to objects universally,' ibid.), in independence of *all* conditions of our sensibility. Hence no simple inspection in pure intuition can justify their application, and a transcendental deduction of the categories is absolutely necessary.

Moreover, Kant holds, the necessity of such a deduction and the problems posed by that deduction seem even more crucial when one realizes the full implications of the fact that the categories are generated wholly by the understanding and that the understanding as a faculty is distinct from sensibility. The implications are that – it certainly seems – objects may be *given* to us in sensibility in independence of the requirements imposed by the understanding, including category-application to those objects (A89/B122 ff.). Hence it is not clear why or how the categories can apply with necessity, as Kant will argue that they do, to at least the intuition-given objects of our knowledge. And so the transcendental deduction of the categories faces the special and fundamental difficulty of explaining 'how *subjective conditions of thought* can have *objective validity*, that is, can furnish conditions of the possibility of all knowledge of objects' (A89–90/B122).

The importance of explaining how such subjective conditions of thought can have objective validity in fact goes well beyond that of the basic problem, internal to Kant's picture of knowledge, of relating the understanding and its categories to the distinct faculty of sensibility. In the light of Kant's view of the given, the distinction of understanding from sensibility means that through sensibility we might be confronted simply with a manifold of atomic, isolated sensations and intuition-elements. Our knowledge might then take the form merely of disconnected, separate acts of awareness of each of these individual elements, these acts of awareness never yielding any awareness of objects distinct from, and known through, the given sequences of these elements. But

Kant hopes to demonstrate later in the Deduction that all our knowledge is of such objects and a Humean experience of atomic, isolated sensations and intuition-elements is not possible for us.[9] He will do so by showing that category application to the objects of our knowledge requires that those objects be distinct from the individual intuition-elements and given sequences of such elements through which the objects are known. Thus he will ultimately eliminate the possibility of such a Humean experience as he spells out his answer to the difficulty of how the subjective conditions of thought can have objective validity.

The general solution to this difficulty that Kant proposes in the Deduction, and the key to the overall structure of the Deduction, is, as I have intimated earlier, the idea of demonstrating that category application is a condition of the possibility of experience. According to Kant, objects do not make the categories possible, in the sense of generating in our minds representations from which we derive the categories, for then the categories would be *a posteriori*. Rather, and as we will see, the categories make the objects possible, in the sense that a necessary condition of any object's being known at all is that that object should, in the form in which it is known, fall under the categories. ('[O]nly through the representation [and so the category] is it possible to *know* something *as an object*,' A92/B125.) Basically, Kant shows that category application forms such a necessary condition by arguing that, in order for us to know an object through an intuition, we must use the categories to *think*, and so to conceptualize, the object of that intuition as that object is known by us. He develops this point in a succinct form at A92–93/B125–26 ff. (here note Axvii on A92–93 as already 'suffic[ing] by itself' to deduce the categories!), and he sets it out in detail in both the A- and B-Deductions. However the fine points of this development proceed, it is at least clear, he takes it, that through the preceding discussion we have now arrived at the 'principle according to which the whole enquiry' of the Transcendental Deduction 'must be directed.' That principle is that the categories 'must be recognized as *a priori* conditions of the possibility of experience, whether of the intuition that is to be met with in it or of the thought' (A94/B126).[10]

Because this principle is so important to the overall structure of the Deduction, we need to consider in more detail than we so far have some of the terminology that is involved in Kant's application of it: first, his talk of 'objective validity' and, second, his idea of 'conditions of the possibility of experience.' By examining this terminology, we can verify and extend the approach to the Deduction that I have been expounding above; and we can show, through appeal to the texts, how this approach is reflected in the details of the Deduction itself.

In regard to the first piece of terminology, Kant's use in the Deduction of the notion of the objective validity or objective reality of a concept is tied quite specifically to his goal of proving the objective validity of the categories.[11] Thus by the Deduction's notion of the objective validity of a concept Kant does not mean, as he sometimes does elsewhere, the mere possibility of knowing a object to which the concept applies.[12] Rather, and as my preceding exposition has been meant to suggest, in the Deduction he is concerned with the objective validity of a concept *with respect to* an object or a set of objects. And he supposes that a concept is objectively valid with respect to that object or set of objects just in case the object in question, or each of the objects in the set, falls under (or, as he at A93/B126 and elsewhere says, 'conforms to') that concept.

Given this account of the Transcendental Deduction's notion of objective validity, it should now be clear that Kant means the Deduction to prove the objective validity of the categories with respect to the *same* group of objects as the Transcendental Aesthetic is supposed to have established the objective validity of space and time – namely, with respect to the set of phenomenal objects, the set of objects that are appearances or objects as they appear via sensible intuitions. Our earlier expositions should have made this point at least implicitly evident. And the point also is evident from the texts. Thus the Aesthetic is said to demonstrate 'the *reality*, that is, the objective validity' (A28/B44) of space and time with respect to the set of objects as objects appear to us (here note also A34/B51). But then, and with similar consequences, the goal of the Deduction is to demonstrate the objective validity of the categories precisely with respect to the set of *all objects whatsoever, insofar as* those objects appear via our intuitions (or are, as appearances, identical to those intuitions). That is, the categories will indeed have, as they purport to have, a *universal* validity with respect to objects. But it will be a universal validity that holds with respect to precisely the same set of objects as does the objective validity of space and time – namely, with respect to phenomenal objects, the objects that we do or can know.

To cite texts supporting these last comments, note that in the Introduction to the first *Critique* Kant speaks of determining 'the *valid employ-*

ment of such concepts [and here he has in mind such *a priori* concepts as the categories] in regard to the objects of *all* knowledge in general' (B23–24, my italics). In the A-Deduction, he writes that the synthetic unity of nature could not be established *a priori* if the subjective grounds of such unity – the categories – 'inasmuch as they are grounds of the possibility of knowing *any object whatsoever in experience*, were not at the same time *objectively valid*' (A125–26, my italics). In the next-but-last section of the B-Deduction, he asserts that 'the categories are conditions of the possibility of experience, and are *therefore valid* for *all objects of experience*' (B161, my italics). And he says or implies many similar things elsewhere, for example in the § 14, A92–93/B125–26 ff., text that we discussed above in connection with his statement of the basic principle of the entire Deduction. All of these texts support the view of the categories' objective validity that we have developed above.

In regard to the second piece of terminology mentioned above, we have already observed that, according to Kant, the objective validity of the categories is to be established by a proof from the possibility of experience – by arguing that that objective validity is a condition of the possibility of experience. All three of the notions introduced here – of experience, of its possibility, and of a condition for that possibility – need discussion. I will consider these notions in the order just given.

First, and as we have seen in Chapter Two, by the term 'experience' Kant sometimes means, in a Lockean sense, 'the raw material of sensible impressions' that our understanding develops into our knowledge of objects. At other times, however, he means simply that (empirical) knowledge of objects itself. Now at A93/B125–26, in the midst of the § 14 discussion leading up to his A94/B126 statement of the principle of the Deduction, Kant writes:

> The question now arises whether *a priori* concepts do not also serve as antecedent conditions under which alone anything can be, if not intuited, yet thought as object in general. In that case *all empirical knowledge of objects* would necessarily conform to such concepts, because only as thus presupposing them is anything possible as *object of experience* [first italics mine].

And it seems clear from this quotation and other evidence that in arguing for the objective validity of the categories as a condition of the possibility of experience, he has in mind this second sense of 'experience.' As I have suggested earlier, he means to show that that objective validity is a condition of the possibility of our having (empirical) knowledge of objects.[13]

Second, in what sense, however, does Kant take such objective validity to be a *condition* of the *possibility* of our having that knowledge? Sometimes he speaks simply of the categories (or of their validity) as *making* experience *possible*; and when he speaks in this way, he often means just to suggest a part of the Deduction's explanation of how such experience comes about. And that part of the explanation is that the categories make our experience possible in the sense that they, or their applications to objects, are indispensable constituents in the mental states that amount to that experience or knowledge.[14] But at other times, and in connection specifically with the Deduction's proof of category application, he speaks in the way that is our immediate concern here – namely, of the categories (or their objective validity) as *a priori* conditions of the possibility of experience. And when he speaks in this way, while he no doubt still has in mind the Deduction's preceding explanation, he also intends a more strictly logical meaning. What he then means is that a *necessary condition* of any *actual* or *possible* mental state's being a case of knowledge, for a being like us, is that the categories should apply to the object of that mental state, as that object is known through that mental state.

This last point is central enough to the structure of the Deduction to deserve a more careful statement and a textual and philosophical defense. To this end, note that what I have just taken Kant to mean can be put more exactly as the claim that

> it is necessary that, for any mental state, if that mental state is a case of knowledge, for a being like us, then the object of that mental state, as that object is known through that mental state, falls under the categories

or, in symbols,

> (A) It is necessary that [(s)(s is a mental state & s is knowledge, for a being like us $\supset$ the object of s, as that object is known through s, falls under the categories)]

And the fact that it is (A) that gives a proper account (as far as it goes) of Kant's meaning can be seen by noting important features of (A) and by appeal to the texts.[15]

First, Kant is evidently interested in the categories as conditions on *all* knowledge, actual *or possible*, that belongs to a being that is like us in possessing a passive, intuitive sensibility and an active, discursive understanding.[16] Thus suppose that in talking of the categories as conditions of such knowledge, he means to consider them as straightforward necessary conditions (a point to which we return below). Then (A) is so far in order, for (A) evidently takes the categories (or their applications) as such necessary conditions. Moreover, because of the way in which it is prefixed by the 'it is necessary that' operator, (A) guarantees that *anything*, actual or possible, that is a mental state and knowledge for a being like us has the categories applying to its object. And, for the reason just noted, that result is exactly what we want from an account of the categories as conditions for the possibility of experience. So, given the supposition about necessary conditions that we have just made, (A) offers a proper such account.

Certainly given that supposition it is (A) that we want rather than, for example, the claim

(B) (*s*) it is necessary that (*s* is a mental state & *s* is knowledge, for a being like us ⊃ the object of *s*, as that object is known through *s*, falls under the categories)

('for any given thing *s*, it is necessary that if *s* is a mental state and is knowledge, for a being like us, then the object of *s*, as that object is known through *s*, falls under the categories'). We want (A) rather than (B) simply because, as logical reflection shows, (B) applies only to all *actual s* (all of which actual *s* (B) then takes to have the necessary feature specified by the conditional embedded in (B)). And hence (B) says nothing at all about possible states of knowledge that belong to a being like us.

Second, if we continue to make the assumption about necessary conditions that we have noted, then it is clear also that we want (A) rather than a claim like

(C) (*s*) [it is possible that (*s* is a mental state & *s* is knowledge, for a being like us) ⊃ the object of *s*, as that object is known through *s*, falls under the categories]

('for any given thing *s*, if it is possible that *s* is a mental state and is knowledge, for a being like us, then the object of *s*, as that object is known through *s*, falls under the categories'). A claim along the lines of (C) does seem to capture one conceivable sense of the expression 'condition for the *possibility* of experience.' But, unlike (A), (C) is again open to the objection that (C) concerns only *actual s* and not also possible *s*. Furthermore, the sense of 'condition for the possibility of experience' that (C) gives can be seen from the texts to be inaccurate in comparison with the sense of that expression that is given by (A).

To see this last point, continue the assumption that conditions of the possibility of experience are, however they are to be understood in detail, necessary conditions. Then were (C) correct rather than (A), one would expect to find Kant writing that a *necessary condition* of *its being possible* that certain mental states constitute knowledge is that the categories apply to the objects of those mental states. But he does not in general write in this way in the first *Critique*. Instead, and in harmony with (A), he usually proceeds as though such category application is a necessary condition for any mental state, *actual or possible*, to constitute knowledge.

For example, in concluding the B-Deduction at § 26, Kant notes that

> All synthesis, ... even that which renders perception possible, is subject to the categories; and since experience is knowledge by means of connected perceptions, the categories are conditions of the possibility of experience, and are therefore valid *a priori* for all objects of experience. (B161)

To say that the categories are conditions of the possibility of experience is here evidently to say only that the categories are necessary for the synthesis which makes even ordinary actual perceptual knowledge possible. And to make that claim is to hold simply that the categories (or their use and application in the course of synthesis) are a necessary condition of, and necessary constituents in, *any* case of such knowledge, actual or possible. Again, at A93/B126 of the § 14 discussion that we commented on earlier, Kant asserts flatly that 'all experience *does* indeed contain a *concept* of an object [that is, a category]' (first italics mine). And here category application clearly is claimed to be a necessary condition of *experience*, actual or possible, and not a necessary condition of *its being possible that there is experience* or that *some mental state counts as experience*. The same conclusion follows also from the B218–19 sketch of the argument of the Analogies, for that sketch indicates that a representation of a necessary connection of perceptions (and hence category application) is necessary *for experience* (and not simply necessary for its *being possible* that there is experience).[17]

Finally, in the important methodological remarks at A736–37 = B764–65 Kant states that a principle of reason or understanding – for example, the principle of causality based on category application – is established

> always only indirectly through relation of these concepts [the pure concepts of the understanding that are involved] to something altogether contingent, namely, *possible experience*. When such experience (that is, something as object of possible experiences) is presupposed, these principles are indeed apodeictically certain; but in themselves, directly, they can never be known *a priori* .... [Such a principle] should be entitled a *principle*, not a *theorem*, because it has the peculiar character that it makes possible the very experience that is its own ground of proof, and that in this experience it must always itself be presupposed.

Like the texts that we have already cited, this quotation clearly makes the truth of the relevant principles (and thus category application) a necessary condition of there being any experience, or knowledge of objects, at all. It does not make the truth of those principles a necessary condition of *its being possible that* there is any experience. (An indication of this fact is that Kant writes 'when such experience... is presupposed' and *not* 'when the possibility of such experience... is presupposed.') Moreover, it is clear also, from the overall context, that in the above text a principle 'makes this experience possible' simply in the sense of expressing a category-governed factor (say a relation between the elements of the experience) that must obtain if that experience is really to amount to *experience*, a genuine knowledge of objects.

Many other texts could be cited to the same effect as those just quoted, including some that we noticed earlier. Although Kant does not always speak as clearly in them as I take him to in the texts that I have just been discussing, the evidence that is provided by all these texts shows that (A) rather than (C) gives the more accurate account of his notion of a 'condition of the possibility of experience.' And so it seems clear, since no other equally plausible candidates for that account come to mind, that it is (A) – instead of, say, (C) or (B) – that we should take as our rendering of that notion. Or, rather, that conclusion seems clear as long as we retain our assumption that by a condition of the possibility of experience Kant means some sort of necessary condition.

But should we retain that assumption? Various commentators have suggested that since the condition in question (and thus category application) is obviously not any sort of sufficient condition for our having knowledge of objects, it can only be a necessary condition of some kind or other.[18] As it happens, I agree. But (and to consider the issues here just in the Transcendental Deduction context and without reference to the overall structure of Kant's treatment of the possibility of synthetic knowledge *a priori*) it should be observed that modern work in logic suggests at least one other relevant type of condition – namely, a presuppositional condition.[19] The basic idea of such a condition is familiar. A statement *A* (say the standard 'the present King of France is bald') presupposes a statement *B* (say 'the present King of France exists') just in case the truth of *B* is a necessary condition of *A*'s having *any* truth-value (and so the truth-value *true* or the truth-value *false*) at all. Such an idea thus abandons the view that all statements or propositions are bivalent – are always either true or false. The presupposition relation that this idea introduces is clearly different from the standard necessary-condition relation.

Applied to Kant, an approach via presupposition would presumably take any statement that does or can express a knowledge-claim (say the statement 'the tree is conical') to presuppose, in the above sense, a statement that the categories apply to the object of that putative knowledge (say the statement 'the tree is a spatial quantum and a substance'). Or, to make a slightly different proposal, such an approach might suggest that the first sort of statement presupposes the second sort of statement, in the sense that a statement that does or can express a knowledge-claim has a truth-value *for us* only if it is true that we can know the object that that statement concerns and so only if it is true that the categories apply to that object.

Now it cannot be denied that there are attractions to such an approach to Kant's idea of a condition of the possibility of experience. In places this approach may tease out ideas latent in Kant's work, or at least ideas that are for modern purposes stimulatingly attributed to Kant. Moreover, much of the interpretation that I offer below could be adapted to a presuppositional approach. Nevertheless I think that such approaches are not really satisfactory as accounts of Kant's own position.

For one thing, Kant's term 'presuppose' (the usual way – and Kemp Smith's way – of translating '*voraussetzen*') is really used by him only to mean 'assume' and so is used without anything like the sophisticated meaning suggested by the above approach. (Thus note the A736–37 = B764–65 text quoted above, as well as the A93/B126 text quoted earlier, where the phrase 'only as thus presupposing them [the *a priori* concepts of the understanding]' evidently means 'only as thus assuming them.') For another thing, where Kant uses 'presuppose,' one can in general interpret him without loss as speaking in the traditional language of necessary or sufficient conditions. Thus we have already argued such an interpretation to be appropriate to the A736–37 = B764–65 text; and at A93/B126 the phrase 'only as thus presupposing them,' which I have just read as 'only as thus assuming them,' seems clearly to mean simply '*only if* [in the necessary-condition sense] they [the *a priori* concepts of the understanding] thus apply to objects.' Furthermore, and as has been indicated in connection with (A), one clearly can, in a straightforward way, interpret Kant's overall talk of conditions of the possibility of experience in terms of necessary conditions.[20]

For yet another thing, the presupposition relation, as applied to Kant in either of the ways suggested above, does violence to his actual position concerning the truth-value of statements that do or can express knowledge-claims. According to the first application that was suggested above, Kant must take a claim like 'God exists' to be without truth-value, for (on the usual interpretation of his position) it is not the case that the categories apply to God if God exists. And, according to the second application, he must regard that claim as without truth-value *for us*, just because on his overall position we cannot know that God exists (and so of course we cannot know that the categories apply to God). Yet, as is well known, Kant's actual position (or one prominent strand in it) is that the claim 'God exists' *has* a truth-value and that, for both moral and other reasons, beings like us *must regard it as having* a truth-value. In fact, he supposes that we must regard it as having the truth-value *true*, even though no being like us can *know* its truth-value.[21] So neither of the two preceding suggestions for treating Kant's ideas presuppositionally really does justice to his view of the truth-value of statements that can express knowledge claims.

Given the reasons that I have just outlined, we do best to read Kant's talk of the categories as conditions of the possibility of experience in straightforward necessary-condition terms. Account (A) therefore is the most satisfactory way to capture that talk. It should be mentioned, in closing the present discussion, that a further reason for accepting (A) and this necessary-condition reading lies in the harmony of (A) and this reading with Kant's distinction, in *Prolegomena*, § 4 and § 5, between the progressive (or synthetic) mode of exposition and the regressive (or analytic) mode of exposition.

The progressive mode of exposition, Kant holds, is the method proper to, and used within, the first *Critique*. The progressive mode looks within pure reason for the elements and laws of pure reason and 'thus tries to develop [*a priori*] knowledge out of its original seeds without seeking the support of any fact.' Such knowledge has 'to be deduced [*abgeleitet*] wholly *in abstracto* out of concepts.' In contrast, the regressive mode is appropriate to 'preliminary exercises' like the *Prolegomena*. It begins with certain pieces of knowledge that are accepted as given and 'ascend[s] to the sources which are not yet known and which, when discovered' both will explain what we knew already and will 'exhibit a large extent of [*a priori*] knowledge which springs exclusively from these same sources.'[22]

The exact interpretation of the progressive-regressive distinction is not altogether clear, as various commentators have shown.[23] But the work of these commentators and a careful reading of the relevant *Prolegomena* and first-*Critique* texts suggest that Kant means to contrast the first-*Critique's* sort of basically deductive, necessary-condition-treatment of the categories (and of synthetic truths known *a priori*) with the *Prolegomena's* nondeductive treatment of such matters. Thus, and as I have urged earlier, the first *Critique* begins with the assumption that an arbitrary piece of knowledge is had by a being like us. And from that assumption, together with further claims about the operations of that being's cognitive faculties, that *Critique* deductively infers the objective validity of the categories (and the reality of certain cases of synthetic truths known *a priori*).[24] In contrast, and however the details are to be interpreted, the *Prolegomena* begins (in the order of logic) by assuming the objective validity of the categories, as well as various further points about *a priori* knowledge. The *Prolegomena* then seeks to show that this assumption is a sufficient condition (whether deductively or in some other explanatory way) for the existence of certain examples of synthetic knowledge *a priori*.

It should be clear that, on the above interpretation, the progressive mode of exposition, as it is practiced in the first *Critique*, is of the necessary-condition, deductive sort that we have associated with account (A). Hence it is plausible, again, to read Kant's remarks on the categories as conditions of the possibility of experience along the lines of (A). We will consequently adopt that reading below. In doing so we will amplify and refine claims like (A) as is required, particularly in regard to the sort of necessity that (A) and similar Kantian claims involve, a sort that we have not yet discussed.

Finally, while adopting the (A)-reading – and while accepting the view that the overall argument of the Deduction is a proof from the possibility of experience – we will not suppose that the same is necessarily true of all the significant subarguments of the Deduction. (For example, it can be seen from our later discussion that such a characterization is not true of central moves in the Deduction's inference from unity of apperception to category-governed synthesis of the manifold.) Nor need we assume, to mention another topic of interest, that every argument of the sort called transcendental, whether Kantian or not, must be a proof from the possibility of experience in the sense explained in this chapter.

4. THE OVERALL SHAPE OF THE TRANSCENDENTAL DEDUCTION; THE A- AND B-DEDUCTIONS

With these last remarks, we conclude our exposition of Kant's views on the objective validity of the categories and allied matters, as those views affect the preliminary discussion of the Deduction in § 13 and § 14 of the first *Critique*. We are now able to see the overall shape of the Deduction's argument, a shape that we will bring more sharply into focus in the remainder of this book. In order to present that shape most faithfully, I should note at once that in speaking of *the* overall shape of the argument, I am of course to some extent idealizing Kant's actual position, for he was occupied with problems bearing on the Deduction at least from the period shortly after the 1770 publication of the Inaugural Dissertation until the end of his working life. And, as one would expect, his thoughts about these problems show an intellectual development during this time.

This development can be seen in texts that begin with Kant's well-known letter of February 21, 1772, to Marcus Herz, run through unpublished notes of the later 1770s, and reach the 1781 edition of the first *Critique*. Kant's views on the Deduction in the 1781 edition then receive further refinement in an important footnote to the 1786 *Metaphysical Foundations of Natural Science*.[25] And, relying in part on the ideas in this footnote, the Deduction achieves what is perhaps its clearest, most comprehensive statement in the 1787 edition of the first *Critique*, a statement that is then repeated in more succinct forms in later works like the *Fortschritte der Metaphysik* (*What Real Progress Has Metaphysics Made in Germany since the Time of Leibniz and Wolff?*), written in the 1790s and published in 1804.[26] However, although there is a genuine development present here in Kant's thought, a development that has been studied in detail by earlier scholars, I think that it is impossible to deny that there is also a great consistency to that thought throughout his career, at least in regard to the large-scale structure and concerns of the Deduction.[27]

As our goal in this book is not to trace the history of the Deduction in Kant's reflections, we must ignore much of the detailed evidence for this last point. But support for it is present in a multitude of passages. As a sample, note, for instance, that the main problem of the 1772 letter to Herz ('how a representation [an intellectual, *a priori* representation like a category]... relates to an object [which may be a sensible object] without being in any way affected by it') is close to the problem of the first *Critique*, § 13, of how the categories, as 'subjective conditions of thought,' can apply to sensible objects.[28] Again, Kant says in the 1786 *Metaphysical Foundations* footnote that the suggestions in that footnote about the Transcendental Deduction affect the 'mode of presentation' of the 1781 Deduction but not 'the ground of explanation, which is already given correctly there.'[29] In the B-edition (whose version of the Deduction is in close agreement with the *Metaphysical Foundations* suggestions), he likewise asserts that although he has changed the mode of exposition of the A-Deduction (and of other A-texts), he has not altered 'the propositions themselves and their proofs' (Bxxxvii; see also Bxxxviii and Bxlii); and he invites us to consult the A-edition for material that he has omitted from the B-text (Bxlii).

As I will suggest below, Kant is right to hold that essentially the same basic pattern of argument (with the exception of various points of detail) is given in the B-Deduction as in the A-Deduction. And it seems clear that later Critical texts like the *Fortschritte der Metaphysik* do not really alter that pattern of argument. Hence it does not seem misleading to present, in summary form, a brief, idealized account of the Deduction's structure here, particularly since in succeeding chapters we turn immediately to the detailed development of that account in the B-Deduction itself.

The idealized account that I propose of the Deduction's overall shape is of the sort that Section 3 and our earlier discussions make plausible. We are to consider an arbitrary being *H* like us, who has – and who can acquire knowledge only through – a passive sensibility and an active but discursive understanding. Like us, this being *H* possesses an apperceptive understanding, and Kant in fact usually refers to this being in the first person ('I [or we] think such-and-such'). As possessing (and needing to use, in order to gain knowledge) both sensibility and understanding, this being, like us, is to know through the joint operation of intuitions and concepts.

As I have already suggested several times over, the Transcendental Deduction then begins with the assumption that *H* has an arbitrary case of experience, or empirical knowledge, through a given, arbitrary sensible intuition. For convenience we will hereafter call this intuition '*i*.' The Deduction makes this assumption in the minimum form noted earlier, for it must not be assumed at the start of the argument that the object known through *i* satisfies the categories. A main goal of the Deduction then is to show deductively that the categories apply – and apply with necessity – to the object known through *i*. In attempting to achieve this goal, the Deduction uses various assumptions about the operation of *H*'s cognitive faculties, as we have already noted. In particular, Kant emphasizes in his presentations from 1781 on that *i* (or the manifold of *i*) is subject to unity of apperception, a subjection that he takes to follow from various of his claims about *H*'s apperceptive understanding.[30] Through the use of such claims and assumptions, he believes that he can demonstrate that the categories indeed apply (with necessity) to *i*'s object. Because *H*'s knowledge through *i* is achieved through an arbitrarily selected intuition, it follows that the categories apply to the object of any sensible intuition through which *H* knows. And to Kant's mind this point can be used to establish (A) (or a point equivalent to (A)) and so the necessity that the object of any mental state through which a being like us knows, as that object is known through that mental state, falls under the categories.[31]

We also will see later a problem, already referred to in Chapter Two, about Kant's goal of showing, through the above reasoning, category application to all the objects that a being like us does or can know. But because this problem looms large only in Chapter Eight, I postpone it until then.

The above, idealized account gives the main structure of Kant's most important line of thought in all versions of the Deduction from 1781 on. But there are of course significant details of these versions – and significant differences between them – that it does not indicate. Because the *Metaphysical Foundations* footnote largely previews the B-Deduction – and because the *Fortschritte* material develops directly out of the B-Deduction line of thought – we can illustrate these details and differences adequately for our present purposes simply by noting the main differences that are now relevant between the A- and B-Deductions. In both those Deductions the subjection of intuition *i* (or the equivalent intuition) to unity of apperception is the key idea, and both Deductions see this subjection as requiring a synthesis of *i* that in turn leads to category application to *i*'s object. But, despite various passages that suggest that Kant's intentions may have been to the contrary, the actual text at least of the official A-Deduction (A115–30) sets no further conditions, or no very clear further conditions, on *i* or on any other intuitions that the argument of the A-Deduction concerns.[32] And, beyond taking the categories as rules for synthesis, the A-Deduction does not make it especially explicit why the synthesis of *i* should be regarded as yielding category application to the object of *i*. In the B-Deduction, however, various additional conditions are set explicitly on the intuitions that the argument considers. And an explicit, detailed connection is made between the categories and judgment in a way that makes clear the connection between synthesis and category application.

The conditions that the B-Deduction imposes on the intuitions that it concerns are presented in the course of the argument's development. In fact (and to continue to refer to our idealized account of the Deduction), the B-Deduction takes intuition *i* to be, specifically, an arbitrary sensible 'intuition in general.'[33] That is, the B-Deduction abstracts completely from all assumptions about the 'mode' in which *i* is sensibly given to *H*; and so the B-Deduction abstracts, for example, from questions whether *i* is structured by specifically human forms of sensibility like space and time and whether *i* is given through one of the specific human senses. So proceeding, the B-Deduction thus does regard *i* simply as an arbitrary 'intuition in general' – an arbitrary intuition in the most general sense of 'intuition.' About this arbitrary intuition in general, the only assumption that the B-Deduction makes is that *i* is passively given to *H*, in the form of a manifold, through *H*'s sensibility. And the B-Deduction first establishes category application to the object of *i*. Then, generalizing, the B-Deduction argues that the categories therefore apply to the object of any sensible intuition in general through which any being like us does or can know.

However, the B-Deduction, like all versions of the Deduction, is of course concerned specifically to demonstrate category application to the objects of our human empirical knowledge. And so Kant must somehow pass from the conclusion at the end of the last paragraph to the result that the categories apply to the objects of our actual or possible human such knowledge. One can imagine different ways of reaching that result, for example by modifying the above line of argument so as to proceed from the start as though *i* were a human empirical intuition. But Kant's own way of proceeding is rather different and, I think, reflects the fact that in the Transcendental Deduction he is engaged in an *a priori*, 'transcendental' investigation of the scope and limits of our own *a priori* knowledge.[34] According to Kant, such an investigation must consider precisely the *a priori* elements of our knowledge, and for this reason he does not seek to modify the above line of argument as it stands, with its conclusion about the application of the categories to the object of any sensible intuition in general.[35] Rather, and seeking to apply that conclusion to the case of the objects of our human empirical knowledge, he now focuses attention on the main results of the Transcendental Aesthetic.

The Transcendental Aesthetic has, after all, already argued that to our human sensibility there belong the *a priori* forms of space and time. The Aesthetic has also argued that all the objects of which we do or can have empirical knowledge are necessarily structured by these forms. And at A76–77/B102 and A78–79/B104, in the Metaphysical Deduction of the Categories, Kant has previously indicated that the manifold of *a priori* sensibility or of such forms offers material on which the categories are to work. In the B-Deduction he now takes up these ideas. He argues roughly that the application of the categories to the object of any sensible intuition in general implies the application of the categories to space and time as the *a priori* forms of our sensibility. And he concludes that, because all the objects of our actual or possible human empirical knowledge are spatiotemporally structured, this last result shows that the categories apply to all such objects.

The connection, mentioned above, that the B-Deduction makes between the categories and judgment is of the sort that one would expect from our Chapter Three sketch of Kant's views of the manifold. Without trying to anticipate complex issues discussed in Chapter Ten, we need note here only that the basic idea of the B-Deduction is this.

Intuition *i* – and so any passively given, sensible intuition in general that yields knowledge – has a manifold. Following just our Chapter Three comments, we would expect the elements of this manifold initially to be presented to the mind as occurring in a potential time order. And, depending on which specific reading of the Chapter Three minimum Deduction assumption we adopt, (i) (on the strong reading) we would take the elements of this manifold, as they are presented to the mind, to put before the mind – in what is not assumed to be a category-involving manner – unspecified features, aspects, or general properties of the object that is known through *i*. Or (ii) (on the weak reading) we would take the elements of this manifold to be presented to the mind; but we would make no claim that those elements, as they are so presented, put any features, aspects, or general properties before the mind.

However, *i* is regarded in the B-Deduction simply as an intuition in general. So in the B-Deduction we regard the elements of *i*, taken in either of the above ways, as being initially presented in some sort of (potentially) sequential fashion; but we do *not* assume that these elements are presented in any specifically *temporal* sequential fashion. Otherwise, however, we take the elements of *i* in just the way noted above. But then (and here I put aside complications) we note – if we take the elements of *i* in the strong way – that *i* has a conceptual manifold. Or we argue – if we take the elements of *i* in the weak way – that *i* has such a manifold. And – on either way of proceeding – we then argue that that conceptual manifold is involved in a judgment. In that judgment, moreover, the conceptual manifold – and so the conceptual manifold of any such intuition in general – must be structured in a manner that involves *a priori* synthesis by the logical functions of thought in judgment. And this *a priori* structuring, Kant holds, requires a parallel structuring of the object of *i* (and so of the object of any such intuition in general), a parallel structuring that yields category application to that object.[36]

In connection with the comments on *i* in the last paragraph, notice that when we suppose that minimum-Deduction-assumption intuition-elements are initially presented in a potential time order through inner sense, nothing stops us from supposing, further, that insofar as those intuition-elements yield knowledge, they occur before the mind in an actual time order (through imagination and other synthetic operations). And this point of course applies to the elements of *i*. Although in the B-Deduction we regard those elements as initially presented in a potentially sequential (although not necessarily in a temporally sequential) fashion, we may still take those elements, insofar as *H* knows through them, to occur before *H*'s mind in an actual sequential order (although not necessarily in a temporal fashion) through imagination and other synthetic operations. In Chapter Five we will, indeed, see strong reasons for taking *i*'s elements in such a way. However, for the present we may ignore this fact.

By making explicit the above two points about conditions on intuition and about the connection between the categories and judgment, the B-Deduction refines or supplements ideas already present in the A-Deduction. As a result, the argument of the B-Deduction is in general much clearer than that of the A-Deduction. But, despite the B-Deduction's acceptance of the above two points and its (relative) clarity – differences that do distinguish the B-Deduction from its predecessor – at a very fundamental level the overall argument of the B-Deduction is essentially the same as that of the A-Deduction. In both versions of the Deduction the central idea is to show that the intuition *i*, or any equivalent intuition, is subject to unity of apperception, that subjection then requiring a synthesis of the manifold of intuition that leads to category application to the object of the intuition. That fact illustrates sharply the great consistency that I have noted earlier in Kant's thought about the problems bearing on the Deduction. So also does the presence of the same central idea (as well as other B-Deduction ideas) in post-1787 writings like the *Forschritte der Metaphysik*.

Because the B-Deduction makes clearer than does the A-Deduction the overall argument of the Transcendental Deduction, the B-Deduction makes it easier than does the A-Deduction to approach the philosophically central core of the Transcendental Deduction. And just because the B-text incorporates the key, unity-of-apperception-involving line of thought of the A-text, one can always appeal to the relevant parts of the A-Deduction in order to elucidate the B-Deduction.[37] Consequently I will focus below largely on the B-Deduction, and in particular on what I think is its philosophically more interesting first half, § 15 to § 20. I will, however, have something to say about the Deduction in general (and so about various parts of the A-text). And I will also comment, below, on the second half of the B-Deduction, § 21 to § 27.

In view of the central role of the B-Deduction in the remainder of this book, it will be useful to have a brief summary, here, of the B-Deduction's proof of the objective validity of the categories. As I have already suggested, this proof falls into two parts. These parts correspond approximately to the B-Deduction's treatment first of category application to the object of intuition in general *i* and then of category application to the objects of our own empirical knowledge.[38]

In the first part of the proof, in § 15 to § 20 of the text, Kant begins, as I will argue in detail in Chapter Five, by assuming in § 15 that the arbitrary being like us, *H*, has an arbitrary piece of knowledge through an arbitrary, passively given sensible intuition in general *i*. In § 16 he then argues for the subjection of the manifold of *i* to unity of apperception, and he urges that that subjection requires a synthetic combination of the elements of *i* within *H*'s mind. He next examines, in § 17, the unitary relation that the elements of *i* have together, in virtue of that synthetic combination (and thus ultimately in virtue of the holding of unity of apperception with respect to them). He claims that that unitary relation *is* or *amounts to* the unitary relation that those elements have together in virtue of their being all related to the object that is known via *i*. To Kant's mind this result establishes that the subjection of *i*'s elements to unity of apperception is tantamount to, and the ultimate source of, the relation of intuition-elements to the object in knowledge. And so, as he concludes in § 18, the unity that such elements receive through the unity of apperception is an objective unity. It is not a subjective unity that is based, for example, on the existence of accidental associative bonds that might hold among those elements in *H*'s own particular mind.

Yet, as an objective unity that gives knowledge of an object, this unity among *i*'s elements is or involves a judgment, as Kant now implicitly holds in § 19. And he holds also that the logical form of any judgment – for example, a judgment's having a singular, affirmative, subject-predicate, assertoric form – simply consists in or arises entirely through the subjection of the concepts in that judgment to unity of apperception. However, he argues in § 20, because the unity among *i*'s elements is or involves a judgment with a logical form that is imposed on those elements through the unity of apperception, the conceptual elements in *i*'s manifold are themselves structured according to the logical functions of thought that determine that logical form. Moreover, he indicates, the categories are concepts under which the elements of *i* (or of *i*'s object) are subsumed in ways that correspond to those logical functions. It therefore follows that the object of *i* is subject to the categories. Hence, generalizing on this last result, he can conclude that the categories apply to the object of any sensible intuition in general through which any being like us does or can know. And with this conclusion he completes the first part of the B-Deduction's proof of the objective validity of the categories.[39]

In the second part of the proof, in § 21 to § 27, Kant begins in § 21 by noting that because the previous, § 15 to § 20 argument has concerned only the objects of sensible intuitions in general, it has not shown that the categories apply to the objects of our own human empirical knowledge – the objects of our own sensible, empirical intuitions. Yet, as he notes in § 22 and § 23, we have at least seen that the categories can function in knowledge only (at best) to structure the manifolds of sensibly given intuitions. So it is at least clear from the § 15 to § 20 argument that, in the case of our human knowledge, category application, if it exists at all, can extend no further than to the objects of our sensible intuitions. Given this point, he argues in § 24 (and on the basis of his § 20 conclusion about the application of the categories to the object of any sensible intuition in general) that the categories apply, through the efforts of our imagination, to our *a priori* intuitions – that is, to space and especially time. After considering issues about inner sense and apperception in the remainder of § 24 and in § 25, he then concludes the main argument of the B-Deduction in § 26. He recalls from the Aesthetic that all our actual or possible empirical intuitions are structured by our *a priori* intuitions. He therefore holds that the application of the categories to space and time yields also the application of the categories to all our actual or possible empirical intuitions or their objects.[40] Hence, Kant takes it, the categories apply to all the objects that we do or can know by means of our empirical intuitions. This conclusion is that toward which the overall proof in the B-Deduction has been directed, and with it the main argument of the B-Deduction ends. § 27 then summarizes the course of the argument and considers briefly some further issues that concern the B-Deduction reasoning.

It is the above line of argument, and especially its crucial first part, that we will develop in the remaining chapters of this book. I will show that this argument is indeed a proof from the possibility of experience. I will also defend the other parts of the interpretation that I have sketched above. Before we turn to the final preparations for our detailed discussion of the B-Deduction, it is, however, useful to notice one last piece of evidence for treating the Deduction, and in particular the B-Deduction, as a proof from the possibility of experience.

The evidence in question arises as follows. If the B-Deduction (or indeed the A-Deduction) is such a proof, then it will begin with the asumption that *H*, the knower, arrives at an arbitrary piece of knowledge via *one* given intuition *i*. And this one intuition will be an intuition that the subsequent Transcendental Deduction argument will hold must be unified by *H*'s mind in order that it can function *for* that mind *as one* single, unified intuition that represents to *H* a single object. (Recall that intuitions, by definition, represent single objects.) One would expect that Kant might want to emphasize that *H* is assumed to know through one such intuition, particularly because in the argument of the Deduction he emphasizes so much the synthetic processes that make the intuition's elements into *one* intuition representing *one* object *for* the mind.[41] And this expectation is strengthened by the fact that, in German, '*ein*' is ambiguous in many contexts as between the indefinite article that means 'a' or 'an' and the numerical adjective that means 'one.' Given that ambiguity, if Kant is indeed assuming that *H* is to know through *one* intuition that functions in the way just noted, then he has compelling reasons for indicating in the text that when he speaks, in the relevant places, of, say, '*eine Anschauung*,' he means *one* intuition so functioning for *H*'s mind and not simply *an* intuition (some intuition or other) that exists, unified or not, within *H*'s mind.

Now means for indicating this fact exist in German – namely, one capitalizes the first letter of '*ein*' or else spaces that entire word, in order to indicate that *one* and not *a* or *an* is what is meant. Moreover, Kant employs such means in the B-Deduction (and also in the A-Deduction). And, in the manner of the above proof from the possibility of experience, his use of these means seems designed to show that we are to assume *H* to know through one intuition that functions for *H*'s mind as one unified intuition representing one object. For example, in B-Deduction § 20, at B143, the fact that we are to make such an assumption is, I think, emphasized by the capitalization in the expression '*in Einer empirischen Anschauung*' – 'in One empirical intuition.'[42] Again, in B-Deduction § 16, at B135, this same fact seems to be pressed home by Kant's lower-case, spaced '*e i n e*.'[43] Other pieces of his usage also support this fact. And so there is evidence, in his pattern of emphases involving '*ein*,' that favors the treatment of the Deduction that I am sponsoring in this book.

In interpreting Kant's argument and these emphases in the above way, I am disagreeing with central parts of Dieter Henrich's important essay on the proof structure of the B-Deduction.[44] In that essay, Henrich notes that the B-Deduction falls into the two halves that we have been discussing. He suggests that this division is to be explained as follows. In the first half, Kant shows that all intuitions, insofar as they already contain unity, fall under the categories. In the second half, he then shows that because space and time are intuitions that (by the Aesthetic) contain unity and include all the objects of our empirical knowledge, space and time and so all those objects fall under the categories. Henrich takes Kant's B-

Deduction emphasis on '*ein*' to direct attention to the restriction of the first half to intuitions already containing unity. And he suggests that one reason Kant does not make the above structure of the B-Deduction completely clear is that he is misled by a fallacious § 16 argument about unity of apperception into thinking – with some uneasiness – that already in § 16 he has a means of showing that all our intuitions whatsoever are subject to the categories.

In a further paper, Henrich deepens his discussion by investigating the legal background to Kant's term '*Deduktion.*' Within Kant's tradition, a juridical deduction was an attempt to justify the possession or use of something by tracing that possession or use back to its origins, in such a way as to show its legitimacy.[45] Henrich argues that while his earlier account of the structure of the B-Deduction can be maintained and Kant does there offer a two-part, deductive proof of category application, Kant's main reason for dividing the B-Deduction into two parts is the distinctive contribution of each part to our understanding of the origins and so the justification of category application.[46]

I have learned a great deal from Henrich's discussions. I believe, however, that at various points my interpretation is closer to the text than is his.[47] As I see it, Henrich is right to insist that the two halves of the B-Deduction must be understood as parts of a single proof; he is right to observe the crucial importance, in that proof, of unity of intuition; and he is right also to see a fallacy in Kant's reasoning in § 16. Henrich's pioneering remarks about the juridical background of the notion of a deduction also throw much light on Kant's strategy in the Deduction, and I think that Henrich has made it very plausible that, at a minimum, Kant had firmly in mind the nature of a legal deduction and modeled much of his discussion on it.

However, rather than accepting Henrich's views about the first half of the B-Deduction and intuitions that already contain unity, I believe, and have argued above, that in § 15 to § 20 Kant begins by arguing for the general, abstract point that every sensible intuition in general through which a being like us knows – and therefore every such intuition that functions for us as one unified intuition representing one object – falls under the categories. As I have suggested, this fact adequately explains the B-Deduction emphasis on '*ein.*'[48]

When Kant then moves to space and time in the second half of the B-Deduction, he does so, first, because (as I have suggested) he is conducting a transcendental investigation of the scope and limits of our *a priori* knowledge, and, having made his basic, general point about the categories as applying to all sensible intuitions in general, he now narrows the investigation to our human case and considers the relation of the categories specifically to the *a priori* formal elements of our sensibility.[49] Second, Kant moves to space and time because in the Transcendental Aesthetic he already has argued that space and time are single, unified intuitions that structure all the objects of our empirical intuitions. So he has an argumentative means of demonstrating, in § 26, category application to such objects by showing that the categories apply to space and time (and to individual, determinate spaces and times) and are responsible for their unity. But in so arguing Kant is not moving, as Henrich suggests, from intuitions already containing unity to space and time as specific examples of such intuitions. Rather, he is moving from sensible intuitions (in general) through which we know, as category-determined unities, to space and time as unities whose unification, he argues, also is due to the application of the categories. Nor, I believe, does Kant confuse himself about the structure of the B-Deduction by an uneasy reliance on the undoubtedly fallacious § 16 argument. Rather, and as I argue in Chapter Eight, that argument has an integral part to play in the overall reasoning of the B-Deduction.

Finally, the account of the B-Deduction that I have proposed is compatible with, although it does not rely on, Henrich's discoveries about Kant's legal model for the Deduction. One might still ask how far Kant's use of that model extends beyond the organization and exposition of an argument whose basic content and structure are already determined (it might be urged) by the philosophical requirements, as Kant and the Cartesian tradition see them, of an investigation into the nature and limits of our knowledge.[50] But as such worries, whether or not they are well founded, do not affect the detailed interpretation that I give below, I will not pursue them here.

5. FINAL PRELIMINARIES. THE § 14, A92–93/B125–26 ARGUMENT FOR THE DEDUCTION

The Transcendental Deduction is a text of nearly inexhaustible complexity, and it has many other aspects that we could examine before beginning the details of the B-Deduction. Here, however, we need consider only two: first, Kant's in some respects opaque distinction, in the A-Preface, between the objective and the subjective deductions; and, second, his initial sketch of an argument, in § 14 at A92–93/B125–26, for the objective validity of the categories.

The first of these topics need be noted only briefly, and for the sake of completeness. Kant says at Axvi–xvii that there are two sides to the investigations of the Deduction: (i) an attempt to 'expound and render intelligible' the objective validity of the categories; and (ii) an attempt to study the pure understanding, 'its possibility and the cognitive faculties upon which it rests.' (By this description, he means an attempt to determine, through an investigation of the faculties involved in the operations of our pure understanding, how those operations can themselves occur and lead to category application.) Of these two points, (i) amounts to Kant's objective deduction of the categories and (ii) to his subjective deduction. He says at Axvi–xvii that while the objective deduction is essential to his main purpose in the Deduction, the subjective deduction, although of great importance to that purpose, nevertheless is not essential to it.

One can argue at length about how to understand the objective-subjective distinction in the A-Deduction and the extent to which it is present also in the B-text. Our present purpose is, however, to investigate the details of Kant's proof, as it is presented chiefly in the B-Deduction, of the categories' objective validity. To achieve that purpose, and indeed to evaluate the success of his proof, we need not try to sort these matters out in any detail, a fact that is particularly gratifying just because he himself says nothing further in the first *Critique* about the objective-subjective distinction, except to make one point, which we will mention below, about his § 14, A92–93/B125–26 argument. Thus I will observe merely that, as we have noted earlier, alongside the Deduction's proof of the objective validity of the categories there occurs an explanation, couched in terms of Kant's theory of the mind, of how category application comes about. This general distinction between the Deduction as proof and as explanation seems to correlate roughly – although not exactly – with the Axvi-xvii distinction between the objective and the subjective deductions. And in our further study we thus need only be sure that we do not confuse what are to Kant's mind inessential features of the Deduction as explanation with what he regards as essential features of the Deduction as proof.[51]

The second of the above topics, the § 14, A92–93/B125–26 argument for the objective validity of the categories, is cited by Kant at Axvii as already 'suffic[ing] by itself' to produce the 'complete conviction' that he hopes for from the objective deduction. As we will see, this § 14 argument is really too sketchy to stand as a proof of category application. But it is worth noting here if only to bring out what Kant takes to be fundamental claims in the overall Deduction. And it is also worth noting because its structure confirms the suggestion above that an objective deduction must be a (deductive) proof of the objective validity of the categories.

The § 14 argument runs essentially thus. In § 14 Kant is responding to his § 13 worry about how the 'subjective conditions of thought,' the categories, can have objective validity with respect to objects that we initially encounter through our apparently thought-independent intuitions. His response begins with the suggestion that the categories may well be 'antecedent conditions under which alone anything can be... thought as object in general [*als Gegenstand überhaupt*]' (A93/B125). That is, it may well be that a necessary condition of our thinking of a thing as being an object, in the most general sense of 'object,' is that, in our thought, we take the categories to apply to that thing. However, Kant in effect now notes, our experience, or empirical knowledge of objects, is a compound of intuition and thought. In particular, all our experience contains,

> besides the intuition of the senses through which something is given, a *concept* of an object which is given in the intuition, or appears... (A93/B126)

Or, in other words, to know a given thing through an intuition, we must apply to that thing, via our faculty of thought, not just concepts at random but, specifically, the concept of (being) an *object*.

Yet, Kant has just suggested, if we are to think of a thing as being an object and so as falling under the concept of an object, then we must take the categories to apply to that thing. (The categories in fact turn out to be what he calls 'concepts of an object in general.') Hence any given thing that we do or can know (and so do or can know via intuition) falls under the categories. Or, as Kant says, the categories

> relate of necessity and *a priori* to objects of experience, for the reason that only by means of them can any object whatsoever of experience be thought. (A93/B126)

And so in § 14 he demonstrates the objective validity of the categories.

It is clear that the deductive form of the above argument supports the idea that an objective deduction is a (deductive) proof of that validity. It should also be clear that, despite Kant's remarkable optimism in this regard, the above argument should really convince no one of the categories' objective validity. After all, one might, at least for the sake of argument, grant Kant that our knowledge is a compound of intuition and concept. But, even if one accepts such a point, nowhere in the first *Critique* has he *argued* prior to § 14 for his claim that we have to apply to any object that we know the general concept of being an object. Nor is this claim on its face obvious. Furthermore, he has not argued anywhere earlier that if we are to think of a thing as falling under the concept of being an object, then we have to take the categories to apply to that thing. Nor, to say the least, is *that* point on *its* face obvious. Therefore the above § 14 argument hardly does 'suffice by itself' to prove the objective validity of the categories.

Although Kant's § 14 argument thus fails as a proof of the categories' objective validity, that argument expresses some of the Deduction's basic claims and patterns of thought in a succinct and rather obscure form. For example, one of the Deduction's key lines of reasoning is that knowledge through a given intuition requires a synthetic unification of the manifold of that intuition; and that synthetic unification in turn requires the application, in thought, of the concept of an object in general, and ultimately the categories, to the object of that intuition. At least the outlines of this line of thought are, I think, deliberately anticipated in a very brief form (and without reference to the process of synthetic unification) in the § 14 argument. (Here recall a point made in Section 3 about this A92–93/B125–26 argument and the need, in knowledge, to *think*, and so to conceptualize, the relevant object of intuition.) We may therefore use this argument – and the fact that Kant himself emphasizes its significance in the A-Preface – as evidence for the general interpretation of the Deduction that is developed below. Now, however, we are at last ready to begin, in Chapter Five, our discussion of the official argument of B-Deduction § 15 to § 20 and its ramifications.

6. SUMMARY

We first considered Kant's thesis that combination cannot be given. We saw that, for Kant, in the case of our coming to know a spatiotemporal object like a tree, what is given is outer-sense intuition-elements as those elements are initially (in the order of logic) generated through our outer sense or as those elements are subsequently put before our mind through our inner sense. Again, in the case (at least for the B-Deduction) of the minimum Deduction assumption, what is given is a group of – potentially – sequential (but not necessarily temporally sequential) elements of the manifold. Finally, the combination of a set of entities occurs just when those entities are related in such a way that they make up *one* thing.

We next discussed Kant's grounds for holding the thesis that combination cannot be given. We observed that those grounds are unsatisfactory. In addition, we noted that a number of philosophers have recently questioned the whole notion of givenness. Thus Kant does not adequately support his thesis. For simplicity, however, we chose to proceed as though his views on combination and givenness were without problems. But we noted that we would return, as necessary, to such views.

We turned then to the Transcendental Deduction itself. The Deduction emerges in the first *Critique* from the problem of how the *a priori* categories of our pure understanding, which seem unconnected with our sensibility and its intuitions, can nevertheless be known to apply *a priori* (and so with necessity) to the objects of those intuitions. The Deduction proves deductively that the categories relate in this way to objects; and it also offers an explanation, in terms of Kant's account of the operations of the mind, of how the categories come to relate to objects in this way. We saw that, in its deductive aspect, the Deduction proceeds as a proof from the possibility of experience. And we observed that, as so proceeding, the Deduction aims to establish claim (A) of Section 3 – the claim that, necessarily, any mental state through which a being like us knows is such that the categories apply to the object of that mental state.

Adopting this view of the Deduction and defending it against possible objections, we sketched the progress of the Deduction in terms of its opening assumption that the arbitrary being like us, *H*, knows through the arbitrary given sensible intuition in general *i*. In the B-Deduction, the version of the Deduction on which we will concentrate in this book, Kant first shows category application, through appeal to the subjection of *i* to unity of apperception, in the case in which *i* is a sensible intuition in general. Generalizing from this result, he concludes that the categories apply to – or govern the synthesis of – the manifold of any sensible intuition in general through which any being like us does or can know. Given that conclusion, he then argues that the categories apply also to the manifolds of the pure forms of our human sensible intuition, space and time, and hence apply to all the objects of our experience. And at this point we may turn specifically to B-Deduction § 15.

CHAPTER FIVE

COMBINATION AND INTENSIONALITY: B-DEDUCTION § 15

1. INTRODUCTION

Using the two-part model of the B-Deduction that we have developed in Chapter Four, we now proceed to our intensive study of that version of the Deduction and especially of its first half. In the present chapter, we will focus on § 15 and its use of Kant's thesis that combination – and so the required combination of the manifold of the sensible intuition in general *i* – cannot be given. We will approach these matters through our Chapter Four discussion of Kant's thesis and of the Deduction as a proof from the possibility of experience. While considering the role of § 15 and of that thesis in the argument of the Deduction, I also will introduce an idea that I have mentioned in the Preface – namely, the idea of intensionality. As philosophers know, the idea of the intensionality of, say, claims expressing our thought is very roughly the idea that our thought always grasps its object under some specific characterization, and in such a way that even though this characterization may be coreferential or coextensive with some other characterization, our thought, in grasping the object under the first of these characterizations (say 'iron'), need not grasp it under the second (say 'element with atomic number 26'). In the present chapter we will see the details of this idea, and we will begin to see also why it is of interest in the interpretation of Kant.

2. CLAIMS OF B-DEDUCTION § 15

As I have indicated in Chapter Four, B-Deduction § 15 begins with the proof-from-the-possibility-of-experience minimum assumption that *H*, the arbitrary being like us, has an arbitrary case of experience, or empirical knowledge, through the arbitrary sensible intuition in general *i*. Before we turn to further details of § 15, we need to provide evidence that § 15 begins with that assumption. To that end, I should recall from Chapter Three that Kant's argument in the Deduction ultimately takes different forms depending on which of two readings of the minimum assumption we accept. However, the texts show that Kant does not initially acknowledge these versions in the B-Deduction (or, I think, in the A-Deduction); and reasons for doing so emerge only as we follow the B-Deduction well beyond § 15, into § 17. So, in considering the evidence in question, we can continue to ignore the two readings of the assumption.

That § 15 begins with the above assumption can be seen at once from Kant's A94/B126 statement, in the introductory Deduction section immediately preceding the B-Deduction, of the basic principle that the Deduction is a proof from the possibility of experience. As I suggested in Chapter Four, his assertion of this principle clearly supports the fact that he relies on the proof-from-the-possiblity-of-experience assumption throughout the B-Deduction and hence beginning in § 15. Further evidence that § 15 starts with that assumption can be found in the § 17, B137–38, emphasis on conditions of knowledge; the § 19, B142, comments on knowledge; the § 26, B161, emphasis on the categories as conditions of the possibility of experience; and the similar emphasis in the § 27, B168–69, 'Brief Outline' of the Deduction. (Recall also the B-Deduction's use of the capitalized or spaced '*ein*.') Of course these texts (which were already partly noted in Chapter Four) follow § 15 itself. But their existence supports the conclusion that Kant uses the above assumption throughout the B-Deduction and hence in § 15.

Given that conclusion, the discussion in the previous chapter shows that the assumption thus used in the B-Deduction concerns a sensible intuition in the most general sense of 'intuition,' a sensible intuition 'in general.' (That discussion of course also shows that this assumption must be genuinely minimum in the sense of not assuming the categories to apply to the object known via that intuition.) Our earlier discussion in addition shows that this assumption is about a being like us.[1]

In the light of these points, we may now turn to the explicit content of § 15. § 15 opens abruptly with a number of mostly unargued claims about the combination of a manifold of representations. As both the B129 heading on 'combination in general' (*Verbindung überhaupt*) and the B130 remarks on 'all combination' indicate, Kant here means 'combination' in the most general sense. Such combination (which, as we have seen, occurs when the entities in a set are so related that they make up one thing) includes conscious and unconscious combination and combination both of the manifold of sensible intuition, empirical or not, and of various concepts (B130).

About combination as thus described, Kant's main claim in § 15 is that while the manifold of representations can be given sensibly and can have, as it is given, an *a priori* form (like space or time in the human case), the combination of a manifold in general cannot be given through sensibility and hence cannot be given through the *a priori* form of sensible intuition (B129). This claim Kant defends merely by holding, in addition, that such a combination is (or results from) an act of spontaneity of the understanding (B129–30). He asserts further that such an act is 'originally one, and equipollent [*gleichgeltend*] for all combination' (B130). He calls this act *synthesis*, in order to indicate that through our action we are ourselves the source of all combination and that 'of all representations *combination* is the only one which cannot be given through objects' (B130). And he makes various other claims about combination.

It seems clear, given the lack of argument for them, that most of Kant's § 15 remarks about combination are meant simply to draw attention to topics that the B-Deduction later discusses in detail. For example, the § 15 description of the act of combination as originally one and equipollent for all combination clearly relates to Kant's § 16, B132, view of unity of apperception and of the synthesis that that unity requires. And the § 15 comments on combination and unity relate not just to § 16 views but also to § 19 points about apperception and judgment.[2] But while in such ways § 15 simply anticipates later parts of the B-Deduction, § 15 also plays a role in the substantive argument of the B-Deduction for category application. The principal such role of § 15 is to introduce into that argument, as a premise, the claim that combination cannot be given. Kant then uses that premise in the central B-Deduction § 16 ff. reasoning that a unity-of-apperception-required act of synthesis is responsible for all combination (including what turns out to be the category-applying combination of intuition *i*).

Besides making the points above, in § 15 Kant makes various other points about combination. Most of these points express views that will become clear later in this book (like the B130 assertion that the act of combination is 'one') or else are based in obvious ways on the main thesis that combination cannot be given (like the B130 remark that analysis presupposes the act of synthesis). I should note here, however, Kant's B130–31 observation that the notion of combination involves not only the notions of the manifold and of the synthesis of the manifold but also the notion of the unity of the manifold. This observation, which looks ahead to B-Deduction § 16, § 17, and § 19, holds that (as we have observed earlier) a combination involves not only a group of elements that are related together but also a group of elements that relate together in such a way that they function as one thing. We will see in subsequent chapters how – as Kant suggests at the end of § 15, in B131 – this unity of the manifold derives from what is, given § 16 and § 19, the synthetic unity of apperception.

3. INTENSIONALITY

The idea of intensionality turns out to be of great importance in interpreting the Transcendental Deduction.[3] In particular, some of the central claims in the Deduction – for example, those having to do with apperception, with the unification of the manifold in synthesis, and with the use of the concept of an object in general and the categories to effect that synthesis – are, I think, most satisfactorily interpreted in terms of matters relating to intensionality. Because in the next chapters we will embark on a detailed study of these claims, it is useful to introduce the notion of intensionality here. As I do so, I will note that intensionality belongs to sentences that, for Kant, express the role of thought in our knowledge. In Section 4 we will then prepare to apply that fact to the central parts of the Deduction. In Chapter Six and in later chapters we will see how appeal to the notion of intensionality helps to illuminate the sorts of Deduction claims that I have remarked above.

In Section 1 I gave a brief explanation of intensionality in terms of the idea that our thought can grasp an object under one characterization without having to grasp that object under another, coreferential or coextensive, characterization. That explanation was, however, only a preliminary one and a more exact account is now needed. To that end, I will note that, as is well known, intensionality is, strictly, a logical phenomenon that belongs to sentences, predicates, and other linguistic entities. In particular, intensionality may be said to belong to a linguistic claim, or sentence (the only case that need now concern us), just in case either (I) this sentence exhibits what logicians call referential opacity, and thus the truth-value of this sentence changes when for some singular term occurring in this sentence there is substituted a coreferential singular term; or else (II) this sentence exhibits what may be called extensional opacity, and thus the truth-value of this sentence changes when for some predicate or sentence occurring in this sentence there is substituted a coextensive predicate or sentence.[4]

Intensionality as thus characterized is well known to belong to linguistic claims expressing various of our propositional attitudes. For example, the sentence 'Mary thinks that the brown object is a wooden table' may be true on one standard reading of that sentence. It may also happen that the singular term 'the brown object' is coreferential with, and so names the same thing as, the singular term 'the largest object in the room'; and it may happen that the predicate 'is a wooden table' is coextensive with, and so applies to exactly the same objects as, the predicate 'is a thing that has at some time been climbed simultaneously by seven ants.' But, despite the coreferentiality of the singular terms in question, Mary may not realize the identity of the brown object with the largest object in the room. And so, on its relevant reading, the sentence 'Mary thinks that the largest object in the room is a wooden table' may be false. Again, despite the coextensiveness of the predicates in question, Mary may not realize the identity of the set of wooden tables with the set of things that have at some time been climbed simultaneously by seven ants. And so, on its relevant reading, the sentence 'Mary thinks that the brown object is a thing that has at some time been climbed simultaneously by seven ants' may be false. Hence our original sentence is intensional by both (I) and (II) above.

Facts of the above sort are familiar to all contemporary philosophers. But it is much less well known that similar facts show intensionality to belong to sentences that, for Kant, express the role of thought in our knowledge. To see this point, suppose, along the lines of one of our earlier examples, that *H* knows the object before *H* to be a conical red spruce; that *H* achieves this knowledge through the manifold of the given intuition *j*; and that this knowledge involves *H*'s having the following concept-utilizing thought that concerns the object that *H* knows through *j*:

(T) *H* thinks that (the object before *H* has the property of being conical)

Then (T) is evidently a sentence expressing the role of thought in *H*'s knowledge through *j*.[5]

Now as reflection on the Mary example shows, the above account of intensionality is meant to be applied in such a way that the truth-value of the relevant sentence is evaluated with respect to the *same* world or state of affairs with respect to which the coreferentiality (or coextensiveness) of the relevant singular terms (or of the relevant predicates or sentences) is evaluated. In the case of a sentence like (T), and as we have seen in Chapter Two, there are *two* distinct worlds with respect to which such a sentence can be evaluated: the world *W* of objects as they exist in themselves in nonspatiotemporal forms; and the world *W′* of phenomenal objects, the world of empirically real, spatiotemporal things that *H* knows through *H*'s empirical intuitions.[6] However, and as we noted in Chapter Two, there are good reasons to take the minds that know, their states of knowledge, and so on, to exist in the world *W*. Yet – and as we also observed – a sentence like (T) should hold true at the world in which exists the thought that is expressed in that sentence. So sentence (T), which expresses a part of what is thus *H*'s world-*W*-occurring state of knowledge, should hold true at *W*.[7] Given that (T) is taken in this way, there are now at least two ways to show the presence of intensionality in a Kantian claim like (T); and there is also a third, indirect, consideration that supports that presence.

First, imagine that (in *W*) the being *J* thinks, about the property of being conical, that that property is interesting, but that *H* does not realize that *J* so thinks. Then at *W* it is the case that

> the property of being conical = the property which is such that *J* thinks that that property is interesting

Yet despite the resulting coreferentiality at *W* of the singular terms in this last identity claim, the truth of (T) at *W* is not preserved if for the first of these singular terms we substitute the second.[8] And thus (T) exhibits intensionality.

Second, Kant's account of clarity and distinctness at B414–415 note and *Logik*, Introduction, § V (Ak. 9, 34–35) shows that a representation – and, in particular, a Kantian concept – can be a clear object of one of *H*'s acts of thought without being a distinct object of that act of thought. That is, within some specific act of thought *H* may be conscious of that representation (or concept) as being just that representation; but *H* may not thereby be conscious specifically of the individual elements of the manifold contained within that representation as being contained within that representation.[9] However, and as can be seen from Chapter Three, for present purposes we can identify a Kantian concept with a (mental presentation of a) general property. So in the context of our present discussion Kant's account implies that the following identity claim may well hold at *W*:

> the property of being conical = the conjunctive property *P*-and-*Q*

where *P* and *Q* are properties that amount to what Kant would call the manifold of properties contained in the property of being conical. Yet, despite the fact that *H* (in *W*) grasps the property of being conical as being just that property, *H* (in *W*) may fail to grasp the manifold of properties that is contained in that property. So *H* (in *W*) may fail to grasp – and *H* may even deny – the above identity. Hence despite the coreferentiality at *W* of the singular terms in the above identity claim, the truth of (T) at *W* need not be preserved if for the first of these singular terms there is substituted the second.[10] And thus, again, (T) exhibits intensionality.[11]

Third, many examples show the familiar fact that claims like '*H* knows that (the object before *H* is conical)' involve intensionality. Since, for Kant, the truth of this latter sort of claim, at *W*, always involves the truth of a claim like (T) at *W*, one evidently satisfying way of accounting for the presence of intensionality in this latter sort of claim is simply to take that presence to derive from the logically antecedent presence of that phenomenon in (T). And thus we have another, indirect ground for supposing that (T) exhibits intensionality.

Further arguments could be given to show the intensionality of claims like (T) that express the role, for Kant, of thought in *H*'s knowledge; but the three arguments above should already make the presence of intensionality in such claims quite plausible. Of course Kant does not himself make use of the modern logical notion of intensionality, but such arguments show that his picture of knowledge has active within itself ideas that imply the presence of intensionality in those claims. Moreover, these ideas do not turn on accidental features of that picture. Rather, they reflect Kant's emphasis on our ability, for example, to have clear but not distinct thoughts and concepts, an ability which is one reflection of the finiteness that he sees as fundamental to all human thought. (Thus, for Kant, our thought grasps objects only through finite sets of conceptually presented general properties, none of which properties, or of their presenting concepts, need always be fully distinct to us.[12] Moreover, our thought may not grasp the coextensiveness of various conjunctions of these properties; and so on. All such facts can be seen to lead naturally to the conclusion, expressed in modern terms, that (T) and related Kantian sentences possess intensionality.) It is thus proper to apply considerations about intensionality in studying Kant's picture of knowledge; and we will see, below, numerous aspects of the argument of the Deduction that are illuminated by such applications.

We should note one last point before we return to our discussion of the Deduction. A number of philosophers have raised questions about the meaningfulness or significance of various sorts of intensional claims. One might think that these questions would seriously damage any attempt to appeal to matters of intensionality in interpreting Kant. In fact, however, I believe that such questions have been satisfactorily answered, at least as far as they impinge on the sorts of appeal that I make in later chapters. Moreover, even if such questions were unanswered, the fact that fundamental parts of Kant's own views lead naturally to the conclusion that claims like (T) are intensional would certainly justify appealing to intensionality in any attempt to understand the Transcendental Deduction in his own conception of it.

4. THE ASSUMPTION THAT *H* KNOWS THROUGH *i*

Given that claims like (T) are properly interpreted as intensional, we need now to state in more detail, and with more textual support than we have so far given, the basic Deduction assumption that *H* knows through the intuition in general *i*. To that end, we will continue to abstract from points that depend on which of the two Chapter Three readings of that assumption we take Kant in the Deduction eventually to consider. And then we should recall that, officially, *i* is an arbitrary sensible intuition in general that belongs to a being like us, *H*, and yields *H* some piece of knowledge (that knowledge being minimally described).

Because *i* is merely a sensible intuition in the most general sense of 'intuition,' it is clear that Kant should make no further assumptions (or very few further assumptions) about *i* itself in the B-Deduction. (This point is independent of the need to understand the opening Deduction assumption as a minimum one.) And the fact that Kant makes no – or very few – further such assumptions can be seen from passages like § 21, B144–45; § 21, B145 (where he indicates that in the B-Deduction § 15-to-§ 20 proof he has abstracted from all features of the relevant intuition in general save from the fact that its manifold is given prior to and independently of the synthesis of the understanding); § 23, B148 (where the relevant intuition in general is held to include any sensible intuition and not just our human, spatiotemporal such intuitions); § 24, B150, B151, and B154; § 25, B157 and B158 (where the apperception-determined synthesis of the manifold of intuition in general is contrasted with the synthesis of the inner-sense manifold); and § 26, B159, B160, and B161. Furthermore, when Kant does refer to space and time or outer and inner sense in the B-Deduction, he is arguing specifically for category application to the objects of our human sensible intuitions, given the already established (in § 15 to § 20) application of the categories to the object of any sensible intuition in general through which a being like us knows.[13] Or else he is simply indicating consequences of his overall conclusions about sensible intuitions in general for our human outer- and inner-sense intuitions.[14]

I will therefore follow the above way of treating *i*. I will make no specific assumptions about *i* beyond the supposition that *i* is indeed an arbitrary intuition that is given to *H*, for *H*'s knowledge, through the affection of *H*'s passive faculty of sensibility in a way that is independent of the operation of *H*'s thought. As part of this supposition, I also will take *i* to be given in the form of a manifold which I will describe, where necessary, as consisting of elements that (as they are given) are presented before the mind as occurring in a potentially sequential fashion. But for the present I will ignore the specific treatments that, depending on what reading of the minimum Deduction assumption we are considering, we saw in Chapter Three to be appropriate for those elements. In addition, I will not assume that *i* is some sort of outer-sense or inner-sense intuition or that *i* (or *H*'s faculty of sensibility) must have an *a priori* form of space or time.

As noted several times in Chapters Three and Four, I will, however, build into the basic assumption that *H* knows through *i* the claim that the elements of *i*, insofar as they yield *H* knowledge, occur before *H*'s mind in an actual sequential order through the operation of *H*'s imagination and of other factors in synthesis. Given other parts of Kant's picture of knowledge, the upshot of this claim is as follows. Each element of *i*, as that element is initially presented before *H*'s mind for the operations of *H*'s thought, has the feature of being such that that element is able to occur in an actual sequence of such elements; but that element of *i*, as it is initially presented, does not occur in any such actual sequence. When, however, *H*'s imagination and other factors in synthesis operate on the elements of *i*, each element acquires the feature of occurring at a certain point in such an actual sequence.[15]

It is clear that the mere minimum Deduction assumption that *H* knows through *i* does not imply the above claim that the elements of *i*, insofar as they yield *H* knowledge, occur before *H*'s mind in an actual sequential order. Nevertheless there are good reasons to see Kant as building this claim into the minimum assumption. For instance, his discussion of examples of knowledge in B137–38, B139, and B142 shows that in the B-

Deduction he thinks of the knowledge belonging to a being like us as proceeding in terms of actually sequential elements of the manifold. (And this point is even clearer in the A-Deduction, where, in texts like A99ff., A115–16, and A120–21, he emphasizes the actual sequential – and indeed temporal – ordering of the intuition-elements on which he there focuses.) Moreover, while it is allowed by the minimum Deduction assumption taken by itself, the view that *H* could somehow know through entities that occur within the mind in only a potentially sequential order can seem very puzzling. Indeed, aside from expressing worries about the nature of such an occurrence, many philosophers will surely argue – I think correctly – that such a view is not of major importance to the overall Transcendental Deduction. What is of real importance, it will be urged, is the case in which *H* is assumed to know through an actual sequence of intuition-elements (or perceptions), and the Deduction then seeks to show that what *H* knows must be a category-subsumed object that is distinct from those elements rather than simply being those elements themselves, taken one by one in isolation or taken together in the actual sequence in which they occur before the mind. After all, such a case – and such a goal of the Deduction – is suggested both by Kant's preliminary Deduction worries about how the subjective conditions of thought can have objective validity and by his related and evident goal of showing that the objects of our knowledge can never be mere isolated Humean atomic perceptions or mere sequences of such perceptions. Hence the view that *H* might know through merely potentially sequential mental entities (a view that, furthermore, corresponds to nothing in our own human experience) is not of major significance to the Deduction.

For the above reasons, we will focus, in this book, just on that understanding of the minimum Deduction assumption that takes *i*'s elements, insofar as they yield *H* knowledge, to occur before *H*'s mind in an actual sequential order. We will of course also suppose that, for Kant, these elements have come to occur in this actual sequential order through processing by *H*'s imagination and other synthetic operations. Our decision to understand the minimum Deduction assumption in this way is especially reasonable given that, as we will see later, several of the Kantian arguments for the holding of unity of apperception with respect to *i* depend on supposing that *i*'s elements have an actual sequential order as they occur before *H*'s mind.

It is important not to misconstrue the above way of understanding the minimum Deduction assumption. One misconstrual arises as follows. As I have just noted, in the Deduction Kant ultimately needs to establish in a non-question-begging manner the conclusion that what *H* knows through *i*'s elements is a category-subsumed object distinct from those elements as they are presented to the mind. Hence, at the start of the Deduction, the minimum Deduction assumption by itself (or at least its Chapter Three weak reading) must not immediately rule out the possibility that, as *i*'s elements occur before *H*'s mind, *H* knows no more than those elements themselves, taken one by one, in separate acts of thought. (And then the rest of the Deduction must demonstrate the above conclusion and so eliminate that possibility.) Now given our decision to take the elements of *i* to occur in an actual sequential order insofar as they yield *H* knowledge, this last fact about the minimum assumption implies the following: We must regard Kant, at the start of the Deduction, as not having eliminated the possibility that although those elements occur in *H*'s mind in an actual, sequential order, *H* nevertheless does not grasp the actual sequential occurrence of those elements in thought but at best grasps the elements only one by one in separate, disconnected acts of thought. However – and here begins the misconstrual – this implication may now seem to cast doubt on our decision to take *i*'s elements to occur in an actual sequential order before *H*'s mind insofar as they yield *H* knowledge. After all (one may correctly note), Kant's official position by the end of the Deduction is that such elements occur in such an order only if, through synthesis, they are taken together in one act of thought to occur in that order. Yet, given the preceding implication (the misconstrual says), our above decision has led us to suppose that it really is possible that *i*'s elements occur in an actual sequential order even though *H* does not grasp those elements in one act of thought. So our decision leads to a supposition contrary to Kant's own official position.

The nature of the preceding misconstrual should be obvious. The above way of understanding the minimum Deduction assumption certainly has the implication just noted. That is, it certainly implies that at the start of the Deduction the minimum assumption does not *by itself* rule out the possibility that the elements of *i* occur in an actual sequential order before *H*'s mind but *H* does not grasp those elements together in a single act of thought. However, the fact that this apparent possibility is *not ruled out* at the *start* of the Deduction by the minimum assumption does not imply – and must not be misconstrued to imply – that Kant supposes that such a situation *really is* possible. And, as noted above, Kant's official position – and one of the Deduction's basic conclusions –

is that, given further points in the Deduction, no such situation really can occur.[16]

Our present understanding of the minimum Deduction assumption will be amplified in Chapter Eight. Here I will note simply that, besides treating *i* and its elements in the ways that I have described above, we must also regard the combination-cannot-be-given thesis as applying to *i* as *i* has just been described. Now, as can be observed from our discussion in Chapter Four, Kant's arguments for that thesis really show that combination cannot be given in the case of *i* described as being initially presented before the mind through inner sense as occurring in a potential time order. However, we can take Kant, in speaking of *i* in our present, official way – as not necessarily temporal or presented through inner sense – to proceed as follows. He will hold that since *i* belongs to a being like us and the combination-cannot-be-given thesis applies to our own intuitions, the Chapter Three arguments for that thesis can be generalized so as to apply to *i* as here officially described. (Or else the thesis can simply be stipulated to hold with respect to *i*.)

We should also note that, as we see especially in Chapters Six and Eight, in the Deduction Kant wants to argue that the combination that *i*'s elements have when they are all, taken together, accompanied by the representation *I think* does not belong to them when they are considered merely as occurring in an actual sequential order before *H*'s mind. And Kant wants further to hold that the combination that *i*'s elements have when they function together to represent a single object of knowledge does not belong to them when they are considered merely in such a way. Now these points cannot be shown by appeal to the original combination-cannot-be-given thesis, for (as we noted in Chapter Four) *i*'s elements, when they are considered merely as occurring in an actual sequential order, make up one sequence, hence form a combination, and so no longer can be counted as given.

However, just because *i*'s elements form such a combination when they are considered merely as occurring in such an order, it does not follow that they also form a combination of either the *I think*-accompaniment or of the above object-representing sort. Indeed, accepting the original combination-cannot-be-given thesis, we can already conclude that (necessarily) such an *I think*-accompaniment or object-representing combination (or any other relevant sort of non-single-sequence combination) is not present in *i*'s elements as those elements are given and is not then retained by those elements when they are considered merely as occurring in an actual sequential order. And then we can widen the original thesis to include the claim that (necessarily) such further sorts of combination do not come to exist in *i*'s elements just because of the mere fact, taken by itself, that those elements occur in such an order. Hereafter I will understand the combination-cannot-be-given thesis so as to include the conclusion just noted and this last claim.

For convenience (and without losing any generality), I will make the above points about *i* specific by taking *i*, as given, to consist of the two elements i_1 and i_2 which are initially presented before *H*'s mind in the – potential – order (i_1, i_2) (an order which is not required to be temporal). I will suppose further that, insofar as *H* is taken in the minimum Deduction assumption to know through *i*, i_1 and i_2 are in fact regarded by Kant as occurring before *H*'s mind in the actual order (i_1, i_2). Following Kant, I will suppose, in addition, that this order has been made actual through an operation of *H*'s imagination and of other factors in synthesis. It will therefore be in terms of the elements i_1 and i_2, as here described, that we regard the B-Deduction § 15-to-§ 20 argument as proceeding.

5. SUMMARY

We first gave evidence that B-Deduction § 15 begins with the minimum Deduction assumption that the arbitrary being like us, *H*, knows through the arbitrary, given, sensible intuition in general *i*. We noted also that the principal argumentative role of § 15 in the B-Deduction is to introduce, as a premise, the claim that combination cannot be given.

We then turned to the notion of intensionality, which we explained, in the case of sentences, in the usual way in terms of the lack of truth-value-preserving substitutivity of coreferential or coextensive terms or predicates. We saw that intensionality belongs to sentences that, for Kant, express the role of thought in our knowledge. Because such sentences are important to Kant's picture of knowledge and the Deduction, the idea of intensionality will play a significant role in the remainder of our discussion.

Lastly we considered additional points about the assumption that *H* knows through *i*. We supposed *i* to be given to *H* in the form of the manifold of elements i_1 and i_2, and we assumed those elements to occur before *H*'s mind in the actual sequential order (i_1, i_2) insofar as they yield *H* knowledge.

CHAPTER SIX

APPERCEPTION: B-DEDUCTION § 16

1. INTRODUCTION

In B-Deduction § 16 Kant now begins the formal Deduction reasoning that starts from the assumption that *H* knows through *i* and proceeds by way of the holding of transcendental unity of apperception to the conclusion that the object of *i* falls under the categories. In particular, he argues that the elements of *i*, i_1 and i_2, are subject to unity of apperception. Hence not only must *H* be able to accompany each of i_1 and i_2 by the act of apperceptive thought that Kant calls the *I think*.[1] But, also, *H* must be able to accompany both of i_1 and i_2 simultaneously by the *I think* in one act of mind. So *H* must be able to think the combined thought 'I think (i_1 and i_2)' that holds i_1 and i_2 together before *H*'s thought-consciousness as one combined set of intuition-elements. However, by § 15, combination cannot be given, and therefore (because there is no other possible source, in the case of a being like us) this combination of i_1 and i_2 must be due to an act of synthesis by *H*'s mind. So the subjection of i_1 and i_2 to unity of apperception requires a synthetic combination of those elements before *H*'s thought-consciousness – a synthetic combination that Kant argues, in § 16, to be the source of *any* (knowable or recognizable) combination that those elements have. In B-Deduction § 16, Kant thus shows, if his reasoning is correct, that the subjection of *i* to transcendental unity of apperception yields a combination of *i*'s manfold from which all other *a priori* forms of combination of that manifold follow, including, as we see in later sections of the B-Deduction, the combination that yields the subjection of *i*'s object to the categories.

In the present chapter we will focus on the above line of thought. We will consider (i) the claims that Kant makes in § 16 about apperception; (ii) the exact logical form of such claims and of the basic § 16 argument for unity of apperception (including the logical form of Kant's minimum claim about the necessity of unity of apperception); (iii) various ways in which in § 16 he can justify a central step in this argument; and (iv) and (v) the failure of these ways to justify that step. In Chapter Seven we will complete this examination of § 16 by discussing, among other things, (vi) further aspects of Kant's views on self-awareness and (vii) his additional views on the necessity of unity of apperception.

2. KANT'S VIEW OF APPERCEPTION IN B-DEDUCTION § 16

By 'apperception' Kant means the capacity of a mind for first-person self-awareness. Specifically, 'apperception' names the capacity of our understanding to make various of our acts of thought either contain the first-person representation *I* or *I think* or else have that first-person representation attached to them. Or, in slightly different terms, 'apperception' names our understanding's capacity to ascribe various of our acts of thought to ourself through the use of some thought-equivalent of the first-person pronoun or a similar device ('*I* think that it is raining out,' '*my* current thought is that it is unseasonably warm,' and so on). As it happens, Kant sometimes also means by 'apperception' simply the representation *I think* (or *I*) itself.[2] However, this ambiguity in his use of 'apperception' raises no difficulties in practice, and for convenience I will follow his dominant usage and take 'apperception' to refer to the capacity just noted.

Before we pass to Kant's specific claims about transcendental unity of apperception, I should observe that a series of questions can be raised about the sense in which the *I think* (or *I*) functions as a representation. Most of these questions can be postponed until Chapter Seven. But note that the *I think* (or *I*) is of course a mental entity by means of which we achieve a first-person-style self-awareness. And so Kant should suppose that as long as we have that entity in mind and utilize it, we achieve such self-awareness, even if we happen to have no linguistic device (like the first-person pronoun) that expresses that entity. He also will take first-person thought and awareness, whether or not linguistically expressed by means of a term like 'I,' always to involve the representation *I think* (or *I*). So he will not suppose that we can achieve such thought without utilizing that representation.[3]

We can now turn to the central B-Deduction § 16 line of thought that I have sketched above in Section 1. Kant begins in § 16 with his well-known B131–32 claim that, for reasons that we consider in Section 4, it must be possible for the *I think* to accompany all my representations (or at least all those representations that are of a sort that we note in Section 4.A[4]). So it must be possible for the *I think* to accompany all the elements of any intuition via which I do or can know. Kant next observes that the *I think* is itself an actively and spontaneously generated *a priori* representation of thought, a product of an 'original' apperception (B132) in a sense that we will note in Chapter Eight. He then directs attention to the main topic of § 16, unity of apperception and its required synthesis.

As the texts show, unity or oneness (*Einheit*) of apperception always is a unity with respect to some group of representations, for example the elements of the manifold of *i*. To say that unity of apperception holds with respect to that group of representations is to say that the *one*, selfsame representation *I think* accompanies (or can accompany) all the representations, taken together, in that group. The result of this accompaniment is that, in the case in which, for example, the representations constitute the manifold of one of my intuitions, I can represent to myself 'the thoroughgoing identity of apperception [of the representation *I think*] of a manifold which is given in intuition' (B133). Or, as Kant also says, I can 'represent to myself the *identity of the consciousness in these representations* [that is, the fact that it is one and the same *I think* that accompanies all these representations]' (B133).

Because Kant has just argued at the beginning of § 16 that the *I think* must be able to accompany all my representations, unity of apperception will hold with respect to *any* group of my representations – and will hold also with respect to any group of representations that belongs to any being like me. And although unity of apperception is defined in terms of the (possible) accompaniment of various representations by the *I think*, we can of course also speak of a unity of apperception that is defined in terms of the capacity of apperception itself. For example, all my representations belong to my *one* apperceptive self-consciousness (here compare B132), in the sense that all those representations, taken together, can be accompanied by the representation *I think* that that self-consciousness generates. Kant calls the unity of apperception a *transcendental* unity (B132), in order, as he says, to indicate that *a priori* knowledge may be yielded by it (or by the fact that it holds with respect to the relevant representations). From the holding of unity of apperception with respect to the manifold of *i*, he now wishes to argue for the further main result of § 16, the synthesis of that manifold by *H*'s understanding in such a way that the elements of *i* form a combination, and so *one* single group, before *H*'s thought-consciousness. In order to present this result, he states (at B132–33) a principle which his strict argument for the result does not really require, but which nevertheless serves to emphasize the importance of the holding of transcendental unity of apperception. That principle is what (by his later comments) he calls the fundamental 'principle of the necessary unity of apperception.'[5] Kant identifies that principle (at B135 and B138) with what he takes to be the following analytic proposition: All my representations, and so all the elements of the manifold of any of my intuitions, must satisfy that condition, whatever it may be, that representations must satisfy in order that they can be subject to transcendental unity of apperception – that is, in order that I can accompany them all by the *I think* and so can take them all to be *my* representations.[6] Whether this principle, as so stated, really is analytic, it certainly is obviously and trivially true, given that unity of apperception holds with respect to all my representations.[7] And, in any case, Kant does not use this principle as a *premise* in the further argument of the Deduction. Rather, he uses (or needs to use) this principle simply to draw attention to the fact that, as he now argues in § 16, there is indeed an important specific condition to which all my representations of the relevant sort (and so all of the elements of i) must conform in view of their subjection to unity of apperception.

As Kant makes clear at B133, B134, and B135 of § 16 (and also at B136–37 and B138–39 of § 17), that condition is that the relevant representations must be synthesized or held together by my mind. He urges (at B133 and again at B134) that the holding of unity of apperception with respect to those representations implies that I must be able to represent to myself the fact that the one *I think* does (or can) accompany all those representations. So, to take the case of i_1 and i_2, I do not proceed merely by 'accompanying each representation with consciousness' (B133) and so merely by thinking, say, the two separate thoughts 'I think i_1' and 'I think i_2.' Rather, I must be able to represent to myself, in one thought, both of i_1 and i_2, taken together, as accompanied by the *I think*. When I think this thought, however, i_1 and i_2, as jointly accompanied by my *I think*, form one single group before my thought-consciousness. (Compare B135: 'I call them one and all *my* representations, which constitute *one* [representation or intuition].') So i_1 and i_2 then form a combination before my thought-consciousness. However, combination cannot be given (as Kant recalls at B134 of § 16), so my mind must have synthesized i_1 and i_2. And that synthesis must have been performed in such a way that its result, the joint accompaniment of i_1 and i_2 by the *I think*, *is* something of which I can become conscious in thought.

We thus see that, by the above § 16 argument, the holding of unity of apperception with respect to the manifold of *i* requires the synthesis of the elements of that manifold in such a way that that *H* can become conscious of those elements as accompanied, together, by the *I think*. Kant summarizes this result – the major result of § 16 – by asserting at B133 that the 'analytic unity' of apperception (the fact that, by a process of analysis, the *I think* can be found to occur in, or to accompany, all my representations) is possible only on the assumption of a synthetic (or synthesis-created) unity of apperception. We will return briefly to these notions of analytic and synthetic unity in Section 5 below.

Kant repeats his argument for the main § 16 result at several places in § 16. He also recalls this result in § 17. We can ignore these repetitions here. But it is worth noting his final observation, at B135 at the conclusion of § 16, about both the principle of the necessary unity of apperception and the claim that the holding of unity of apperception with respect to a given manifold requires a synthesis.[8] That observation is that both the principle and the claim apply only to beings of a certain sort – namely, only to beings whose apperception can generate the single representation *I think* (or, as he says at B138, the single representation *I am*), but whose apperception cannot by itself thereby also supply a manifold which is accompanied by (and so is unified through) that single representation. After all, were apperception by itself to supply such a manifold, then the mere generation of the *I think* would yield unity of apperception with respect to the manifold. And thus, contrary to the principle and claim just noted, no special act of synthesis of an independently given manifold of intuition would be required. But, Kant observes, our understanding (and its capacity of apperception) is nonintuitive and does not operate in such a way.[9] For us, the holding of unity of apperception with respect to a given manifold requires such an act of synthesis.

3. THE BASIC STRUCTURE OF THE § 16 ARGUMENT ABOUT APPERCEPTION; THE PROBLEM OF VALIDATING KANT'S CLAIM IN § 16

In order to understand how successful the preceding § 16 line of thought is, we must now turn to its details and logical structure. Kant's argument in effect proceeds as follows. At the start of the Deduction, as we have seen, Kant makes the knowledge assumption that

(K) *H*, a being like us with a passive sensibility and an active, discursive (and so nonintuitive) understanding, knows through

> the sensible intuition in general *i*, which is given to *H* in the form of the manifold consisting of i_1 and i_2

In (K), *H* is of course an arbitrary being satisfying the conditions that (K) lays down; and *i* is an arbitrary sensible intuition in general that satisfies those conditions.[10] Moreover, in taking Kant to assume (K) I follow Chapter Four in supposing that while i_1 and i_2 are initially presented before *H*'s mind in only a potentially sequential order, i_1 and i_2 nevertheless occur before *H*'s mind in an actual sequential order insofar as *H* is taken – as in (K) – to know through *i*. Furthermore, in writing that '*H* ... knows through the sensible intuition in general *i*,' I of course mean that *H* knows an object via *i*. And I understand the claim that *H* knows an object via *i* according to one or the other of the two readings of the minimum Deduction assumption that were indicated in Chapter Three. I will note some details of these readings in Sections 4 and 5 below. However, for the most part we can ignore these readings as we consider Kant's § 16 reasoning.

As we have noted above, in B-Deduction § 16 Kant wishes to show on the basis of (K) and previously established premises that *i* and its elements are subject to transcendental unity of apperception. Now Kant has several slightly different ways of expressing that subjection. His dominant form of expression is the one that we have observed many times earlier, in which *H* is supposed to be conscious, or to be able to be conscious, in thought, that the *I think* accompanies all of *i*'s elements taken together. But, besides using this dominant form, he sometimes expresses what he takes to be the same subjection in at least two other ways. To see these other ways, consider the remark that *H* is conscious, in thought, that *he himself* (or that *she herself*) thinks a given element. Take this remark to express, in the third-person, the thought of *H*'s that *H* would express in the first-person by saying 'I think the element,' 'the element is thought by me' (and so on).[11] Then, first, besides using the dominant form, Kant also expresses the subjection of *i* to unity of apperception by saying simply that *H* is or can become conscious, in thought, that the representation *I* by itself accompanies all of *i*'s elements taken together. Second, he in addition sometimes expresses that subjection by speaking, not of *H's* consciousness that the *representation I accompanies* all those elements taken together, but rather of the fact that *H* is or can become conscious, in thought, that *he himself* (or that *she herself*) *has* or *possesses* all those elements taken together.[12]

These various expressions of *i*'s subjection to unity of apperception are not logically equivalent. Each expression allows one to set out the major argument of B-Deduction § 16 in the way that I have developed that argument above, however. Thus for simplicity we may proceed in terms of the dominant, *I think*-accompaniment form, and I will draw attention to the other forms only as is needed to understand Kant's § 16 reasoning. Adopting the dominant form, we therefore see that in order to show *i*'s subjection to unity of apperception, Kant clearly must demonstrate, from (K) and previously established premises, *at least* the weak unity-of-apperception claim

(W) Each element of the manifold of *i* is such that *H* is or can become conscious, in thought, that the *I think* accompanies that element

or, formally:

$(y)(y$ is an element of $i \supset H$ is or can become conscious in thought that the *I think* accompanies $y)$

But, just as clearly, to show that subjection Kant must in fact demonstrate not merely (W) but, also, the strong unity-of-apperception claim[13]

(S) All of the elements of the manifold of *i* are such that *H* is or can become conscious, in thought, that all of those elements, taken together, are accompanied by the *I think*

or, formally:

$(y)(z) \ldots [y, z,$ and $\ldots$ are the elements of $i \supset H$ is or can become conscious in thought that the *I think* accompanies $(y$ and z and $\ldots)]$

(S) is the central claim of the § 16 argument, for if Kant can show that *H* is conscious, in thought, that the elements of *i* are all accompanied, together, by the *I think*, then those elements, as thus *all* belonging to *one* group of elements standing before *H*'s thought-consciousness accompanied by the *I think*, will form a combination. Hence, applying his § 15 combination-cannot-be-given premise (and his general view of the faculties of *H*'s mind), Kant can argue that *H*'s mind must synthesize the elements of *i* in such a way that this last situation can arise. And he can attempt to show category application to the object of *i* and the additional results of the Transcendental Deduction. We consequently need to focus carefully below on the question of whether Kant can indeed demonstrate (S), given, to work with, only (K) and the general picture of knowledge that has been presented up to this point in the first *Critique*. And, since it may seem that if he can demonstrate a claim like (W), he can then argue from (W) to (S), we also will attend to Kantian means of arguing for (W). As the above discussion has shown, if (S) – or something like (S) – cannot be established, then the argument of the Deduction, as Kant presents that argument, simply fails.

In following sections of this chapter we will examine various means that Kant has available to argue for (S) and (W). Before we turn to those means, however, I should note, for future reference, two important questions that one can raise about the Deduction argument that follows on – and is itself based on – Kant's attempt to demonstrate (S). I also should indicate one way that Kant should *not* attempt to establish a claim like (S) (or (W)). In addition, I should say something briefly about a topic I referred to in Section 1 but have not returned to later: namely, the question of the necessity that Kant in B-Deduction § 16 means to attribute to unity of apperception – and the relation of this necessity to the overall logical structure of the Deduction.

The first of the two questions that I have mentioned is important because it points to an obvious, but little-noticed, issue about Kant's attempt to prove category application on the basis of (S). This question is posed by his view that all the elements of *i* are such that the arbitrary being like us, *H*, is *or can become* conscious, in thought, that the *I think* accompanies all of those elements taken together. On the one hand, such a view seems proper for Kant to hold. In the absence of strong reasons to the contrary, it is implausible to claim that in every case of knowledge the knower *in fact* is conscious, in thought, that 'the *I think* now accompanies *this* representation and this *other* representation and this *further* representation' or else *in fact* is conscious (in less directly Kantian-theoretical terms) that '*I* now think *this* thing – say *this* feature or aspect [represented to me, whether or not I realize the fact, by an intuition-element] – and this *other* thing and this *further* thing.' Yet, on the other hand, Kant's view here obviously leads to a serious difficulty for the basic Deduction argument that passes from the holding of unity of apperception with respect to *i* to the combination of *i* and thence to the actual, category-governed synthesis of the manifold of *i*.

In holding the view in question, after all, Kant is considering the arbitrary human or human-like knower *H* as such. So, as assumed in (K), he takes *H* to be equipped with just the usual cognitive apparatus of passive sensibility and active, discursive understanding. And his claim is that in gaining knowledge, *H*, exactly so equipped, either is conscious, in thought, that the *I think* accompanies the relevant representations or else is in principle able to use his or her cognitive apparatus, in the usual manner, in order to become conscious of that *I think*-accompaniment. It seems clear, however, that, given just this last claim of Kant's, in the above Deduction argument he can at best infer that, when *H* knows through *i*, the manifold of *i* either is or else *can* be an *I think*-governed combination before *H*'s thought-consciousness. But in order to use the above argument to show that the categories *do* apply to the object of *i*, he needs to show that, when *H* knows though *i*, the manifold of *i is* an *I think*-governed combination. Thus we arrive at the serious difficulty that I have noted. I know of no easy, plausible resolution of this difficulty within Kant's framework, and for the present I will ignore it, along with the complications that are posed by phrases like 'the knower *can* become conscious, in thought, that such-and-such.'[14] But the difficulty should be remarked here just because of its relation to the 'is or can become conscious' expression in claims like (S) and (W).

The second of the two questions that can be raised about the Deduction argument arises when one examines the reasons that that argument provides for supposing that all the elements of *i* must be held together by *H*'s mind. As I have indicated above, these reasons turn on the § 15 claim that combination cannot be given. In Chapter Four, however, we saw grounds for rejecting that claim (and Kant's reasons for it). Moreover, the discussion in that chapter suggests that, once that claim is rejected, there is no bar, within Kant's system, to supposing that those elements, as they are given to *H*, do stand in a combination – and, indeed, do stand in the precise sort of combination that is required by the holding of unity of apperception with respect to them.[15]

This last suggestion agrees with the views of various recent philosophers who have considered Kant's work. Such philosophers argue that the holding of unity of apperception with respect to a group of representations requires only that those representations should somehow occur together before the mind. These philosophers (who in general also entertain strong doubts about Kant's arguments for his idealism) then go on to suggest that the Deduction really at best establishes the nonsynthesizing (and non-idealist) conclusion that the holding of unity of apperception with respect to our cognitive states requires merely that a certain categorial order should hold in and among the objects of those cognitive states – those objects that, for these philosophers, are presumably mind-independent and are surely not synthesized by the knower's mind.[16] We need not deal with such suggestions now (or with the suggestion in the last paragraph), for our present interest is in the basic Deduction argument and its § 16 beginnings as Kant himself presents that argument. But it will be important to return to such suggestions later, and they are therefore worth bearing in mind.

As I have noted above, a further point that we need to make concerns one way in which Kant in § 16 should *not* attempt to establish claims like (S). That way would be as follows. As Kant will suppose, (a) the act of thought by means of which *H* knows through *i* is an act that grasps, as among its immediate objects, *i*'s elements. However, suppose then that he holds that (b) this act of thought actually or potentially involves the representation *I think* (or *I*) and does or can reflect upon itself, in every case in which it operates, in such a way as to recognize that involvement. Since this reflection reveals the *I think* as accompanying all of *i*'s elements as those elements are grasped by that act of thought, *i*'s elements are such that *H* is or can become conscious, in thought, that the *I think* accompanies all of those elements taken together. So we demonstrate (S) from the knowledge assumption (K) and points (a) and (b) of Kant's general account of knowledge.

The trouble with this demonstration of (S) does not lie in the assumptions (a) and (b), which Kant will clearly accept.[17] Rather, the trouble lies in the fact that, given the overall goal of the Deduction, he has no right to assume (b) without argument at the start of the Deduction. The Deduction argues for the initial result that the categories apply to the object of *any* sensible intuition in general through which *any* being like us does or can know; and the Deduction then uses that result in order to establish the final conclusion that the categories apply to *all* the objects that we do or can know through our human empirical intuitions. Now (S) (which is a claim about the *arbitrary* intuition in general *i* and the *arbitrary* being like us, *H*) is part of the Deduction's overall proof, by way of the above initial result, for this final conclusion. Hence for Kant to assume (b) without argument in order to prove (S) is for him to assume that there cannot be any cases of human or humanlike empirical knowledge that do not actually or potentially involve the *I think* and so do not (for all that he knows) have the categories applicable to their objects. But Kant cannot make this assumption without defending it.

If he does so, an opponent of the Deduction can simply ask him: 'Even if we grant, for the sake of argument, that the categories apply to those objects our knowledge of which involves first-person intellectual self-awareness through the *I think*, how do you know that *all* objects of our knowledge are like this? How do you know that there may not be genuine, actual cases of knowledge that, to use Strawson's terminology, the knower cannot self-ascribe using the *I think*? Or how do you know that there may not be cases of human or humanlike empirical knowledge that the knower actually can self-ascribe in such a way but that the knower *could* still possess in the absence of the ability to self-ascribe them in that way? (For example, suppose that the knower can now self-ascribe his or her knowledge of the mere presence of sensory qualities – of there being redness, roundness, and rubberiness here or there. Might not the knower possess such knowledge even after his or her intelligence was reduced to a level where such self-ascription was no longer possible? Or might not the knower possess such knowledge even before his or her intelligence was raised to a level where such self-ascription was possible?) But if you do not know these things, then how do you know that the categories apply to the objects of all the above cases of knowledge? Certainly not on the basis merely of an undefended assertion of (b).'

For the Deduction to succeed, Kant's demonstration of (S) must therefore do something more than merely assume (b). And, we should note, that something more must be something more than a mere appeal to the plausibility of (a). Following Kant at least for the sake of argument, we might well accept (a) and so accept that all the representations that are involved in *H*'s knowledge through *i* are or can be taken up into what is *H*'s 'one thought-consciousness' (to adapt some of the language of A116).[18] Yet, in accepting that point, we are not compelled by logic to suppose, also, with (b), that that one thought-consciousness is actually or potentially a first-person, *I think*-involving *self*-consciousness.

The defense of (S) by means of (K), (a), and (b) is thus nothing that the author of the Deduction should intend. Moreover, for similar reasons Kant cannot defend (S) by asserting it as a generalization evident in itself. Since *i* is supposed to be an arbitrary sensible intuition in general, and *H* is supposed to be an arbitrary being like us, if Kant asserts (S) as a self-evident truth, his opponent can simply argue: 'Even if we grant, for the sake of argument, that something like (S) holds for some specific intuitions – and, for some of those intuitions at least, may indeed hold in some way necessarily – how do you know that (S) holds for the *arbitrary* sensible intuition in general and the *arbitrary* being like us, *H*? If you do not know this, and if your proof that the categories apply to *all* actual and possible objects of our knowledge depends on accepting (S) in the case in which *i is* an arbitrary sensible intuition in general and *H is* an arbitrary being like us, then your proof fails.'

Undefended assumptions about knowledge's always being first-person self-ascribable or about the self-evidence of claims like (S) (or, for that matter, (W)) therefore are not means that Kant should use in § 16 to defend such claims. Moreover, the fact that Kant cannot demonstrate (S) by such a means surely agrees with what one would independently want to say about knowledge and mental states. It surely is not obvious – at least on the surface – that every bit of knowledge (and, in particular, every bit of sensible, empirical knowledge) by a being like us actually can be self-ascribed in a first-person or in some equivalent way. Nor is it obvious that such a being actually can self-ascribe, in some first-person way, all of his or her mental states or representations. And, as I hope the reader will agree after our discussion (below and in Chapter Seven) of the necessity of unity of apperception, it is no more obvious that, necessarily, such a being should be able to self-ascribe, in some first-person way, all of his or her knowledge, mental states, or representations. (Of course further reasoning might show some of these not-obvious points to be true. But I know of no such reasoning that seems thoroughly convincing.[19])

Before we turn to Kant's means of demonstrating (S) (or (W)), I should fulfill my final promise above and consider the necessity that in § 16 (and elsewhere) he means to attribute to unity of apperception. I also should comment on the bearing of this necessity on the overall logical structure of the B-Deduction. In fact, at B135 of § 16, as well as later at B142 and B144, Kant describes unity of apperception as a 'necessary unity'; and, in what is clearly a related way, at the end of B135 he describes the synthesis required by unity of apperception as a 'necessary synthesis' (compare also earlier in B135, as well as B151 and B162). Kant's ideas about these necessities involve a tangle of different views, but one of his basic lines of thought is that since it *must* be possible for the *I think* to accompany all my representations, unity of apperception holds necessarily with respect to the elements of any intuition through which I know. (Here note B131–32.) And from this fact Kant derives the necessity of the synthesis of those elements required by the holding of unity of apperception. These results then bear on the conclusion of the first half of the B-Deduction that, necessarily, the object of any sensible intuition in general through which any being like us knows is subject to the categories. And they bear also on the final conclusion of the second half of the B-Deduction, the conclusion that any objects which we human beings do *or can* know are subject to the categories.

Even the above line of thought about the necessity of unity of apperception involves many complications. It therefore seems best to present Kant's means of demonstrating (S) as far as possible in independence of his views on necessity. We may then return to such views below in Chapter Seven as a separate topic. For future reference, I should note at once, however, that our present discussion at least lets us set minimum conditions on the sorts of necessity that Kant should seek to arrive at in the first half of the B-Deduction. Thus recall from Chapter Four that, in speaking of the categories as necessary conditions of the possibility of experience, Kant is seeking to prove what I called claim (A) – namely, the claim that 'it is necessary that, for any mental state, if that mental state is a case of knowledge for a being like us, then the object of that mental state, as that object is known through that mental state, falls under the categories.' But then given his views about states of knowledge as always involving intuitions whose objects are the objects of those mental states, the overall goal of the Transcendental Deduction is to establish the claim that 'it is necessary that, for any intuition, if that intuition is an intuition through which a being like us knows, then the object of that intuition, as that object is known through that intuition, falls under the categories.' And indeed, given such Kantian views, (A) and this last claim can be shown to be equivalent.

Consequently, and in the light of our earlier discussions of the specific structure of the B-Deduction itself, the final conclusion (noted in the next-to-last paragraph) that the second half of the B-Deduction seeks to demonstrate should be the following: 'it is necessary that, for any intuition, if that intuition is a sensible, empirical intuition via which a human being knows, then the object of that intuition, as that object is known through that intuition, falls under the categories.' Since this conclusion is arrived at by applying to the case of space and time the B-Deduction first-half conclusion that I have just recalled (in the next-to-last paragraph), it is clear that that first-half conclusion, with its contained expression of necessity, is itself to be understood as making, at a minimum, the necessary-category-application claim that can be rendered precisely as[20]

(NCA) It is necessary that, for any sensible intuition in general and for any being like us, if that being knows through that intuition, then the object of that intuition, as that object is known through that intuition, falls under the categories

or, formally:

> It is necessary that $(v)(w)(v$ is a sensible intuition in general & w is a being like us & w knows through $v \supset$ the object of v, as that object is known through v, falls under the categories)

Moreover (and as our discussion in later chapters will show), this last claim (NCA) itself is ultimately arrived at on the basis of Kant's § 16 results about the necessity of unity of apperception (and about the necessity of the synthesis required by unity of apperception). Hence in holding unity of apperception in § 16 to be necessary, Kant must be making, *at a minimum*, a claim, structurally analogous to claim (NCA), that says that it is necessary that if a sensible intuition in general yields knowledge to a being like us, then the elements of the manifold of that intuition in general are subject to unity of apperception. Or, more precisely, he is making, at a minimum, the necessity-of-unity-of-apperception claim

(NUA) It is necessary that, for any sensible intuition in general and for any being like us, if that being knows through that intuition, then all of the elements of that intuition are such that that being is or can become conscious, in thought, that all of those elements, taken together, are accompanied by the *I think*

or, formally:

> It is necessary that $(v)(w)[v$ is a sensible intuition in general & w is a being like us & w knows through $v \supset (y)(z) \ldots [y, z$ and $\ldots$ are the elements of $v \supset w$ is or can become conscious in thought that the *I think* accompanies $(y$ and z and $\ldots$)]]

The necessity of category-application and the necessity of unity of apperception, as those necessities are arrived at in the first half of the B-Deduction, should therefore be taken by Kant to amount *at least* to the sorts of necessities that are expressed in (NCA) and (NUA) above. We can confirm this fact and further understand those necessities by noting that it is just such necessity-involving conclusions as (NUA) and (NCA) that he can reach in that half of the B-Deduction by arguing deductively from his above proof-from-the-possibility-of-experience assumption (K) (in conjunction, perhaps, with already-established results from his picture of knowledge).

That is, suppose that Kant shows that (S) can be validly deduced from (K) (with the aid, perhaps, of already-established results). Then the conditional claim is logically valid whose antecedent is (K) ((K) taken in conjunction with such results) and whose consequent is (S). Since '*H*' and '*i*' in this claim are arbitrary names for a being like us and a sensible intuition in general, Kant can generalize and infer the claim that, 'for any sensible intuition in general and for any being like us, if that being knows via that intuition (and if such already-established results hold), then all of the elements of that intuition are such that that being is or can become conscious, in thought, that all of those elements, taken together, are accompanied by the *I think*.' As following validly from a logically valid claim, this latter claim is, however, itself logically valid. Hence it is necessary, and so it can be prefixed by 'it is necessary that.'[21] But the result of that prefixing is (NUA) itself – or a version of (NUA) that includes an 'if such already-established results hold' clause.

We thus see that in the first half of the B-Deduction, and starting from (K), Kant can indeed infer (NUA) or a 'already-established-results' version of (NUA). But nothing in our preceding discussion or in the Chapter Four reflections about claims like (A) suggests that such a version of (NUA) (or a similar version of (A)) should not satisfy Kant in the Deduction rather than the simpler (NUA) (or (A)) by itself. Thus if, in the B-Deduction's first half, Kant deduces (S) from the proof-from-the-possibility-of-experience assumption (K), that deduction will lead him precisely to the requisite sort of necessity. Hence our view that he should take the necessity of unity of apperception to amount to at least the (NUA) sort is confirmed by the type of conclusion that he can argue for from (K) in that half of the B-Deduction.

Similarly, we can see that, in the overall argument of the first half of the B-Deduction, Kant will try to deduce from (K) (or from (K) in conjunction with already-established results) the claim 'the object of *i*, as that object is known through *i*, falls under the categories.' (Of course this deduction will itself involve deducing (S) from (K).) But then, proceeding exactly analogously to the above argument for (NUA) (or for its 'already-established-results' version), Kant can infer, as logically valid, the conditional claim whose antecedent is (K) ((K) in conjunction with already-established results) and whose consequent is this last claim. Generalizing on this conditional claim, and taking the generalization (as following validly from a logically valid claim) itself to be logically valid, he can then arrive at the necessity-involving claim that is (NCA) itself or an 'already-established-results' version of (NCA). But then, again, nothing in our earlier discussion indicates that he will not be satisfied by such a version of (NCA) in the Deduction. And so our view that the B-Deduction-first-half necessity of category application should amount to at least the (NCA) sort of necessity is also confirmed by what Kant can argue for from (K) in that half of the B-Deduction.

We will see below and in later chapters that the first half of the B-Deduction is very plausibly interpreted as following the above pattern of argument from (K) to claims like (NUA) and (NCA). The minimum that Kant requires of the necessity of unity of apperception and of category application should thus be the sort of necessity that such claims express. That sort of necessity is at bottom the necessity that belongs to a conditional claim that is logically valid because its consequent follows validly from its antecedent (or that belongs to the universal generalization of such a logically valid conditional claim). For just that reason, however, in the remainder of this chapter we will not have to focus specifically on the necessity of unity of apperception; as the preceding discussion has just shown, if Kant can validly derive (S) from (K), then he can successfully argue to (NUA) and so to the (NUA) necessity of unity of apperception. So in considering whether he can validly deduce (S) from (K), we will in effect be focusing on the necessity of unity of apperception in the minimum sense above. Thus no further discussion of that necessity is required in the following parts of this chapter.

In Chapter Seven, we will resume our comments on the necessity of unity of apperception and of category application. We will see there and in later chapters how Kant's present views relate to the additional claims about necessity that he makes in the B-Deduction. But now we are ready to turn to Kant's attempts to demonstrate the strong unity-of-apperception-expressing claim (S) itself.

4. THREE ULTIMATELY INADEQUATE KANTIAN ATTEMPTS TO VALIDATE UNITY-OF-APPERCEPTION CLAIMS LIKE (S)

(S), the reader will recall, is the claim that all of *i*'s elements are such that *H* is or can become conscious, in thought, that all those elements, taken together, are accompanied by the *I think*. As we have seen, in order to develop the argument of the B-Deduction, Kant must show that (S) can indeed be validly deduced from the proof-from-the-possibility-of-experience assumption (K). It seems clear that he can establish such a point either by making heavy use of the idea in (K) that *H knows* through *i* or else by ignoring that idea and trying to show, independently of the detailed content of (K), that (S) somehow itself follows from the fact (assumed in (K)) that *i* is a representation belonging to *H*. As I see it, of the three arguments that the text of the B-Deduction suggests for (S), the first and second represent attempts to argue validly to (S) from the idea that *H* knows through *i* (and from certain of Kant's already-established results about knowledge). And the third represents an attempt to demonstrate, without appeal to that idea, that (S) can be established immediately from the fact that *i* is one of *H*'s representations.[22]

The first of these arguments is found in the opening sentence of § 16 and the second in Kant's § 17 (and also § 16) considerations about *i*'s elements as functioning in knowledge as *one* intuition for *H*. The third is a piece of reasoning, present both in § 16 and elsewhere, which proceeds from what I will label the possibility of my calling all my representations mine. I will consider these three arguments in order. We will see that none of them succeeds in demonstrating (S) in a way suitable for the purposes of the Deduction. In Section 5 we will then briefly examine a further, fourth argument for (S) that can be developed on the basis of A-Deduction views about synthesis and knowledge.

Before turning to these various arguments, I should note that evidence exists that Kant on occasion either (a) confuses or does not bother to discriminate between an *all*-elements-of-*i*-concerning claim like (S) and an *each*-element-of-*i*-concerning claim like the weak unity-of-apperception claim (W) ('each of *i*'s elements is such that *H* is or can become conscious, in thought, that the *I think* accompanies that element') or else (b) supposes that one can pass rapidly and without difficulties from a claim like (W) to a claim like (S).[23] The fact that Kant may well do (a) or (b) creates numerous complications for the interpretation of his ways of proving (S). (W) and (S) are logically different claims (here see Section 4.A immediately below); and yet, given the possibilities noted in (a) and (b), one can wonder whether Kant is arguing directly for (S) or else arguing directly for (W), which he then takes somehow to yield (S). To avoid such complications, I will simply state what I think is the most straightforward version of each of Kant's arguments for (S), whether that version proceeds directly to (S) or to (S) by way of (W). Except in one case where it is important, I will then leave any further (W)- or (S)-style variations on these arguments for the reader to consider.[24]

Finally, I should observe that, as I have suggested above, the second of the arguments below for (S) depends on the general idea that the elements of *i* function for *H* as *one* intuition – and, more specifically, on the idea that those elements (because they so function for *H*) stand together before a single act of thought through which *H* knows. As we will see later, the further, fourth argument for (S) in Section 5 also depends on the general one-intuition-for-*H* idea. Given such dependencies, it is worth noting that if those arguments for (S) are to succeed, the claim '*H* ... knows through the sensible intuition in general *i*' in (K) must be understood according only to some of the readings of the minimum Deduction assumption that we remarked in Chapter Three and will consider in further detail in Chapter Eight.

In particular, in the second and fourth arguments this claim clearly can be understood according to the first, strong reading of that assumption. On that reading, as one can see from Chapter Eight, this claim should mean simply that *H* knows a single object through i_1 and i_2 (an object that is distinct from i_1 and i_2 and from the actual sequence (i_1, i_2)); and so *i*'s elements will have to function together before a single act of *H*'s thought as a single intuition that represents that single object. Hence that reading will require the operation of the above sort of ideas in the second and fourth arguments.

Again, and although the following fact is not made obvious by Kant's treatment of unity of apperception in the texts, one can see that in the second and fourth arguments the claim '*H* ... knows through the sensible intuition in general *i*' in (K) can be understood according to some specifications of the second, weak reading of the minimum assumption. To amplify the statement in Chapter Three, on that reading the above claim in (K) means that an object is known through i_1 and i_2. But (as far as (K) by itself goes) no specific claims are made about the nature of that object, and it is allowed that what *H* knows may amount simply to the actual (i_1, i_2) sequence occurring before an act of *H*'s thought-consciousness or to i_1 and i_2 taken separately.

Suppose, now, that we take the specification of this weak reading on which *H*'s knowledge through *i* is held to involve at least the occurrence of the (i_1, i_2) sequence before a single act of thought-consciousness through which *H* knows. Then such a specification of the weak reading evidently requires the operation of the one-intuition-for-*H* idea, interpreted (as turns out to be possible on both the second and fourth arguments) as meaning merely that the elements of *i* function before a single, knowledge-yielding act of *H*'s thought as *one* mental entity – here as the one actual sequence (i_1, i_2). So this sort of specification, too, requires the operation of the above sort of one-intuition-for-*H* ideas. However, in general the weak reading will not require the operation of such ideas. In its full generality that reading simply takes *H* to know through *i* and allows that (as far as (K) by itself goes) *H*'s knowledge might consist in no more than *H*'s knowing, in separate, disconnected acts of thought, first the occurrence of i_1 and then the occurrence of i_2, *H* not thereby being aware of any relations as holding between i_1 and i_2.

In considering the second and fourth arguments below, I will recall, as necessary, that they require certain readings of the claim '*H* ... knows through the sensible intuition in general *i*' in (K). We need not, however, examine the dependence of those arguments on such readings in further detail here. (Moreover, that dependence does not in any case apply to the first and third arguments for (S), which are independent of which version we consider of the claim in (K).)

4.A. (S) Demonstrated by appeal to the opening sentence of § 16

In this sentence, Kant writes:

> It must be possible for the *I think* to accompany *all* my representations; for otherwise something would be represented in me that could not be thought at all, and that is equivalent to saying that the representation would be impossible, or at least would be nothing to me. (B131–32)

As the talk here of '*the* representation' shows, this sentence at its end seems to make a (W)-style point. But two sentences later Kant makes a point ambiguous between (W) and (S) ('All the manifold ... has ... a necessary relation to the *I think*,' B132); and, shortly after that, without indicating any change in his opinion, he affirms the (S)-style view that an intuition's elements '*can* stand together in one universal self-consciousness' (B132). So I take Kant in the above § 16 opening sentence to be arguing for (W), from which he takes (S) itself immediately to follow.

This reasoning for (S) can be seen in more detail by noting first that Kant supposes that animals have representations without having the capacity for self-awareness through the *I think*.[25] So in interpreting the opening § 16 sentence we can disregard the clause 'the representation would be impossible.'[26] With this clause disregarded, the sentence then makes the (W)-style claim that, necessarily, any representation of mine that is not 'nothing to me' is such that my *I think* can accompany it. In § 16 Kant does not say under what conditions a representation is not nothing to me. But he surely thinks that those representations through which I know are not nothing to me.[27] So in the opening § 16 sentence Kant is asserting, in (W)-style, that it is necessary that any representation of mine that is not nothing to me, including any representation through which I know, is such that my *I think* can accompany it.

Given this (W)-style assertion, Kant now can be interpreted as arguing to (S) as follows. Since, by (K), *H* knows through *i*, the elements of *i* are representations of *H*'s that occur before *H*'s thought-consciousness; and, precisely as these representations so occur and play a role in *H*'s knowledge, they are not nothing to *H*. Consequently, by the (W)-style assertion, which of course applies to *H* and the elements of *i*, *H*'s *I think* can accompany each of these representations as it so occurs before *H*'s thought-consciousness. So we reach the conclusion, (W) itself, that each element of *i* is indeed such that *H* is or can become conscious, in thought, that the *I think* accompanies that element. And hence, given that Kant here regards (S) as immediately following from (W), (S) is itself demonstrated.

The preceding argument for (S) clearly hews to the Kantian texts. But it has two fatal defects. First, in order to establish (W) it assumes the above (W)-style assertion and hence assumes, in effect, that since each of *i*'s elements is a representation through which *H* knows, the *I think* can accompany that element. But because *H* is any being like us and *i* is any sensible intuition in general, to assume this last thing is tantamount to assuming that any representation or intuition-element through which any being like us knows is such that the *I think* can accompany that element. Yet evidently this assumption, which Kant makes without justification, is subject to the same sorts of questions as is his assumption (b) in our rejected Section 3 argument to (S) from (K), (a), and (b). ((b) held that the act of thought through which *H* knows actually or potentially involves the *I think*.) So the present argument does not establish (W) satisfactorily, let alone (S).

Second, the present argument takes (S) to follow immediately from (W). But this position is mistaken. It is true that (W) follows from (S). However, (W) by itself does not imply (S). Suppose (W) is true in the case in which *H* is aware in separate acts of thought of the *I think* as accompanying i_1 and of the *I think* as accompanying i_2. Then the truth of (W) by itself clearly allows that *H* may *not* actually be aware, in *one single* act of thought, of the *I think* as accompanying both i_1 and i_2 taken together. The truth of (W) by itself also gives no reason to suppose that *H*, the *arbitrary* being like us, even has the *ability* to be aware, in one single act of thought, of the *I think* as so accompanying i_1 and i_2. (To see this point graphically, imagine that not 2 but 200 intuition-elements are in question.[28]) Hence in this case (W) is true but not (S). So (S) does not follow from (W) when (W) is taken by itself. And thus the present argument does not derive (S) satisfactorily from (W).

4.B. (S) Demonstrated Through the 'One Intuition for H' Idea

This second argument for (S) can be reached from § 17 remarks at B138 and, for example, from A354. (Thus note B138 on each intuition as having to stand under synthetic unity of consciousness 'in order *to become an object for me*' and the A354 remark that 'I think (the manifold in a [or: in one] representation).' Compare also B132 and B135 of § 16.) For reasons indicated at the beginning of Section 4, this argument succeeds only if the claim '*H* ... knows through the sensible intuition in general *i*' in (K) is understood according to one of the readings of the minimum assumption that we noted there. In fact, texts like those just cited suggest taking this claim according to the first, strong reading, according to which a single object distinct from i_1 and i_2 and the (i_1, i_2) sequence is known through *i*. However, and as noted, this claim can be read also according to the sort of specification of the second, weak reading that we observed – the specification on which *H*'s knowledge through *i* involves at least the occurrence of the (i_1, i_2) sequence before a single act of thought-consciousness through which *H* knows. I will assume that this claim is read in one of these ways.

Put succinctly, the second argument says that since, by (K), *H* knows through *i*, *i*'s elements function together before *H*'s thought-consciousness as a single intuition. However, in order for these elements to function in such a way, they must occur before the single act of thought through which *H* knows and there form one thing. Yet this single act of thought is an act of *H*'s apperceptive thought-consciousness. So it does or can involve within itself the *I think*, and it does or can reflect upon itself so as to recognize that involvement. (Here recall the similar-sounding (b) from our Section 3 argument for (S) from (K), (a), and (b).) Therefore through this single act of thought *H* is or can become conscious that all of *i*'s elements, taken together, are accompanied by the *I think*. And hence (S) holds.

The difficulty for this second argument for (S) should be obvious. Like the Section 3 argument for (S) from (K), (a), and (b), it assumes without justification that *H*'s knowledge through *i* involves a single act of self-reflective, *I think*-involving thought-consciousness, a single act before which all the elements of *i* somehow occur together. But, as we have seen in discussing the Section 3 argument, Kant cannot make such an assumption without justification if his demonstration of (S) is to do the work that the Deduction requires. Therefore the present argument for (S) fails. We could of course defend this argument by adopting precisely such an assumption; and one could, indeed, defend the first argument for (S) or the argument from (K), (a), and (b) by making a similar assumption or simply by assuming (b) itself. The effect of so proceeding would be to weaken the Deduction by restricting it to the conclusion that the categories apply to any object that any being like us does or can know through an act of thought-consciousness that actually or potentially involves the *I think* (and is directed to a sensible intuition). Some such weakening may be forced on us if no satisfactory argument for (S) emerges. (Here see Chapter Seven.) But it is crucial to see now that such a weakening abandons the Deduction's original, strong goal of demonstrating that the categories apply to *all* objects that any being like us does or can know. If Kant were to abandon that goal, he would have to admit that the Deduction yields no proof that *every* object of a spatiotemporal intuition, actual or possible, falls under the categories. Such an admission would require major changes in the remainder of the Deduction and of the first *Critique*. Kant himself could hardly accept these changes with equanimity. So such a weakening of the Deduction cannot be anything that he would be at all eager to accept.

4.C. (S) Demonstrated by Appeal to the Possibility of My Calling All My Representations Mine

As noted above, this third argument for (S) focuses on the fact (assumed in (K)) that *i* is one of *H*'s representations, while ignoring the idea in (K) that *H* knows through *i*. Moreover, and as we will see, this argument in the end establishes not the *I think*-accompaniment form of unity of apperception that is expressed in (S) but, rather, the form that says that *H* is or can become conscious that *he himself* (or that *she herself*) *possesses* all of the elements of *i* taken together. At the beginning of Section 4 we observed that such a possession form of unity of apperception is not equivalent to (S) itself. But because Kant can conduct the remainder of the B-Deduction argument in terms of this possession form – and because he himself does not bother to distinguish the possession from the *I think*-accompaniment form – I will proceed in terms of the possession form. I should also note that the statement of the third argument is complicated by the question of whether Kant intends this argument to reason directly to (S) (or to a possession form of (S)) or only to (S) by way of (W). In order to deal with these complications, I will eventually consider both such versions of the argument.

The third argument can be found in three places in B-Deduction § 16 and also appears in the A-Deduction, for example at A129. Here are the relevant passages from § 16:

As *my* representations ((q) even if I am not conscious of them as such) (r) they must conform to the condition under which alone (s) they *can* stand together in one universal self-consciousness, because otherwise (t) they would not all without exception [*insgesamt*] belong to *me*. (B132–33, with emphasis altered and letters inserted)

The thought that the representations given in intuition one and all [*insgesamt*] belong to *me*, is therefore equivalent to [*heißt demnach soviel, als*] the thought that I unite them in one self-consciousness, or can at least so unite them ... In other words, only insofar as I can grasp the manifold of the representations in one consciousness, do I call them one and all *mine*. (B134)

I am conscious of the self as identical in respect of the manifold of representations that are given to me in an intuition, because I call them one and all *my* representations, and so apprehend them as constituting *one* intuition. (B135)

And here is the A-Deduction:

Now this very idea – that all these appearances, and consequently all objects with which we can occupy ourselves, are one and all in me, that is, are determinations of my identical self – expresses as necessary a complete unity of these same appearances and objects in one and the same apperception. (A129; compare A122)

In its direct-to-(S) version, the argument that Kant has in mind in these passages can be illustrated by explicating the first quotation, from B132–33. Since all my representations belong to me (by (t)),[29] he is arguing, they all can stand together in a universal self-consciousness ((s)), even if I am not in fact conscious of them as being my representations ((q)). But (s) here, the claim that all my representations can stand together in a universal self-consciousness, will yield (S), given the fact that the elements of *i* are my representations. So, using that fact (itself contained in (K)), Kant thus argues to (S) via the obvious or trivial point that all of my representations are mine.[30] Moreover, he clearly can argue in a similar way to (S) by way of (W). Thus (to appeal again to the B132–33 quotation) since all of my representations belong to me (by (t)), we see once more that (s) holds – that all of my representations can stand together in a universal self-consciousness. But this last claim (s) can be read as equivalent to (or as implying) the result that *each* of my representations can occur before my apperceptive self-consciousness. (Here note, also, the B134 and B135 quotations above.) This result then yields (W), given that *i*'s elements are representations of mine. And from (W) Kant can attempt to argue to (S).[31]

These last two versions of the third argument for (S) can profitably be gone over in slow motion. In the case of the first version, Kant in effect proceeds thus. He takes (c)

(c) All of *H*'s representations are *H*'s

in the first-person form (d)

(d) All of my representations are mine

as being trivial or analytic (and as in any case embodying a necessary truth).[32] Regarding me as *H*, he then supposes that I am or can become conscious, in first-person thought, of the truth stated in (d), whence we get (e):

(e) I am or can become conscious in thought that (all of my representations are mine)

But now he regards 'all of my representations' in (e) (and in (d)) as meaning 'the sum total of my representations.' So my consciousness in (e) is really the consciousness expressed in

(f) I am or can become conscious in thought that (the sum total of my representations is mine)

But, he seems to think, 'the sum total of my representations = the individual representations *r*, *s*, *t*, and so on.' Hence from (f) he concludes that

(g) I am or can become conscious in thought that (the individual representations *r*, *s*, *t*, and so on are mine).

And since the elements of *i*, i_1 and i_2, are of course my representations and so are supposed to be listed alongside *r*, *s*, *t*, in (g), from (g) a possession form of (S) follows: I am or can become conscious in thought that I possess i_1 and i_2 (the elements of *i*) taken together.[33]

In the case of the second version of the above argument for (S), Kant in effect argues just as does the first version down to (e). But then he takes 'all of my representations' in (e) (and (d)) to mean 'each of my representations' rather than to mean 'the sum total of my representations.' My consciousness in (e) thus is now expressed in

(h) I am or can become conscious in thought that (each of my representations is mine)

From (h) Kant then takes it to follow that

(i) Each of my representations is such that I am or can become conscious in thought, of it, that it is mine

But since i_1 and i_2 are my representations, from (i) a possession form of (W) follows: Each of *i*'s elements is such that I am or can become conscious, in thought, that I possess that element. And Kant will take the possession form of (S) noted at the end of the last paragraph to follow immediately from this possession form of (W).

Unhappily for Kant's discussion in B-Deduction § 16 and elsewhere, there are many problems with the above two versions of our third argument for (S). For one thing, in each version it is assumed that I, in my role as *H*, must be able to recognize and accept, in first-person thought, the truth (d), in such a way that (e) becomes true ((e) read according to the relevant construal of 'all my representations,' of course). But *H*, as an arbitrary being like us, cannot be assumed *not* to be very young, ignorant, or unintelligent. Yet Kant does not supply, and I myself do not see, any good reason to suppose that such a being *has* to be assumed to accept – or to be able to accept – the specific truth in (d). Thus from (K) by itself we cannot deduce, on either construal of 'all of my representations' in (d), that I (as *H*) am or can become conscious in the way expressed in (e).[34]

For another thing, and crucially, each version of the argument is plagued by fallacious inferences having to do with intensionality. To see this point, recall from Chapter Five that sentences like (T) of that chapter ('*H* thinks that the object before *H* has the property of being conical'), which contain expressions like '*H* thinks that,' exhibit intensionality. But then the same thing is evidently true of sentences above, like (e) to (i) (or like (W) and (S)), that contain expressions like '*H* is [I am] or can become conscious in thought that.' And indeed all such expressions are intensional operators, in the sense that prefixing such an expression to a given sentence yields a new sentence that exhibits intensionality.

Consider now, however, the first version of the third argument. According to that version, Kant wishes to deduce (g) from (f), given the identifying fact that 'the sum total of my representations = the individual representations *r*, *s*, *t*, and so on.' But this deduction cannot succeed. Suppose that I (*H*) do not know that identifying fact.[35] Then while I can be conscious of the general truth, contained in (f), that the sum total of my representations is mine, I certainly need not be conscious of the specific truth, contained in (g), that the *individual*, *particular* representations *r*, *s*, *t*, and so on are mine. Indeed, (f), prefixed as it is by an intensional operator, clearly possesses intensionality; and, as we have seen in Chapter Five, substitution of coreferential expressions in intensional sentences does not, in general, preserve truth-value. So it will be a clear intensional fallacy to infer (g) from (f) on the ground that, because of the above identifying fact, the term 'the sum total of my representations' in (f) is coextensive with the term 'the individual representations *r*, *s*, *t*, and so on.' Moreover, it is hard to see how Kant can hope to pass from (f) to (g) without committing this fallacy.

Consider next the second version of the third argument. In supposing that if a possession form of (W) can be derived, then a possession form of (S) follows immediately, that version is of course already in trouble for nonintensional reasons of the sort indicated in our Section 4.A discussion of the first argument. But that version also faces a difficulty turning on intensionality. In taking (i) above to follow from (h), Kant falls victim to an intensional-operator-shift fallacy.

To explain most clearly how this operator-shift fallacy arises, I need to digress for a moment and comment on the *de re-de dicto* distinction. Some intensional sentences are *de re*, in the sense that they express the thought (or belief, knowledge, hope, fear, and so on), *of* or *about* some individual, particular object, that that individual, particular object is such-and-such. Other intensional sentences are *de dicto*, in the sense that they express the thought that some purely general proposition is true, so that that thought does not concern any individual, particular object. *De re* sentences include such claims as (on their usual readings) 'Jane thinks (or believes, knows, and so on), *of* the World Trade Center, that it is high' or (to recall our discussion in Chapter Five) '*H* thinks, *of* the object before *H*, that it conical.' *De dicto* sentences are given by such examples as (on their usual readings) 'George thinks (or believes, knows, and so on) that there is a high building' (where George simply accepts the general, existential proposition that is expressed by the that-clause in this last statement and concerns himself with no individual, particular object).

The *de re-de dicto* distinction is best understood as turning on the scopes of the intensional operators that occur in sentences like those above. By introducing quantifiers and other logical tools, one can easily capture these matters of scope in precise terms. Our above *de re* sentence about Jane, for example, may be read as (equivalent to) a claim that asserts that[36]

> There is a certain particular, individual thing that is the World Trade Center and is such that Jane thinks (or believes, knows, and so on) that (that thing is high)

or, formally:

> $(\exists x)(x$ = the World Trade Center & Jane thinks (or believes, knows, and so on) that (x is high))

And our above *de dicto* sentence about George may be read as (equivalent to) a claim that asserts that

> George thinks (or believes, knows, and so on) that (there is a certain thing [which is not here specified as being any one particular, individual thing] which is such that that thing is a high building)

or, formally:

> George thinks (or believes, knows, and so on) that $(\exists x)(x$ is a high building)

By proceeding along the general lines above, one can satisfactorily understand any given *de re* or *de dicto* claim. And this point now returns us to the intensional-operator-shift fallacy that I have mentioned. Suppose (as on the second version of the third argument for (S)) that (h) is got through (e) and so through *H*'s being aware (or being able to be aware), in thought, of the trivial truth stated in (d) (with 'all of my representations' in (d) and (e) construed as 'each of my representations'). Then (h) can only be the *de dicto* claim that

(h*) I am or can become conscious in thought that (for each thing, if that thing is a representation of mine, then that thing is mine)

or, formally:

I am or can become conscious in thought that $(x)(x$ is a representation of mine $\supset x$ is mine)

But (i), if (i) is to yield a possession form of (W) and then such a form of (S), must be the *de re* claim

(i*) For each thing, if that thing is a representation of mine, then I am or can become conscious in thought that (that thing is mine)

or, formally:

$(x)(x$ is a representation of mine $\supset$ I am or can become conscious in thought that x is mine)

Yet to infer (i*) from (h*) is fallaciously to move the intensional operator 'I am or can become conscious in thought that' across the quantifier and the implication sign of (h*) and is thereby also fallaciously to transform a *de dicto* into a *de re* claim. This shift is clearly fallacious, for my consciousness of the general truth that each of my representations is mine of course does not require my consciousness, *of* each particular one of my representations, that *that* particular representation is mine.

We thus see that, besides its other difficulties, the second version of our third argument for (S), like the first version, is undercut by a problem of intensionality. It is worth noting that the intensional problems of both versions are very similar. While the problem for the first version does not involve any operator-shift fallacy, that problem is like the problem for the second version in making what is obviously an erroneous transition from a *de dicto* to a *de re* claim.[37]

Kant, of course, does not himself explicitly offer each step either of the above (c)-to-(g)-to-(S) first-version argument for (S) or of the above (c)-to-(W) second-version argument for (S). But the first three passages quoted above strongly suggest that something like the first-version reasoning is going on in B-Deduction § 16. And the last, A129, passage suggests that the same sort of reasoning operates also in the A-Deduction. Moreover, if we read in (W)-style a claim like 'I am conscious of the self as identical in respect of the manifold of representations that are given to me' in the above B135 passage, then it is clear how this B135 passage (and, indeed, the other three passages quoted) can suggest that something like the second-version reasoning is present in B-Deduction § 16 (and, by A129, is present also in the A-Deduction).

Furthermore, suppose that we read in (S)-style the parts of the above B134 and A129 quotations that concern the idea that my representations all occur in or before my self-consciousness. Then these quotations clearly show Kant identifying (or very nearly identifying) statements on the order of (f) with statements on the order of (g). (Note, especially, the first sentence in the B134 quotation.) Or, again, suppose that we read in (W)- rather than in (S)-style the parts of these B134 and A129 quotations that concern the idea just mentioned. Then these quotations clearly show Kant identifying (or very nearly identifying) statements on the order of (h*) with statements on the order of (i*). (Note, again, the first sentence of the B134 quotation.)

It also is worth observing that such fallacious identifications were not uncommon in Kant's time. Berkeley's notorious argument in *Principles of Human Knowledge*, Part I, § 23, that one cannot, in logic, conceive of an unperceived object is plausibly interpreted as resting, in part, on an identification similar to the (h*)-(i*) one. It is thus not exegetically or historically surprising to find Kant arguing fallaciously, in the ways just sketched, insofar as he offers one or the other version of his defense of (S) from the possibility of my calling all my representations mine. But once we recognize the fallacies, we have no choice but to abandon these ways of defending (S).

Since this defense of (S) is, along with the other two arguments noted above, the only way of establishing (S) that Kant offers in the B-Deduction, we are forced to look to other possible defenses of (S). In particular, we will find it helpful to look briefly at Kant's A-Deduction account of the synthesis of the manifold of intuition in knowledge.

Before turning to that account, let me note finally that in his well-known paper on the proof-structure of the B-Deduction, Dieter Henrich argues in effect that in § 16 Kant commits a fallacy of ambiguity by passing from the claim that the representations given to me are mine (mine = in my sensibility and only available to be taken up into my consciousness) to the claim that the representations given to me are mine (mine = occurring or capable of occurring as an object of my consciousness). If we read in an (S)-style the idea that the representations given to me occur or are capable of occurring as objects of my consciousness (that is, if we read in an (S)-style what Henrich regards as the second sense of 'mine' here), then I take this fallacy to be, in fact, the illicit *de dicto*-to-*de re* transition from (f) to (g) that we have seen above in the first version of our third argument for (S). Again, if we read in a (W)-style the idea in question, then I take this fallacy to be the (h*)-to-(i*) *de dicto*-to-*de re* operator-shift fallacy that we have just seen in the second version of our third argument for (S).[38]

5. CAN (S) BE VALIDATED BY KANT'S ACCOUNT OF SYNTHESIS? A FOURTH ARGUMENT FOR (S)

The upshot of the preceding discussion is that none of the B-Deduction § 16 arguments allows Kant to demonstrate (S) (or, indeed, (W)) in a way that is satisfactory for the goals of the Transcendental Deduction. It is worth noting that (S) – or perhaps (W) – will evidently express the situation that obtains in *H*'s mind when, through a process of analysis, *H* discovers (or is in a position to discover) that the *I think* does or can accompany all of *H*'s representations. So a claim like (S), or perhaps (W), will express the holding of what at B133 of § 16 Kant calls analytic unity of apperception (with respect to the elements of *i*).[39] Thus the preceding discussion shows that in § 16 he does not establish that holding in a satisfactory way. Hence he is in no position to argue from the holding of analytic unity of apperception to the further, central § 16 result that a synthetic unity of apperception is required with respect to *i*'s elements – that is, a synthesis, by *H*'s mind, of those elements must occur in such a way that (S) comes to be true. Kant's failure in § 16 to establish (S) therefore brings the basic argument of the Deduction to a halt almost before it is begun. Moreover, this failure of course also halts any attempt to reconstruct the Deduction that ignores his views on synthesis but still aims to demonstrate a claim like (S).

Kant is consequently in difficulties in § 16. I myself see no escape from these difficulties that is in the end completely satisfactory. However, before we come (in Chapter Seven and following chapters) to the rather unhappy consequences of this situation, I want to notice a final, fourth line of argument for (S), a line of argument based on Kant's A-Deduction account of synthesis. This line of argument itself fails. But it is worth considering here both for the sake of completeness and because – as we note in Chapter Seven – one of its subconclusions, which can be related to further Kantian views on apperception, suggests a way that Kant (although not we) might think it really possible to derive (S), or a claim akin to (S), from (K) and already-established Kantian views.

As observed at the start of Section 4, this fourth line of argument requires the claim '*H* ... knows through the sensible intuition in general *i*' in (K) to be taken according to one of the readings of the minimum Deduction assumption that we noted there. As with the second argument for (S), the texts suggest taking this claim according to the strong reading (according to which *i*'s elements are assumed to function for *H* as one intuition representing a single object). Yet, as we remarked, this claim can be taken also according to the sort of specification of the weak reading that we noted (according to which the (i_1, i_2) sequence occurs before a single act of thought through which *H* knows). For simplicity I will present the fourth argument simply in terms of the strong reading. It should then be clear how a similar argument could be developed by appeal to the specification of the weak reading.[40] I should note also that, contrary to my practice below, it would be possible to detach the basic fourth-argument reasoning from any appeal to Kant's views on synthesis. However – and while to my knowledge no form of the fourth argument is ever explicitly presented by Kant – some of his discussions in the A-Deduction threefold-synthesis passage come close to the form of the fourth argument that I will now develop. That form illustrates also Kant's well-known Deduction inference from unity of object (or intuition) to unity of apperception. So I focus on that form below.

Presented as simply as possible, the relevant form of the fourth line of argument runs as follows.[41] Since, by (K), *H* knows through *i* (this claim taken on the strong reading), *i*'s elements function for *H* as one intuition representing a single object. For these elements to function in this way, however, they must be synthesized by *H*. Moreover, since these elements are presented before *H*'s mind in a fleeting, sequential order, this synthesis must itself take place in a sequential fashion. At each of its stages it must retain (through the reproductive power of *H*'s imagination) the element then presented to *H* along with the previously presented elements.[42]

In addition, knowledge through *i* involves concept application. (This concept application is of course to the object known through *i* – an object which we are not here assuming to be subject to the categories – but we can ignore the details of that point now.) Indeed – and given Kant's account of concepts – the sequential synthesis of *i*'s elements must involve *H*'s thinking that the various sequentially presented features that are put before the mind by these elements do make up – or otherwise relate to – a general concept that is here applied.

For Kant, however, we arrive at knowledge in what is in general an ongoing, cumulative fashion. And that point about cumulativeness we may take to imply that at each later stage of this last synthesis, and at its conclusion, *H* can recall the earlier stages. *H* can consider the individual thoughts (about sequentially presented features, and so on) that those earlier stages have involved, and *H* can take those individual thoughts together to have yielded one overall piece of knowledge (for example, the knowledge that such features make up a concept). However, this one piece of knowledge is itself expressed in a single thought. And – as we may hold the above point about cumulativeness to imply – *H* can take the individual thoughts in question to have yielded this single thought. Hence when *i* is synthesized, *H* can take there to be a single thought – which by the foregoing comments (we can argue) *H* can realize to involve i_1 and i_2 – that has yielded knowledge.

For Kant, however, each (act of) thought belongs to a single thinker; and the present line of argument supposes that *H* can regard this Kantian position as true.[43] Hence *H* can take the single thought just noted, which *H* can realize to involve i_1 and i_2, to belong to a single thinker. So (K) and Kant's views on synthesis imply that

(j) *H* is or can become conscious in thought that there is a single thing that has both i_1 and i_2

,slp -2 or, formally:

> *H* is or can become conscious in thought that $(\exists! u)(u$ has both i_1 and $i_2)$

Suppose, however, that from (*j*) we can derive the first-person

(k) *H* is or can become conscious in thought that *he himself* (or that *she herself*) has both i_1 and i_2

Then, because i_1 and i_2 are the elements of *i*, from (k) we can reach (S) (or, in fact, a possession form of (S)).[44]

As we will see in Chapter Seven, particularly in its claim (j) this fourth argument makes assertions of great interest for Kant's further account of apperception. But here I want only to observe that, despite its thoroughly Kantian nature, this argument fails, for at least two reasons.[45]

First, if *H* is conscious, in a genuinely first-person way, that *he himself* (or that *she herself*) does *F*, then *H* is *de re* conscious, *of* the entity that *H* in fact is, that *that* entity does *F*.[46] So a genuinely first-person claim like (k) will imply that

> *Of* the entity that is in fact *H*, *H* is or can become conscious in thought that (that entity has both i_1 and i_2)

or, formally:

> $(\exists x)[x = H$ & *H* is or can become conscious in thought that (x has both i_1 and i_2)]

But a *de dicto* claim like (j) clearly implies no such *de re* claim. Hence (j) fails to imply (k), and therefore the fourth argument cannot reach (S) from (j) in its desired way.

Second, perhaps this first difficulty with the fourth argument could be evaded. (Here note Chapter Seven, on the 'purely existential form of (S).') However, a further problem remains. The fourth argument does not establish (j) itself in a way satisfactory for the Deduction.

To see this further problem, note that Kant must establish (S), and hence (j), in such a way as to show that, for *any* being like us, actual or possible, and for *any* intuition through which that being knows, that being is or can become conscious that the *I think* accompanies the elements of that intuition. Consequently in demonstrating (j) the fourth argument should assume nothing about *H*'s knowledge that does not apply to the knowledge belonging to any actual or possible being like us. But the fourth argument does make such assumptions, both in its view that knowledge is always arrived at in an ongoing, cumulative fashion and in its view that *H* can take each thought to belong to a single thinker.

Thus the process of our arriving at most of our actual knowledge may well be cumulative, in the sense that in the later stages of this process we can recall the earlier stages and can take the thoughts that they involve to have yielded knowledge. But it also seems that beings like us might have much knowledge that they subsequently could *not*, even in principle, recall in such a way as to take the thoughts that that knowledge involves to have yielded knowledge. (Stretches of knowledge of individual instances of sense-qualities – say of individual red or green patches, and so on – might provide an example.) Again, it seems very implausible to claim that every actual or possible being like us regards as true – or must be able to regard as true – the position that each thought belongs to a single thinker. Even if that position proves true and is accepted as true by trained Kantian philosophers, it is hard to see that there could not be – or that there *are* not – unexceptional human knowers who can know many objects like sticks, stones, bricks, and stars yet who simply cannot grasp that position, let alone regard it as true.[47]

For the reasons just cited, the fourth argument does not succeed. At the cost of weakening the Deduction, we could of course simply assume as true the two conditions that *H*'s process of arriving at knowledge is cumulative and that *H* does or can accept the position that each thought belongs to a single thinker. So proceeding, we could use the fourth argument to reach (j). And by means of further points that we see in Chapter Seven, we could then argue that (j) itself – or a purely existential, (j)-like form of (S) – expresses a result adequate to show the holding of unity of apperception with respect to *i*.[48]

Because an argument along these lines exists – and because Kant (although not we) might really think it possible to defend a purely existential, (j)-like form of (S) by appeal to such an argument – we will briefly consider the idea of such a form of (S) further in Chapter Seven. It should be evident at once, however, that the appeal to any such argument would seriously weaken the Deduction. Instead of showing that the categories apply to all objects that are or can be known by beings like us, the Deduction would then show only that the categories apply to all objects – that are or can be known by beings like us – that happen *also* to satisfy the above two conditions. That conclusion is far from Kant's own desired result in the Deduction. Thus it is clear already that no such reasoning will attain the original, strong goals of the Deduction. So, although it will be worth considering the purely existential form of (S) in

Chapter Seven, no such reasoning yields that form of (S) and the holding of unity of apperception in a satisfactory way. Moreover, it seems clear that if the fourth argument cannot be satisfactorily defended even by such reasoning, then we have no ground to suppose that the fourth argument can be satisfactorily maintained by any means.

With the failure of that argument, there collapses the last hope that I see, within the original framework of the Deduction, of actually demonstrating (S) and so the holding of unity of apperception with respect to *i*. Postponing further comments on this situation and its rather unhappy consequences until Chapter Seven, I will remark here simply that Kant's failure to prove (S) is serious and perhaps surprising. The sense that one gets from the Deduction and from many of the commentators is, after all, that while much argument is needed to establish category-application on the basis of the holding of unity of apperception, that holding itself is easily demonstrated. But if the above discussion is correct, Kant has no good argument for (S) and so for unity of apperception. Consequently B-Deduction § 16 – and the main line of thought in both the A- and B-Deductions – begins with a much less certain claim about that unity than is often realized.[49]

6. SUMMARY

We examined the opening stage of the B-Deduction § 16 line of argument that passes from the assumption (K) that *H*, the arbitrary being like us, knows through sensible intuition in general *i* to the subjection of the manifold of *i* to unity of apperception and thence to the synthesis of that manifold by *H*. In this stage, Kant assumes (K) (and already-established Kantian results) and attempts to deduce the strong unity-of-apperception-expressing claim (S) – the conclusion that *H* is or can become conscious in thought that the *I think* accompanies all of the elements of *i* taken together. If Kant can show (S), then he can infer that *i*'s elements form a combination before *H*'s thought-consciousness. And, using the § 15 claim that combination cannot be given, he can conclude that those elements must therefore have been synthesized by *H*.

Before considering the possible ways of demonstrating (S), we noted a difficulty about the phrase 'or can become conscious in thought' in (S); we remarked that (S) (and the combination of *i*) can still be accepted even if one rejects the § 15 idea that combination cannot be given; and we observed that Kant should not try to show (S) simply by assuming that the act of thought that grasps *i* always somehow involves the *I think*. We saw also that, in speaking of the necessity of the unity of apperception, Kant must mean at least the claim (NUA) – the claim that, necessarily, for any sensible intuition in general and for any being like us, if that being knows through that intuition, then that being is or can become conscious in thought that the *I think* accompanies the elements of that intuition taken together.

We then considered three proofs that the B-Deduction suggests for (S). These proofs were the argument from the opening sentence of § 16; the argument from the idea that *i*'s elements must function as *one* intuition for *H*; and the argument from the possibility of my calling all my representations mine. We saw these three arguments to fail. The first and second fail because they assume without proof that all of *H*'s knowledge by means of intuition-elements involves *I think*-accompaniment. The third fails because (among other things) it commits one or another logical fallacy involving intensionality. Lastly, we examined a fourth argument, which tries to show (S) on the basis of claims about synthesis and the Kantian view that each thought belongs to a single thinker. This fourth argument raises points about apperception and a possible way of deriving (S) that we will examine further in Chapter Seven. But it itself fails, for it fails to demonstrate its subconclusions properly. Thus we concluded that the B-Deduction (and the A-Deduction) in fact has no satisfactory argument for (S).

CHAPTER SEVEN

TRANSCENDENTAL UNITY OF APPERCEPTION AND ITS NECESSITY

1. INTRODUCTION

As we saw in Chapter Six, (S) is the strong unity-of-apperception claim that all the elements of *i* are such that *H* is or can become conscious, in thought, that the *I think* accompanies all those elements taken together. Given the failure of our Chapter Six arguments for (S), it seems impossible for Kant to prove that *i*'s elements form a synthesis-established (and necessary) unity within *H*'s mind in a way that leads to category application to the object of *i*. So, also, he cannot generalize to the main, B-Deduction first-half conclusion that, necessarily, the object of any sensible intuition in general through which a being like us knows is subject to the categories. And hence in the second half of the B-Deduction he cannot apply that conclusion to the human *a priori* intuitions of space and time in such a way as to reach the final B-Deduction result that, necessarily, the object of any sensible, empirical intuition through which we know falls under the categories. Kant's failure to demonstrate (S) thus abruptly halts the argument of the B-Deduction (and of the A-Deduction), and so we must decide how to proceed if something like the Deduction reasoning is to be maintained.

As I will suggest below, in the end it is best to abandon the attempt to *prove* (S) and, instead, simply to *stipulate* that (S) (or some related claim) holds true, so that *H*'s knowledge through *i* is thereby assumed subject to unity of apperception. However, before we come to the details of this stipulation, we should consider the topic, raised at the end of the last chapter, of what I there called the purely existential form of (S). The idea of such a form of (S) bears on the form in which we stipulate (S), and it also is of interest in itself. In Section 2 I examine that idea and the appropriate way to stipulate (S). In Section 3 I then complete the basic parts of our Chapter Six discussion of the necessity of unity of apperception. This discussion is meant to be comprehensive but brief. While Kant emphasizes points connected with the necessity of unity of apperception at various places in the Deduction, it turns out that in Chapter Six we have already seen the claims about that necessity for which he can hope to make a reasonable case.

2. STIPULATING (S) AND UNITY OF APPERCEPTION

Claim (K), as it was introduced in Chapter Six, is the opening deduction assumption that, roughly, *H* knows through the sensible intuition in general *i* and its manifold i_1, i_2. As we saw there, if we assume (as I explained) that *H* arrives at knowledge in an ongoing, cumulative way and that *H* does or can accept the Kantian position that each thought has a single thinker, then, using the fourth argument for (S), we can at least infer from (K) and already-established Kantian results that

(j) *H* is or can become conscious in thought that there is a single thing that has both i_1 and i_2

or, formally:

H is or can become conscious in thought $(\exists!u)(u$ has both i_1 and $i_2)$

However, from (j) and the fact that i_1 and i_2 are the elements of *i*, the conclusion clearly follows that

[Purely existential form of (S):]

All of the elements of the manifold of *i* are such that *H* is or can become conscious, in thought, that there is a single thing such that that thing has all of *i*'s elements taken together

or, formally:[1]

$(y)(z) \ldots [y, z,$ and $\ldots$ are the elements of $i \supset H$ is or can become conscious in thought that $(\exists!u)(u$ has $y, z,$ and $\ldots)]$

This conclusion is evidently an impersonal, existentially quantified version of (S) – or, for short, a purely existential form of (S).[2]

In our discussions so far, we have understood unity of apperception with respect to *i* to be expressed in claims like (S) in which *H* is taken to have the genuinely first-person thought that *he himself* (or that *she herself*) has *i*'s elements – or in which *H* has the thought that the first-person representation *I think* accompanies *i*'s elements. It is clear, however, that for purposes of the Transcendental Deduction unity of apperception is equally well expressed by the purely existential form of

(S). If that form of (S) is established, then all of *i*'s elements do or can occur related together before *H*'s thought-consciousness through *H*'s thought, concerning those elements, that there is a single thing that has all of them taken together. As so occurring, those elements form one group of elements for *H* (namely, the group of elements that are taken to belong to the single thing). Hence, too, those elements form a combination for *H*. And from that result the remainder of the Deduction can be developed as before.[3]

Because we have seen reasons to reject the two assumptions about the cumulativeness of *H*'s knowledge and the single-thinker-for-a-thought position, the fourth-argument reasoning for the purely existential form of (S) fails. And doubts like those raised in Chapter Six about the truth of (S) apply to that form of (S). However, because, for purposes of the Deduction, the purely existential form of (S) functions as well as does (S) to express unity of apperception with respect to *i*, that form is as good a candidate for stipulation as is (S). Moreover, that form agrees with various aspects of Kant's descriptions of our self-awareness through the *I think*.

To see this last point, note that up to now we have supposed that the *I think*, taken by itself in independence of its relation to sensible intuition, is an act or representation of pure thought that yields us a genuinely first-person, *de re*-like awareness *of* ourself as ourself. On this view, the *I think* represents the self that knows and so ultimately the self in itself.[4] This view is in harmony with our usual first-person, *de re* understanding of claims like 'I think' (*ich denke*, *cogito*), and it is also accepted within Kant's ethical theory (which holds that through the *I think* we are made aware of our morally acting self as it is in itself). In addition, Kant relies heavily on a version of this view in answering Pistorius. As I observed in Chapter Two, Pistorius argues that Kant rules out existent knowledge of appearances by an existent self, for on Kant's theory the category of existence applies only to appearances and so cannot apply to the knowing self (as it is in itself) to which objects appear.[5] Kant's B-text answer is that the *I think* gives us what is in effect a *de re*-expressed awareness (although of course no knowledge) *of* our knowing self as our knowing self is – nonphenomenally – in itself. And the *I think* gives us such an awareness *of* our knowing self as – nonphenomenally – *existing* (as existing in what, Kant holds, is not, strictly, a categorial sense).[6]

The first-person, *de re*-like view of the *I think* evidently fits (S) and our use of (S) to express the holding of unity of apperception with respect to *i*. Yet, and to come now to the point that I claimed above about the purely existential form of (S), there are other strands in Kant's treatment of the *I think* which agree with the use of such a form of (S) to express that holding.

For example, and as we have seen earlier, one of Kant's basic positions is that acts of thought, taken by themselves in independence of sensible intuition, cannot grasp, in a *de re*-like manner, individual, particular things as such. Rather, such acts can at best grasp, in a *de dicto*-like way, some thing or other but no individual thing in particular. Or (as I will say) such acts can at best grasp the fact or situation *that there is a single thing* that does so-and-so. Hence the *I think*, taken by itself as an act of pure thought, can at best yield us merely the *de dicto*-like awareness that there is a single thing of such a sort. A number of texts support this view of our self-awareness through the *I think*.[7] And such a view obviously agrees with the view of unity of apperception that is expressed by the purely existential form of (S).

The fact that the purely existential form of (S) and (S) itself each agrees with a part of Kant's views about the *I think* does not mean that those views are without problems. There are difficulties in integrating each of the above two views into his overall picture of knowledge. Thus – and as just observed – in its first-person, *de re*-like aspects the first view certainly fits our (and what seems to be Kant's) natural understanding of claims like 'I think.' But in taking the *I think* by itself to be a pure act of thought that grasps the knowing self in a *de re*-like way, the first view contradicts the basic Kantian position that pure acts of thought, taken by themselves, cannot grasp individual, particular things as such. Moreover, and while the following fact shows no flaw in the first-person, *de re*-like character, itself, that the first view attributes to the *I think*, Kant's attempted first-view answer to Pistorius is unpersuasive. (How does Kant know – and how can he make it convincing – that through the first-person, *de re*-like *I think* the knower indeed grasps, veridically, the nonphenomenal existence of the knowing self in itself?) Again, the second, *de dicto*-like view clearly fits the basic Kantian position about pure acts of thought. But the *de dicto*-like view hardly agrees with our natural, first-person, *de re*-like understanding of 'I think'; and by itself it suggests no way at all of answering Pistorius. (The mere occurrence of the *thought* that *there is* a single thing that has i_1 and i_2, and so on, obviously does not guarantee the veridicality of that thought and the nonphenomenal existence of that single thing.[8]) Moreover, because the two views make inconsistent claims about the operation of the *I think* taken by itself, these views cannot be held jointly.

I believe that Kant is led into these varying, and mutually inconsistent, views because of conflicting intellectual pressures that he is under. He is under pressure, for example, to defend his basic first-*Critique* position (which underlies his attack on past dogmatic metaphysics) that pure acts of thought do not grasp entities in a *de re*-like way. He is under pressure, further, to defend this position while he also tries to respect the genuine first-person (and *de re*-like) character of 'I think' or 'I' (a character much emphasized by his Cartesian predecessors) and he tries to avoid problems like the one raised by Pistorius.[9] The philosophical and exegetical issues that surround these pressures go exceptionally deep and involve many more points than I have noted here. But because they do not directly affect the main Deduction argument from unity of apperception to category application to the object of *i*, I will not examine them further in this book.

I will note only, for the record, that it is not clear how far Kant recognizes the differences between his two views and how far he thinks he can reconcile them. It is easy to illustrate this lack of clarity from the B-text, where he especially feels the need to utilize the first-person, *de re*-like view to answer Pistorius. Here we can see Kant as first suggesting that the *I think*, taken by itself simply as the representation that accompanies all other representations and knowledge – that is, the *I think* 'taken problematically' (A348/B406)[10] – is a merely *de dicto*-like act of thought. (Thus note, for instance, the B418 implication that in the *I think*, so taken, we 'begin with the *concept* [my emphasis] of a thinking being in general' or the B422 claim that unity of consciousness through the *I think* 'is only unity in *thought*, by which alone no object is given'; and observe A346/B404: 'consciousness in itself is not a representation distinguishing a particular object, but a form of representation in general, that is, of representation insofar as it is to be entitled knowledge.') However, Kant then holds that there is a way in which this purely *de dicto*-like *I think* can become *de re*-like and yield us consciousness *of* the self. And – I believe he supposes – this way allows him to reconcile or to hold together this *de dicto*-like view of the *I think* and his first-person, *de re*-like view.

In offering this reconciling way (as he presents it in the B-text), Kant considers the *I think* taken not simply as accompanying all other representations and knowledge but, in particular, as being applied to the manifold of inner-sense representations so as to yield the specific assertion that *I* (this particular, empirically existing self or person) think.[11] He supposes that when the *I think* is so taken, it expresses the application, to sensible intuition, of the concept – of a 'thinking being in general' – that is expressed by (or thought through) the *de dicto*-like *I think* taken by itself as an act of pure thought. Since this application is to inner-sense sensible intuition, he holds that through this application we are yielded a *de re*-like awareness *of* the particular, empirically existing self or empirical I as having all the various representations presented in inner sense. (Here note the *de re*-like effect of the fact that a sensible intuition is the thing to which a concept is applied; and recall also that it is the empirically existing self that is supposed to appear through inner sense.) However, Kant supposes, once we achieve this *de re*-like awareness *of* the empirically existing self, we can then abstract, in thought, from the empirical nature and existence of this self. And we can thereby form at least the idea, in thought, that this self has an existence in itself.

In this reconciling way, Kant takes it, we can thus move from the use of the merely *de dicto*-like *I think* to a *de re*-like awareness *of* our empirical self. And thence, by the above abstraction, we can pass to the idea just noted. However, through this passage we do not gain any awareness or knowledge *of* our empirical self as existing, in itself, separately from our experience or empirical knowledge of it.[12] Because this sort of passage can be made, we can indeed reconcile the *de dicto*-like view of the *I think* and the first-person, *de re*-like view's claim that through the *I think* we gain a genuinely *de re*-like awareness *of* a self that has an (unknowable) existence in itself.

This subtle way of reconciling the *de dicto*- and the *de re*-like view resembles (and I believe should be regarded as an application of) Kant's Phenomena and Noumena discussion of how, by abstracting in thought from the conditions of sensible intuition, we can come at least to *think* the individual, particular objects of our knowledge to have an unknowable existence in themselves.[13] Regretably, however, this reconciling way is unsuccessful, for it does not account for everything that Kant holds in his original *de re*-like view. In particular, it does not accommodate his claim, in that view, that the *I think*, taken *by itself* as an act of pure thought, grasps the self in a *de re*-like fashion.[14] Moreover – and here we reach Kant's apparent lack of clarity about the differences between his two views of the *I think* – it is not clear how far he recognizes that the reconciling way and its own *de re*-like treatment of the *I think* do not accommodate everything that the original *de re*-like view claims. So, by implication, it also is not clear how far he realizes that there is an unbridgeable distance between his original *de dicto*-like view (which at best yields the above, reconciling *de re*-like treatment) and his original *de re*-like view.

Various B-texts show that it really is not clear how far Kant recognizes that the reconciling way and its own *de re*-like treatment do not accommodate all the claims of his original *de re*-like view. Thus, in such texts, after presenting the reconciling way, he turns to answering Pistorius; and in his answer – and without indicating that a different, and an incompatible, view of the *I think* is now at issue – he offers the first-person, *de re*-like view. For example, immediately after the B426–27 presentation of the reconciling view, quoted in part above, Kant at B428–29 suggests what sounds like a restatement of the *de dicto*-like view. ('Thought, taken by itself, is the mere logical function; ... thought takes no account whatsoever ... of the mode of intuition,' B429.) One therefore expects him to proceed, in reconciling fashion, by noting that only through the application of the *de dicto*-like *I think* to inner-sense intuition do I gain a *de re*-like awareness *of* my (empirical) self. And in fact Kant notes such an application later, at B430 (where he speaks of 'the thought of myself applied to the empirical intuition of myself').

However, what Kant now actually does at B429, after an ambiguous but perhaps still reconciling statement,[15] is suddenly to urge (without any sign that he is introducing an incompatible view) that through the *I think*, taken by itself, I am made aware, in a *de re*-like manner, *of* my self in itself.[16] Indeed, in texts like this one he urges, in a nonreconciling way, that the *I think* is a pure act of thought that can precede *any* experience that determines an object of knowledge (including any experience of our empirical self). And he holds that the *I think*, as such an act of thought, 'designates' (*bezeichnet*) – and so yields a *de re*-like awareness *of* – our self as 'object in itself' (B430), even though the *I think* does not note any properties of our self in itself (in Kant's terms, it does not conceptually and categorically 'determine' our self in itself) in such a way as to yield us knowledge of that self. [17]

As the B-text illustrates, Kant thus does not seem clear about the differences between his *de dicto*-like view (with its associated reconciling, *de re*-like treatment) and his first-person, *de re*-like view of the *I think*. Because of the relations that we have seen between these views, on the one hand, and (S) and the purely existential form of (S), on the other hand, it follows that he also should not be clear about the differences between (S) and the purely existential form of (S). We cannot, however, follow Kant in this lack of clarity but must choose some expression of unity of apperception to stipulate. Because of the differences between (S) and the purely existential form of (S), we cannot assume both of these claims as equivalent expressions of the unity of apperception. Rather, it would seem best, below, to continue our discussion by making, officially, two different and alternative assumptions about unity of apperception: one in the form of (S) and the other in the form of the purely existential form of (S). However, it is unnecessary to follow out both of these alternative assumptions; and we have been emphasizing, earlier, the first-person, *de re*-like view of the *I think* – a view which gives the natural reading of claims like 'I think,' is underlined by Kant in the B-Deduction, and has its own considerable philosophical interest. Hence I will focus, below, just on the assumption of (S) as our expression of the holding of unity of apperception with respect to *i*. But the reader may find it helpful also to consider, where necessary, the effects of assuming the purely existential form of (S).

We should note three final matters in connection with our preceding stipulation of (S). First, the problems that we have remarked about Kant's two views of the *I think* undoubtedly create difficulties for his own use of (S) in the Deduction. However, and as I implied above, these difficulties do not affect the main Deduction argument from unity of apperception to category application to the object of *i*. Thus, for example, in merely arguing on the basis of (S) for category application, we are not forced to accept the specific Kantian points that create these difficulties – the points (i) that we cannot hold the knowing self in itself to exist in a categorial sense (a point which creates Pistorius's problem) or (ii) that our awareness of the knowing self through the first-person, *de re*-like *I think* is achieved *a priori*, without relying on anything that could be called sense experience (a point which creates the conflict between the first-person, *de re*-like view and the position that acts of thought, taken by themselves in independence of sense experience, can never be *de re*-like). And given that we doubt Kant's claim that we cannot know objects in themselves – and given recent ideas about first-person self-awareness and related topics – we have no good reasons to accept such points.[18] Thus while the above difficulties seriously affect Kant's overall picture of knowledge, we need not worry about them in stipulating (S) and continuing with the Deduction itself.

Second, (S) involves the claim that *H* is *or can become* conscious, in thought, that the *I think* accompanies all of *i*'s elements. In our later discussions of the Deduction argument from the holding of unity of apperception to category application to the object of *i*, we will not appeal to the presence, in (S), of this 'or can become' clause. As already noted in Chapter Six, I see no clear, plausible solution to the problem that, given the occurrence of this clause in (S), Kant can at best argue merely that the categories do *or can* apply to the object of *i*. Instead, we will proceed directly from the assumption that *H is* actually conscious in thought of the *I think*-accompaniment of *i*'s elements. Thus our ultimate stipulation will not be of (S) (or of the purely existential form of (S)) but, rather, of (S) with its 'or can become' clause deleted.

I call (S) with that clause deleted the actual-consciousness version of (S); and I similarly call the purely existential form of (S) with that clause deleted the purely existential form of the actual-consciousness version. It will thus ultimately be in terms of the actual-consciousness version of (S) (or, if one prefers, of its purely existential form) that the remainder of our discussion will proceed. However, as we have seen, Kant emphasizes claims like (S); and it is simplest, below, to consider his further views about the necessity of unity of apperception in terms of such claims. So until the end of Section 3 I will not stress our stipulation of the actual-consciousness version of (S) as against a stipulation of (S) itself.

Third, proceeding in the above way in terms of (S) (or of its actual-consciousness version) amounts to stipulating that, besides (K), (S) holds true. For reasons of a sort that I have emphasized earlier, making such a stipulation restricts the conclusion of the Deduction to the claim that the categories apply just to those objects of the knowledge, belonging to a being like us, for which a unity-of-apperception claim like (S) is true. As we have seen, such a restriction considerably weakens Kant's desired conclusion in the Deduction. However, given that he has failed to demonstrate anything like (S), I see no alternative to this weakening here.[19]

3. NECESSITY OF UNITY OF APPERCEPTION

We now turn to the necessity of unity of apperception, focusing on those of Kant's points that go beyond the matters considered in Chapter Six. Here the main thing to note is that, in the Transcendental Deduction, Kant has three basic reasons for wanting to demonstrate that necessity. First, and as we have already seen in Chapter Six, Section 3, what he must at a minimum hope to show, by establishing such a necessity, is the claim (NUA):

> It is necessary that, for any sensible intuition in general and for any being like us, if that being knows through that intuition, then all of the elements of that intuition are such that that being is or can become conscious, in thought, that all of those elements, taken together, are accompanied by the *I think*

(NUA) is of interest to Kant because (among other reasons) in the course of demonstrating that claim (if he could do so), he in effect shows that (S) follows validly from the opening Deduction assumption (K), that *H* knows via *i* ((K) perhaps taken in conjunction with already-established Kantian results). As was suggested in Chapter Six, Kant's showing that (S) follows validly from (K) is one step toward his establishing of such basic Deduction claims as (NCA), the claim that it is necessary that if any being like us knows through a sensible intuition in general, then the object of that intuition, as that object is known through that intuition, falls under the categories. And we have already explained in Chapter One and later why such a claim is important to Kant.

Second, Kant associates the necessity of unity of apperception (understood as noted below) with what he calls the objectivity of that unity.[20] And he relates that objectivity to the fact that (as he holds) because *H*'s knowledge through *i* is subject to necessary unity of apperception, that knowledge concerns an object that satisfies appropriately strong Transcendental Deduction – and, ultimately, categorial – conditions.

Third, and as is well-known, Hume argues against the holding of any sort of necessary connections among the distinct objects of our knowledge. Although Kant's official replies to Hume come much later in the first *Critique*,[21] already in the Deduction Kant takes the holding of necessary unity of apperception with respect to all our knowledge to imply the holding of necessary connections, of sorts that Hume rejects, among the objects that we know.[22]

In Chapter Six we have already considered Kant's first reason for hoping to prove the necessity of unity of apperception. To that discussion, we need only add here that the remarks in Section 2 above evidently show the possibility of expressing his minimum claim of that necessity in the form not of (NUA) itself but of a purely existential form of (NUA). However, since nothing crucial is added by introducing such a form of

(NUA), I will continue below just in terms of (NUA). (And I will ignore the possibility of purely existential forms of other Kantian necessity claims that we note below.[23])

In addition, our interest now is in those of Kant's claims about necessity of unity of apperception that go beyond – or simply differ from – (NUA). So it will be simplest for the moment to assume, contrary to our results in Chapter Six (and Section 2 above), that Kant has validly derived (S) from (K) and thus has succeeded in showing (NUA) through such a derivation. Finally, I should note that since, as seen in Chapter Six, various of his other claims about necessity (for example, (NCA)) parallel (NUA), much of what we say below can be generalized so as to apply to those other claims. But we need consider no such generalizations here.

In the following discussion, I begin by noting Kant's claims about the necessity of unity of apperception that differ from (NUA). Then I consider (as far as is possible at this point) how his views about that necessity, including his points about objectivity and his response to Hume, bear on those claims.

Kant's non-(NUA) claims are conveniently classified into three groups. (I) Sometimes Kant does not accept (or does not accept only) (NUA) itself, with the 'it is necessary that' operator governing the entire conditional claim (or generalized conditional claim) that is contained in (NUA). (Here observe (NUA) as recalled above; and see also the Chapter Six formal version of (NUA).) Instead, Kant offers the different claim to the effect that

(N_1) For any sensible intuition in general and for any being like us, if that being knows through that intuition, then it is necessary that (that being is or can become conscious in thought that the *I think* accompanies all the elements of that intuition taken together)

or the appropriate formal version of (N_1).[24]

In (NUA) itself the necessity is *de dicto*, for 'it is necessary that' governs the whole generalized conditional claim (beginning 'for any sensible intuition in general' and running to the end) that (NUA) contains. And (NUA) simply attributes necessity to that whole claim and asserts nothing about any individual, particular intuitions, intuition-elements, or knowers as having certain properties necessarily. In (N_1), however, the necessity is *de re*, for 'it is necessary that' in (N_1) governs just the consequent of the generalized conditional claim. And (N_1) says that if you pick any individual, particular being like us and any individual, particular sensible intuition in general through which that being knows, then it is necessary, with respect to *that* individual, particular being and *that* individual, particular sensible intuition, that that being does or can think, with regard to the elements of that intuition, in the way just noted.[25]

It is clear that such a *de re* view of the necessity of unity of apperception is logically independent of the *de dicto* claim made by (NUA) itself. Obviously (NUA) by itself does not imply the *de re* (N_1). And (N_1), because (among other things) it concerns just actual beings like us and actual sensible intuitions in general, does not imply (NUA), which given its structure concerns all actual and possible beings like us and all actual and possible sensible intuitions in general.

(II) I will return, shortly, to the further philosophical interest, for Kant, of the (N_1)-style view of the necessity of the unity of apperception. Meanwhile, and to observe the second of Kant's non-(NUA) types of claim about that necessity, I should emphasize again that, insofar as Kant accepts just (NUA), he accepts merely the necessity of the (generalized) conditional claim that, for any sensible intuition in general and for any being like us, *if* that being knows through that intuition in general, *then* that being is or can become conscious in thought that the *I think* accompanies the elements of that intuition taken together. So he does not thereby accept any necessity as belonging to the antecedent of this claim or to its consequent. Yet sometimes (and in a way different from that in (N_1)) Kant also asserts something like the necessity of that consequent. For instance, in his A121–22 remarks on the 'affinity' of the manifold, he denies that 'it is entirely accidental that appearances should fit into a connected whole of human knowledge,' denies further that 'a multitude of perceptions' *could* exist separately in my mind without all together 'belonging to a consciousness of myself [through unity of apperception],' and avers that 'all appearances [and intuition elements] *must* so enter the mind or be apprehended, that they conform to the unity of apperception' (my italics). And in the B-Deduction he asserts what seems to be the same set of views. Such views clearly suggest that Kant here accepts a further claim of *de re* necessity of unity of apperception that differs from (although it structurally resembles) (N_1): namely, the claim (or claims)[26]

(N_2) For any individual, particular intuition-elements (or appearances, representations, and so on) and for any being like us, if those intuition-elements (and so on) belong to that being, then it is necessary that [that being is or can become conscious in thought

> that the *I think* accompanies all those intuition-elements (and so on) taken together]

In (N_2), as in (N_1), necessity attaches to the consequent of the relevant conditional claim. But whereas (N_1) makes the holding of that necessity conditional on the having of knowledge, by a being like us, through the relevant intuition-elements, (N_2) makes the holding of that necessity conditional simply on the possession, by a being like us, of the relevant intuition-elements. Furthermore, like (N_1), (N_2) is evidently itself logically independent of (NUA). We will come back to the interest of (N_2) (which parallels that of (N_1)) below.

(III) Finally, it is not just that, besides (NUA), Kant also accepts *de re* necessity-of-unity-of-apperception claims that have the structure common to (N_1) and (N_2) above – that is, that have the structure of a claim that holds

> For any sensible intuition (or its elements, and so on) and for any being like us, *if* so-and-so is the case with respect to that sensible intuition (or its elements, and so on), then it is *necessary* that [that being is or can become *conscious* that the *I think* accompanies (all the elements, taken together, of) that intuition (and so on)]

In addition, at various points Kant certainly seems to endorse *de re* necessity-of-unity-of-apperception claims in which the places, here, of the necessity and consciousness-in-thought operators are exchanged – that is, claims which in structure are of the form.[27]

> (N_3) For any sensible intuition (or its elements, and so on) and for any being like us, *if* so-and-so is the case with respect to that sensible intuition (or its elements, and so on), then that being is or can become *conscious* in thought that [it is *necessary* that the *I think* accompanies (all the elements, taken together, of) that intuition (and so on)]

For example, in A108, third sentence, Kant first claims, in effect, that if there is to be knowledge through a manifold of intuition-elements, then a 'necessary consciousness' obtains of the identity of self throughout that manifold. But he next speaks of the 'consciousness of an equally necessary unity of ... synthesis' that this necessary consciousness yields. Such a consciousness of a necessary unity of synthesis must, in logic, derive from a consciousness of the necessary identity of self throughout that manifold rather than directly from the 'necessary consciousness' of that identity. So Kant here in effect endorses (or should, in logic, endorse) a claim in the (N_3) form, to the effect that if a being like us knows through a manifold of sensible intuition, then that being is or can become conscious of the necessity that the *I think* accompanies all the elements of that manifold taken together (or, in possession form, that being is or can become conscious of the necessity that *he himself* – or *she herself* – has all of those elements taken together).[28] It is clear that such (N_3)-style claims are logically independent both of (NUA) itself and of claims like (N_1) or (N_2).

There are, I think, two main grounds that Kant has for his defense of the three above groups of non-(NUA) claims in the Deduction. First, such claims yield at least part of his answer to Hume's denial of necessary connections among the distinct objects of our knowledge. Second, Kant links such claims to his Deduction view that necessary unity of apperception is an objective unity. We cannot, in this book, discuss Kant's answer to Hume (whose development goes far beyond the Deduction) in detail. And we consider Kant's views on objectivity further in Chapter Nine. However, brief comments may be helpful here.

First, and in regard to Kant's answer to Hume, observe that the *de dicto* necessity-claim (NUA) (if Kant could establish it) already runs contrary to Humean views. In his discussion of necessary (causal) connections Hume implies that there is no logical or metaphysical impossibility in the occurrence of the extreme situation in which a being like us knows through i_1 and i_2 – and has those individual intuition-elements as the sole objects of its (separate pieces of) knowledge – without thereby recognizing or even being able to recognize any relations (except, perhaps, relations of succession or resemblance) to obtain between those elements.[29] Obviously, however, such a situation is impossible given (NUA) (or given *de re* necessity-claims like (N_1), (N_2), or (N_3)). Moreover, suppose that Kant could show necessary unity of apperception, in the *de dicto* style of (NUA), to hold with respect to the elements of the manifold of a single intuition. Then he could use similar reasoning to show that such a *de dicto*-style necessary unity of apperception holds with respect to all of the different elements, taken together, of any group of intuitions through which a being like us knows distinct objects. And he would take that result to imply the holding, in a *de dicto*-style of necessity, of category-governed connections among those distinct objects. For example (and note the analogy with a claim like the Chapter Six (NCA)), Kant could show that it is necessary that (if a being like us knows distinct objects o_1, o_2, and o_3, then that being takes category-governed relations to hold among those distinct objects).[30] Yet such a result is contrary to Hume's view that a being like us could know various distinct objects separately without having to take anything like Kantian category-governed relations to hold among those objects.

Even *de dicto* Kantian necessity-of-unity-of-apperception claims like (NUA) thus already oppose Hume. But, as the implications of texts like A121–22 (noted above in connection with (N_2)) suggest, for Kant the strongest opposition to Hume's views comes from the non-(NUA) *de re* necessity claims. Thus suppose that Kant could establish one or more of these claims as holding with respect to the elements of the manifold of a single intuition. Then he could use similar reasoning to show that the relevant sort of *de re* necessity of unity of apperception holds also with respect to all of the elements, taken together, of any group of intuitions through which a being like us knows various distinct objects. And that result would then imply, for Kant, the holding, in an anti-Humean, *de re* fashion, of the necessity of category-governed relations among those objects.

Second, and in regard to his views on objectivity, Kant focuses on the sort of objective union that intuition-elements have when (as is required by the holding, with respect to them, of unity of apperception) they represent a category-satisfying object. His basic idea is that such a union is to be distinguished from the mere, accidental association of such elements within their possessor's mind on the ground, in part, that such an objective union is necessary, owing to the necessity of unity of apperception, in a way that such an accidental association is not. As we see in Chapter Nine, many questions can be raised about this idea. However, as far as the idea itself goes, the necessity of unity of apperception that it involves can be any of the sorts expressed by Kant's non-(NUA) claims (as well as the sort expressed by (NUA) itself). (And, corresponding to that necessity, there would then be a similar necessity of the objective union of the relevant intuition-elements.)

When they are pursued in depth, Kant's non-(NUA) claims raise many complicated questions about the interpretation of Kantian points of exceedingly fine detail. However, the basic fact that we now need to notice about these claims does not turn on these complicated questions, in which we must here avoid entanglement. That fact is that these claims are in themselves no more evident than is a claim like (S) or (NUA), so that Kant must provide some argument for them. Yet, given only the apparatus and results of the Deduction, it does not seem that he can provide any more satisfactory argument for those claims than he can offer for (S) or (NUA) themselves.

After all, in Chapter Six we have already noted reasons to doubt that a claim like (S) (and, as can be seen, a claim like (NUA) itself) holds true for all actual and possible cases of knowledge belonging to a being like us. We evidently have just as much reason to doubt that any of the non-(NUA) claims holds true. Suppose that an arbitrary being like us knows through an arbitrary sensible intuition in general. Then is it really *necessary* that that being can become conscious in thought that the *I think* accompanies the elements of that intuition? Or, in other words, is it really true that any individual, particular sensible intuition in general and any individual, particular being like us jointly have the conditional, relational, essential property that (N_1) attributes to them – namely, the property of being such that *if* such a being knows through such an intuition, *then* it is *de re*-necessary, with respect to *that* being and *that* intuition, that that being is able to be conscious in thought of the *I think* as accompanying all the elements of that intuition taken together? Or, in non-Kantian terms, is it really obvious that if a being like us has a certain state of knowledge, then it is *de re*-necessary, with respect to that being and that state of knowledge, that that being can self-ascribe that state of knowledge (or various of its components)?[31] And, similarly, for (N_2)- and (N_3)-style claims.

If, however, Kant's non-(NUA) claims are thus as much open to doubt as are (S) and (NUA) themselves, then Kant clearly should provide some argument for those claims. But no plausible argument exists here, as far as I can tell. *If* Kant could prove (S) from (K), he could derive the *de dicto* claim (NUA). But his own arguments to derive (NUA) (by proving (S) from (K)) fail. And, as reflection shows, he cannot plausibly argue for any of his non-(NUA) claims – all of them independent of (NUA) and all of them involving *de re* necessities – unless he can derive these non-(NUA) claims from already acceptable *de re* assumptions that concern intuition-elements and their *I think*-accompaniment (or else from some further *de re* assumptions that somehow validly yield the non-(NUA) claims). Yet it seems clear that no such *de re* assumptions are available. After all, any such assumptions surely will be as strong as, and as much in need of demonstration as, the non-(NUA) claims themselves. Hence Kant seems unable to provide plausible arguments for any of the non-(NUA) claims. And therefore he seems to have no satisfactory grounds for asserting these claims in the Deduction, despite their undoubted significance for his Deduction views.[32]

We need not examine in detail all of Kant's lines of argument that introduce into the Deduction the non-(NUA) claims. Study shows that they amount to Kant's introduction of the necessity-operator into his four Chapter Six arguments for (S) in various fallacious ways. Through these fallacious ways, Kant moves, apparently without realizing that he does so, from *de dicto* necessity claims structurally like (NUA) – *de dicto* necessity claims that he can reach *if* he can prove (S) – to *de re* claims structurally like the non-(NUA) claims.

For example, (i) Kant often argues in ways that at best prove (NUA) but then states his conclusion in the form of something like (N_1) or (N_2), thus in effect moving the necessity operator into the consequent of (NUA) and so fallaciously changing a *de dicto* into a *de re* claim.[33] Again, (ii) Kant seems sometimes to start with one or the other version of the third argument for (S) (recall Chapter Six, on the two versions of this argument). As he does so, he regards the 'all my representations are mine' (or the 'all of *H*'s representations are *H*'s') claim in that argument (step (c) or (d) of the argument) as being itself necessary since it is for him analytic. Proceeding in this way, he next takes *H* to be able to become conscious in thought of this necessity that all my representations are mine. (Compare step (e) of the third argument.) His reasoning then (on either version of the third argument) continues with one or the other of the intensional fallacies that we noted in Chapter Six. Thus either he now concludes that I (*H*) am able to become conscious in thought that it is necessary that I possess i_1 and i_2 taken together, or else he now concludes that each of my representations is such that I am able to become conscious in thought that it is necessary that that representation is mine. (Compare step (g) or (i) of the third argument.) Owing to the way in which he has introduced the necessity operator, each of these continuations of his reasoning obviously creates a further intensional fallacy (structurally like the one already present in the original continuation in the third argument itself) that involves the necessity operator. The ultimate result of the various fallacies here is an operator-interchanged version of (S) that will eventually yield an operator-interchanged, non-(NUA) claim of the (N_3) sort – namely, the claim that if a being like us knows through a sensible intuition in general, then that being is or can become conscious in thought that it is necessary that the *I think* accompany the elements of that intuition taken together.[34]

Further, (iii), and as our A108 example given above in connection with (N_3) shows, Kant sometimes infers an (N_1)- or (N_2)-style, *non*-operator-interchanged non-(NUA) claim; and from that claim he then fallaciously derives an operator-interchanged, (N_3)-style non-(NUA) claim simply by exchanging the places of the necessity and consciousness-in-thought operators. Finally, (iv) it seems that Kant in places combines various of these fallacious paths to non-(NUA) claims. To take but one example, he seems on occasion both to use way (iii) of reaching an operator-interchanged, (N_3)-style claim and also to introduce an additional necessity operator via way (i). He thus arrives at the doubly *de re*-necessity claim that if a being like us knows through a sensible intuition in general, then (by way (i)) it is necessary that (by way (iii)) [that being is or can become conscious in thought that (it is necessary that *I think* accompanies the elements of that intuition taken together)].[35]

In places Kant shows a clear understanding of what looks like the *de re-de dicto* distinction, and so one might be surprised at the charge that he commits the above fallacies.[36] But of course a general understanding of that distinction can coexist with confusions about its instances. Moreover, both above and in Chapter Six we have already seen that he does commit such fallacies. Furthermore, Kant – like his contemporaries who commit similar fallacies – lacks the sharp logical tools that enable one to deal straightforwardly with these fallacies.[37]

One might also be led by B142 of B-Deduction § 19 to doubt that Kant accepts the non-(NUA) claims and commits the above fallacies. He there claims that the subjection of representations to necessary unity of apperception does not mean that the representations '*necessarily* belong *to one another* in the empirical intuition' but that 'they belong to one another *in virtue of the necessary unity* of apperception in the synthesis of intuitions.' And the example that he gives implies that while the representations of body and heaviness, as they occur together in the contingent judgment that bodies are heavy, are subject to necessary unity of apperception, this fact does not mean that that judgment itself is a necessary one. These B142 claims are not completely clear, but it is easy to read them as denying that any of the non-(NUA) claims properly expresses the subjection of representations to necessary unity of apperception.

For example, one might argue that if that subjection were properly expressed in, say, an (N_3)-style claim, then from the fact that I know through some sensible intuition it would follow that I am or can become conscious that it is necessary that the *I think* accompanies the elements of that intuition taken together. However, given Kant's further Deduction reasoning, as discussed in Chapter Eight, from this last result Kant would infer that I am or can become conscious that it is necessary that there is an object that has the features that those representations put before my mind. Yet such a conclusion conflicts with the implication of the above B142 example that necessary unity of apperception can hold with respect to a group of representations without the knower's having to regard as necessary the judgment in which those representations occur (or to which they are otherwise relevantly related). So, one might hold, we should reject the idea that an (N_3)-style claim – or, indeed, any of the non-(NUA) claims – properly expresses the subjection of representations to necessary unity of apperception.

I think, however, that the above reading of Kant's B142 points is not completely correct. After all, those points do not rule out his expressing the holding of necessary unity of apperception by an (N_1)- or (N_2)-style claim to the effect that if a being like us knows through – or has – a given sensible intuition, then it is *necessary* that that being is or can become *conscious* in thought that the *I think* accompanies the elements of the intuition taken together. Of course, as far as I can see, the B142 points do imply that the holding of necessary unity of apperception should not be expressed in an (N_3)-style claim. And, given the central place of § 19 in the overall B-Deduction, this fact strongly suggests that we should not regard Kant as ultimately endorsing anything like (N_3) as expressing that holding. However, as noted, he in fact makes (N_3)-style claims (and even at B135 in B-Deduction § 16).[38] So I will continue to refer to (N_3)-style claims as is needed, as well as to (N_1)- and (N_2)-style claims. In any case the important point is that Kant appeals to non-(NUA) expressions of the holding of unity of apperception, and the B142 claims do not rule that general appeal out of order.

We could continue this investigation of Kant's views on necessary unity of apperception much further then we have. But for present purposes there is no need to do so, for it should by now be evident that Kant really has no satisfactory way to show any of the non-(NUA) claims that we have noted above. Yet we have seen earlier that he cannot derive (S) itself successfully from (K). And since it seems clear that he can establish (NUA) only by such a derivation, it also seems clear that he is unable to establish even (NUA). Hence, and not to avoid unhappy conclusions, Kant does not seem able to demonstrate any of his views on necessary unity of apperception in a satisfactory way.[39]

What then should we do in our further examination of the Deduction? It is clear that where Kant emphasizes the necessity of unity of apperception we must attend to his claims. But, and as I am afraid that the above discussion may already have begun to illustrate, following out his detailed treatment of that necessity is an intricate, minutely ramifying business. Because he has not demonstrated his basic views – which anyway seem open to doubt – there is no good reason to pursue that business through every nook and cranny of our discussion, either by focusing on every one of his subsequent uses of 'necessary' or by building various specific necessity-references into our Section 2 stipulations. So, and to come back to our conclusions at the end of Section 2, in considering the rest of the Deduction we will stipulate that, besides (K), (S) – or, alternatively, the purely existential form of (S) – holds true with respect to *H* and the elements of *i*.

Or, to put this stipulation with absolute exactness, recall from Section 2 that in our later discussion we will not appeal to the presence, in (S), of the 'or can become' clause but will instead proceed solely in terms of the actual-consciousness version of (S), which lacks that clause. So our precise official stipulation will be that unity of apperception holds true in the form of

> the actual-consciousness version of (S):
>
> All of the elements of the manifold of *i* are such that *H* is conscious, in thought, that all of those elements, taken together, are accompanied by the *I think*
>
> *(or, formally:*
>
> $(y)(z) \ldots [y, z,$ and $\ldots$ are the elements of $i \supset H$ is conscious in thought that the *I think* accompanies (y and z and $\ldots$)])

Or, alternatively, we could stipulate the holding of unity of apperception through the purely existential form of the actual-consciousness version of (S), as that version was explained in Section 2. However, as observed there it is unnecessary to follow out the consequences, in the Deduction, of both these stipulations. And so we will focus, hereafter, just on the stipulation of the actual-consciousness version of (S). We will then introduce Kantian points about necessity independently of that stipulation and only where those points require serious attention.

4. SUMMARY

We focused above on Kantian views about unity of apperception and its necessity that go beyond the views considered in Chapter Six. We first noted the idea of the purely existential form of (S) to the effect that *H* is or can become conscious in thought that there is a single thing that has all of *i*'s elements taken together. We observed that this form of (S) can be derived from the opening Deduction assumption (K) (that *H* knows via *i*) by appeal to the Chapter Six, fourth argument for (S), if we are willing to grant that argument's claims about the cumulativeness of *H*'s arriving at knowledge and about the Kantian position that each thought belongs to a single thinker. ((S) is the strong unity-of-apperception claim that *H* is or can become conscious in thought that the *I think* accompanies all of *i*'s elements taken together.) We saw that, for purposes of the Deduction, the holding of unity of apperception with respect to *i* is expressed as well by this form of (S) as by (S) itself. And we remarked that this form of (S) agrees with various of Kant's own characterizations of the *I think*. Indeed, we saw that, besides the first-person, *de re*-like view of the *I think* that we have followed Kant (and (S)) in accepting up to now, Kant also proposes a different, *de dicto*-like view. This *de dicto*-like view takes the *I think*, as an act of pure thought, to yield only the awareness *that there is* a single thing that does so-and-so. Such a view is in harmony with the purely existential form of (S) and so shows how well that form agrees with various of Kant's claims about the *I think*.

However, we noted that Kant cannot hold both of these views of the *I think* consistently and that what seems to be his attempt to reconcile this inconsistency fails. (Indeed, as we saw, he does not seem clear about the real differences between these two views.) So in the end we reached two different Kantian expressions of the holding of unity of apperception – namely, (S) and the purely existential form of (S). Since in Chapter Six we have already rejected both Kant's argument for (S) and the fourth-argument points about cumulativeness and the single thinker for a thought, we noted that neither expression of unity of apperception is plausibly established by Kant. And, as we observed in Section 2, because of the Chapter Six 'or can become conscious' problem, in our further discussion of the Deduction we in fact must appeal not to (S) (or to the purely existential form of (S)) but, rather, to the actual-consciousness version of (S), which drops the 'or can become' clause. We therefore decided to proceed officially by making two different, and alternative, stipulations: of the actual-consciousness version of (S); and, again, of the purely existential form of that version. But there is no need to pursue both of these stipulations in the Deduction. And because of our earlier emphasis on the first-person, *de re*-like view of the *I think* and the interest of that view, we will focus, hereafter, just on the actual-consciousness version of (S). As we have urged, stipulating that version considerably weakens Kant's desired conclusions in the Deduction.

Finally, we have just observed Kant's views on the necessity of unity of apperception. As we saw in Chapter Six, Kant wants to demonstrate the *de dicto* result (NUA) (the claim that it is necessary that if a being like us knows via a sensible intuition in general, then that being is or can become conscious in thought that the *I think* accompanies all the elements of that intuition taken together). Moreover, as we have now seen, Kant is committed also to at least three other *de re* and non-(NUA) versions of the necessity of unity of apperception. We noted briefly connections that he wants to make between such versions and his claims about the objectivity of unity of apperception and about Hume's denial of necessary connections among the distinct objects of our knowledge. We have just shown, however, that he has failed to establish any of these non-(NUA) versions. Because he also has not succeeded in what seems to be his only plausible way of demonstrating (NUA) itself, Kant has not established any of his claims about the necessity of unity of apperception. Given this result, and in view of the complications that they involve, we thus will not introduce those claims below except where they are of real importance to the Deduction. In particular, I will largely ignore those claims in our discussion, in the next chapter, of Kant's § 17 inference from the holding of unity of apperception with respect to *i* to the union of *i*'s manifold in the concept of an object.

CHAPTER EIGHT

THE UNION OF THE MANIFOLD OF INTUITION IN THE CONCEPT OF AN OBJECT: B-DEDUCTION § 17

1. INTRODUCTION

Given our Chapter Seven stipulation, we have the actual-consciousness version of (S) at our disposal. As we have seen, that version of (S) states that *H* is conscious in thought that the *I think* accompanies all of *i*'s elements taken together. In B-Deduction § 17 Kant wishes to show that from the truth of that version of (S) and the opening Deduction assumption (K) (that *H* knows via *i*) it follows that the manifold of *i* is united in the concept of an object. He also must show that the object in whose concept that manifold is united is the object that *H* knows through *i*. These new conclusions go considerably beyond the claim, which we have noted already, that the holding of unity of apperception with respect to *i* implies that *i*'s elements form a combination and hence require synthesis. These conclusions imply that such a combination is, specifically, a combination of *i*'s elements such that *H* thinks there to be a single object – the object that *H* knows through *i* – that appears through those elements (or that is, as an appearance, identical to the synthesized manifold of those elements). Or, in other terms, the § 17 line of argument amounts to the reasoning that, to the extent that we self-ascribe our knowledge in a first-person way, our knowledge concerns a single object that is distinct from the mental states through which we know.

It is not at all obvious that the knowledge belonging to a being like us must concern a single object of the last sort rather than, say, the individual mental states through which that being knows. Thus by itself the § 17 line of argument is already of much philosophical interest. Moreover, even if Kant has been unable to demonstrate (S), it would still be of much interest for him to show that *if* the actual-consciousness version of (S) holds (and so *if H* does self-ascribes *H*'s knowledge through *i*), *then H*'s knowledge concerns the above sort of object. Furthermore, the § 17 line of argument also is of crucial importance to the overall reasoning of the Transcendental Deduction. In the B-Deduction Kant later uses this line of argument to show that *i*'s manifold forms an objective unity for *H* (§ 18) and so the object known through i is the subject of a judgment (§ 19) and hence falls under the categories (§ 20). Without the above line of argument, this later reasoning collapses.

We therefore need to attend carefully to Kant's B-Deduction § 17 claims about unity of apperception and the union of i's manifold in the concept of an object. In Section 2 we consider his view of what is meant by that union. In Sections 3 and 4 we then examine his § 17 claims and the reasons that Kant might give for the occurrence of such a union. Section 5 discusses some crucial questions that can be raised about Kant's conclusions in § 17.

2. UNITING THE MANIFOLD OF *i*

There are many complications in Kant's attempts to prove that the manifold of i is united in the concept of an object. These complications stem largely from the fact that his proofs take different forms – and yield different specific results – depending on the exact version of the minimum Deduction assumption (K) that he appeals to in these proofs. They also derive from the fact that, as urged in Section 3, he does not initially acknowledge both of the versions that we have noted earlier of that assumption. Before we begin to examine Kant's arguments for the union of i's manifold in the concept of an object, it thus is best to have a clear idea of what such a union amounts to. Since Kant does not make that idea clear in § 17, we must go outside § 17 in order to understand it. In particular, we need to consider Kant's account of the synthesis of the manifold via concepts.

For concreteness, I will present that account in terms of our Chapter Two, Three, and Five example, which I now somewhat amplify, of H's knowledge, by means of intuition j, of the conical, red spruce that is before H. Suppose that through j H knows this spruce to be a conical, thickly needled red spruce and that (to simplify harmlessly) j is given to H in the form of the manifold j_1, j_2, j_3, and j_4. Suppose also that we follow, for the case of j, our Chapter Five assumption that, insofar as H knows through i, i's elements occur before H's mind in an actual, sequential order. Specifically, imagine that j_1 to j_4 occur in the actual, temporal order j_1, j_2, j_3, and j_4, and that, as they so occur, first j_1 puts before H's mind (or is) the property P_1 of being thickly needled, then j_2 puts before H's mind (or is) the property P_2 of being a red spruce, then j_3 puts before H's mind (or is) the spatial part s_1 that is the conical top half of the spruce, and then j_4 puts before *H*'s mind (or is) the spatial part s_2 that is the trunk and bottom half of the spruce. And ignore, for simplicity, the Chapter Three points about the mere potential generality or the mere potential definiteness of P_1 and P_2 or s_1 and s_2 as they are first presented by j_1 to j_4.[1] Then, as our earlier discussions show, Kant holds that, insofar as *H* knows the spruce through the manifold of *j*, *H* must synthesize that manifold in order to arrive at the single, unitary intuition *j* that represents to *H* a single object as having all of P_1, P_2, s_1, and s_2, and so as being the conical, thickly needled red spruce.

The nature of *H*'s synthesis of *j* can be seen from various first-*Critique* passages. Kant's description of that nature involves claims about *H*'s imagination, a matter that I have largely ignored in this book, as Kant's position on imagination does not affect the main discussion. However – to note the relevant points briefly – as well-known texts like A98–104 show, Kant supposes *H* to synthesize *j* by mentally running through *j*'s elements and 'taking them up' into imagination so that they are there reproduced and come to form a single, overall representation that presents all of the properties and spatial parts P_1, P_2, s_1, and s_2.[2] Moreover, this single, overall representation must represent *an object* as *having* all of these properties and spatial parts and so as falling under the concepts of being conical and of being a thickly needled red spruce. And, Kant holds, this representation can function in this way only insofar as *H*'s synthesis is carried out in a conceptually rule-governed fashion.

The specific character of that synthesis is brought out in familiar texts like these:

> It is only when we have produced synthetic unity in the manifold of intuition that we are in a position to say that we know the object ... this unity is impossible if the intuition cannot be generated in accordance with a rule by means of such a function of synthesis as makes the reproduction of the manifold *a priori* necessary, and renders possible a concept in which it is united. Thus we think a triangle as an object, in that we are conscious of the combination of three straight lines according to a rule by which such an intuition can always be represented. This *unity of rule* determines all the manifold, and limits it to conditions which make unity of apperception possible. The concept of this unity is the representation of the object = X, which I think through the predicates, above mentioned, of a triangle. (A105)

> ... a concept is always, as regards its form, something universal that serves as a rule. The concept of body ... , as the unity of the manifold which is thought through it, serves as a rule for intuitions only insofar as it represents in any given appearances the

necessary reproduction of their manifold, and thereby the synthetic unity in our consciousness of them. The concept of body, in the perception of something outside us, necessitates the representation of extension, and therewith representations of impenetrability, shape, etc. (A106).

We can see the implications of such texts most easily by focusing on Kant's view that concepts simply *are* general properties, taken as general. In this connection note also Kant's frequent assertion that knowledge requires not only an intuition but also a concept via which we *think* the object of this intuition.[3] And recall the Chapter Three discussion of concepts and Kant's view that concepts are manifolds of general properties.[4]

Given these points, we can plausibly regard Kant, in the above texts, as implying that the following must occur when *H* synthesizes the manifold of *j*. *H* must take the general properties P_1 and P_2, as they are presented in the above single, overall representation, conjunctively to constitute the general property – and so the concept – of being a thickly needled red spruce. *H* must also take the spatial parts s_1 and s_2, when they are thus presented before *H*'s mind, jointly to specify the presence of the shape-property – and so the presence of the shape-concept – of being conical. In addition (and for reasons of a sort given in Section 3), *H* must use a certain concept of an object in order to take all these properties and spatial parts, as they are thus presented, to belong to a single object. By proceeding in this way, *H* brings it about that the single, overall representation in question functions as a single representation that represents a single object as having all of P_1, P_2, s_1, and s_2. Hence *H* brings it about that that representation functions as the single, unitary intuition *j* that represents a single object as falling under the concepts of being a thickly needled red spruce and of being conical.[5]

The concept that *H* uses, in the above process, to think a single object as possessing all the relevant properties and spatial parts is identified in an important passage at A93/B126 that we have already partly quoted in Chapter Four. Kant writes:

All experience [which by B1, B147, B165, and the next sentence is empirical knowledge of objects] does indeed contain, in addition to the intuition of the senses through which something is given, a *concept* of an object which is given in the intuition, or appears. Concepts of objects in general thus underlie all empirical knowledge as its *a priori* conditions.

It is this concept of an object in general – that is, this concept of an object in the most general sense of 'object' – that Kant supposes H to employ in thinking, in synthesis, of there as being a single object of the relevant sort.[6]

Moreover, Kant takes this concept of an object in general to be *a priori* (as the A93/B126 text shows) and to amount to the generic category, of which each of the twelve individual categories is a realization or specification. The concept of an object in general is *a priori* because it is the concept of *an* (or of *one*) object that is or can be taken to have various sense-presented features. And Kant, in taking combination not to be given, will suppose that sense experience cannot give us the idea of one thing that is or can be related to other things. The concept of an object in general is also *a priori* because it is necessary that if we know an object, then that concept is used in knowing that object. We see the exact relation of the concept of an object in general to the individual Kantian categories in Chapter Ten.

We may set aside until Chapter Ten the details of Kant's claims about the concepts that are (or present) the properties of the spruce. If we do so, then, with a qualification noted below, the above account amounts to the claim that, when H knows through j, H synthesizes the manifold of j in such a way that, by using the concept of an object in general, H comes to think in the following way:[7]

(Tj) j_1, j_2, j_3, and j_4 occur before H's mind in the temporal order j_1, j_2, j_3, j_4 and, as they so occur, put before H's mind (or are) P_1, P_2, s_1, and s_2 and H is conscious in thought that [there is a single object x such that first j_1 puts before H's mind (or is) P_1 and x has P_1 and then j_2 puts before H's mind (or is) P_2 and x has P_2 and then j_3 puts before H's mind (or is) s_1 and x has s_1 and then j_4 puts before H's mind (or is) s_2 and x has s_2 and P_1 and P_2 jointly constitute the property of being a thickly needled red spruce and s_1 and s_2 jointly specify the property of being conical]

or, formally, and with abbreviations:

j_1, j_2, j_3, and j_4 occur before H's mind in the temporal order j_1, j_2, j_3, and j_4 and, as they so occur, pm (or are) P_1, P_2, s_1, and s_2 & H is conscious in thought that $(\exists x)$[x is an object & j_1 pm (or is) P_1 & x has P_1 & then j_2 pm (or is) P_2 & x has P_2 & then j_3 pm (or is) s_1 & x has s_1 & then j_4 pm (or is) s_2 & x has s_2 & P_1 and P_2 jointly constitute the property of being a thickly needled red spruce & s_1 and s_2 jointly specify the property of being conical]

And (*Tj*) will express the act of *H*'s thought that results in *H*'s knowledge, through *j*, of the spruce.

Two comments about (*Tj*) are worth making immediately. First, in formulating *H*'s knowledge-yielding act of thought as I have in (*Tj*), I of course treat that thought as a *de dicto*-like thought to the effect that there is some object or other (but no object in particular) that has P_1, P_2, s_1, and s_2. This treatment is correct insofar as we consider that thought simply as arrived at through *H*'s use of the concept, taken by itself, of an object in general (and we thus ignore the fact that, in that thought, *H* goes on to take the object to have the properties P_1 and P_2 and – especially – to have the spatial parts s_1 and s_2). As we have observed in earlier chapters, Kant holds that through a concept taken by itself we can achieve only *de dicto*-like thoughts. And Kant's own descriptions of the concept of an object in general support this treatment of *H*'s thought.[8] (In Section 3 we then consider how *H*'s *de dicto*-like thought can yield *H*'s actual *de re*-like knowledge of the single, individuated spruce that is before *H*.)

Second, the qualification that I have mentioned above in connection with (*Tj*) arises thus. In accepting (*Tj*) as our formulation of *H*'s knowledge-yielding act of thought, we imply that *H*, as knower (as against us as philosophers reflecting on *H*'s knowledge), must explicitly take the intuition-elements j_1 to j_4 to be those specific intuition-elements and to put before *H*'s mind (or to be) the relevant properties or spatial parts. Yet this implication is highly implausible. There is no reason to suppose that every being like us (including every philosophically unsophisticated such being) does or even can grasp the philosophical concepts necessary to take j_1 to j_4 in these ways.

Kant himself does not bother to distinguish between the thoughts required of *H* as knower and the thoughts required of us as philosophers reflecting on *H*'s knowledge. However, in order to develop his picture of knowledge with the utmost clarity and plausibility – and without, I think, altering his fundamental intentions – we must draw such a distinction. And then we should note that the clauses whose presence in (*Tj*) creates the above implication – various clauses within the scope of the '*H* is conscious in thought that' operator, like 'j_1 puts before *H*'s mind (or is) P_1' – evidently describe *H*'s act of thought from the viewpoint of the Kantian philosophical account of how intuition-elements function in knowledge. They do not describe H's act of thought simply from the viewpoint of what H, as mere knower, nust think through j. So in principle such clauses should be modified in order to escape this difficulty. However, the modification introduces unnecessary complications. And we can achieve its effect by understanding a clause like the one just quoted to claim, roughly, 'an experience is occurring which grasps (or is of) P_1.' The experience here referred to we (but not necessarily H) can then take to be the intuition-element j_1 that is P_1.[9] (Here and later I often simplify phrases like 'j_1 puts before H's mind (or is) P_1' to 'j_1 is P_1.') So, understanding such clauses in that way, I will leave those clauses (and similar clauses in claims below like (Ti)) unmodified.

We can now turn to the case of the union of the manifold of i in the concept of an object. In doing so, we can ignore the differing descriptions of i_1 and i_2 that arise depending on which version of the minimum Deduction assumption we ultimately adopt. Our interest is in the union of i's manifold that, on either version, Kant attempts to establish. Proceeding by analogy with the case of j, we can see that Kant should hold that, insofar as H knows through i_1 and i_2, i_1, and i_2 form the single, overall representation i that represents a single object as having various features and aspects that are – and that H takes to be – experienced through i_1 and i_2. They form that single representation because they are reproduced by H's imagination and synthesized by H in thought using the concept of an object in general.

For Kant, that synthesis involves, specifically, the sequential occurrence of i_1 and i_2 before H's mind in such a way that each is there a feature or aspect. As i_1 and i_2 occur in this way, H in thought grasps those features or aspects, which H takes i_1 and i_2 thus to present. And, using the concept of an object in general, H takes there to be a single object to which all those features and aspects belong. Moreover (and to note a point that we can ignore until Chapter Ten), as H takes the features and aspects to belong to that object, H regards them as constituting or specifying various general properties.

As can be seen from the preceding description, in H's synthesis, H grasps in thought the very features and aspects that i_1 and i_2 present, and H takes *those* features and aspects to belong to the relevant single object. (Of course a similar point holds for the case of j.) In expressing H's synthesis we should therefore in strictness hold that there is a specific feature F such that i_1 puts before H's mind (or is) F and H thinks *of that* feature, in a *de re*-like fashion, that *it* belongs to the relevant object. And we should make a similar supposition about i_2. Hence, given this fact and the preceding points, when H knows through i, H synthesizes i's manifold in such a way that H thinks the thought expressed, strictly, in the following claim – or in its final, 'H is conscious in thought that' conjunct:

(T*i*) There are features or aspects F and G such that i_1 and i_2 occur before H's mind in the sequential order i_1, i_2, and, as they so occur, i_1 puts before H's mind (or is) F and then i_2 puts before H's mind (or is) G and H is conscious in thought that there is a single object x such that first i_1 puts before H's mind (or is) F and x has F and then i_2 puts before H's mind (or is) G and x has G

or, formally:

$(\exists F)(\exists G)$ [i_1 and i_2 occur before H's mind in the sequential order i_1, i_2, and as they so occur, i_1 pm (or is) F and then i_2 pm (or is) G & H is conscious in thought that $(\exists x)$(x is an object & first i_1 pm (or is) F & x has F & then i_2 pm (or is) G & x has G)]

And through the thought expressed in (T*i*) H knows the single object that H thinks, the object of i as that object is represented by i.[10]

The relevant parts of (T*i*) are to be understood in the same way as the corresponding parts of (T*j*). For brevity, instead of speaking of 'the thought expressed in (T*i*),' I will talk hereafter of 'thinking the (T*i*)-thought' or 'thinking (T*i*).' In addition, when H comes to think, in (T*i*), *of* specific features or aspects F and G, H's thinking should be understood in the light of the Kantian view of concepts presented in Chapter Three. So we should suppose that F and G, as they occur before H's mind, are potentially general properties or features. When H knows through i, H in an act of thought grasps those potentially general properties and gives them a form, or generality, while in the same act of thought H takes there to be a single object to which they all belong. And thus H arrives at the full (T*i*) thought.

3. PRELIMINARIES TO B-DEDUCTION § 17

Kant's main goal in B-Deduction § 17 is to demonstrate (T*i*) on the basis of the actual-consciousness version of (S). As we emphasize in Section 4, Kant also needs to show that the (T*i*) thought yields H knowledge. In § 18 to § 20 he will then use (T*i*) and other claims to establish category-application to the object known through *i*. There is, however, a problem about the role of (T*i*) in this line of thought that we should note before turning to the § 17 reasoning itself.

The problem is this. (T*i*) (like (T*j*)) is a *de dicto* claim, for in (T*i*) *H* thinks simply *that there is* an object *x* that has *F* and *G* (and so on). Yet, as we have indicated earlier, (T*i*) expresses the act of thought that unifies the manifold and yields *H*'s knowledge through *i*. And that knowledge, like any knowledge through an intuition, ought to be a *de re*-like knowledge *of* a single, individuated object. So how can (T*i*), although *de dicto*, nevertheless express a thought that yields such knowledge?

In the Deduction, Kant say little that bears directly on this problem. But from the B-Deduction and other Kantian claims one can work out his basic answer to it. To put this answer more explicitly than he does himself, recall that our acts of thought never, by themselves, grasp single, individuated objects as such. Thus for the (T*i*) thought to concern a single, individuated object, there must be something connected with that thought that goes beyond the fact that (as (T*i*) itself indicates) that thought is an act of *thinking* there to be an object that has the relevant features. And, given our earlier discussions, as well as Kant's B-Deduction remarks on the need for intuition in order that our thought may concern specific, 'determinate' (and, I take it, single, individuated) objects, it is clear that this something has to do with intuition.[11]

Precisely what this something is Kant does not make clear. In the case of human knowledge through empirical sensible intuitions, the fact that an act of thought is about a single, individuated object is intimately bound up with the fact that that act grasps its object, through the intuition (and in a direct-object fashion), as occurring at a definite, particular spatial location that the intuition presents.[12] Because *i* is an intuition in general, from whose specific sensible nature we are completely abstracting, we cannot apply this idea to the (T*i*) thought. However, given Kant's overall views, he ought to make the individuating supposition that there is something special about *i* or about i_1's and i_2's presentations of the features *F* and *G*. That something special makes the (T*i*) thought *de re*-like with respect to the object *x* when *H* thinks that *x* has the specific, i_1- and i_2-presented *F* and *G*.[13] By supposing that such a special thing exists, Kant will – if he can justify this supposition – solve the above problem. (I comment in Section 5 on the issue of justifying this supposition.)

In order to understand the B-Deduction § 17 claims about (T*i*), it is important, next, to observe Kant's view that it is the (T*i*) act of thought that relates the elements of *i* to the single object that is known through *i*. At its most plausible, Kant's defense of this view runs as follows.[14]

Grant that through *i* a single object is indeed known by *H* as having features or aspects presented by i_1 and i_2. Grant also that this single object is *the* object known by *H* through *i*. Then the elements of *i* relate to the object known (in the form that that object takes in *H*'s knowledge of it) just in case they function together to represent to *H*, for *H*'s knowledge, that object. For *i*'s elements to function in this way, however, they must form a combination before *H*'s mind. As combination cannot be given, they cannot form this combination simply as they occur in the actual, sequential order i_1, i_2. Nor can their relation to any object independent in existence of i_1, of i_2, and of *H*'s acts of thought, as that object exists in itself, be part of their forming this combination. (Here we ignore issues raised by the Chapter Two problem for Kant.) By the Transcendental Aesthetic, beings like us cannot know objects as they exist in themselves or the relations of such objects to the mind. And, Kant holds, the relation of *i*'s elements to the object known is (like the fact of *H*'s knowledge itself) something that beings like us can know.

The elements of *i* must therefore form a combination – and so relate to the object known – because of something that occurs only after (in the order of logic) they are produced in *H*'s mind by an object as that object exists in itself and they then occur sequentially before *H*'s mind. And, in the light of Kant's account of the mind, it is clear what this thing must be. *H*'s active faculty of understanding, or thought (the only capacity of *H*'s mind able to grasp the combination of *i*'s elements), must hold together *i*'s elements before *H*'s thought-consciousness in such a way that, as those elements occur sequentially before the mind, *H* thinks the (T*i*) thought.

That is, as those elements occur in the way just noted and present features, *H* in thought must take those elements to present those features and *H* must take there to be a single object to which those features belong. (And, of course, although Kant does not indicate the following fact, as *H* proceeds in this way, the above individuating supposition about *i* must be justified, so that *H*'s thought is *de re*-like with respect to the single object in question.) Indeed, when *H* thinks (T*i*) (and when the individuating supposition is justified), *i*'s elements do function together to represent a single, individuated object having features or aspects that i's elements present. And then to maintain that that object is the object of H's knowledge through i, Kant need only observe that, as we have granted above, *the* object known by H though i is a single object having features presented by i's elements; and, by the preceding discussion, only through the (Ti) act of thought can i's element be related to such an object.

By the argument just presented, we establish Kant's claim that the (Ti) thought relates i's elements to the object known through i. Or, put loosely, the relation of i's elements to the object known just *is* (or occurs through) H's synthetic unifying, through the concept of an object in general, of those elements in the (Ti) thought.[15]

It would be impressive if Kant could establish (Ti) – and H's (Ti) thought as yielding H's knowledge through i. Indeed, the establishment of (Ti) (and of further claims like the one about the relation of i's elements to the object known) is a main goal of the Transcendental Deduction. In order to discuss whether Kant can achieve such a goal, we need to note a last point, which concerns what reading of the minimum knowledge-assumption (K) – or, specifically, of the claim 'H ... knows through the sensible intuition in general i' in (K) – the Deduction should be regarded as adopting. (Recall from Chapter Six that (K) says, officially, that H, a being like us with a passive sensibility and an active, discursive understanding, knows via the sensible intuition in general i, which is given in the form of the manifold consisting of i_1 and i_2.)

As we saw in Chapter Three, 'H ... knows through the sensible intuition in general i' in (K) can be given a strong reading, according to which H knows a single object through i, an object distinct from i_1 and i_2 taken separately and from the (i_1, i_2) sequence before H's thought-consciousness. We may now develop this strong reading further by expressing it as the claim that there is a single object x (a single, individuated object), and there are features F and G, such that i_1 and i_2 occur before H's mind in the sequential order i_1, i_2, and, as they so occur, i_1 puts before H's mind (or is) F and then i_2 puts before H's mind (or is) G, and H knows, *of* x and F and G, that x is an object and i_1 puts before H's mind (or is) F and x has F and then i_2 puts before H's mind (or is) G and x has G. And that object x, in the form that it takes as it is thus represented to H, is distinct from i_1 and i_2 taken separately and from the (i_1, i_2) sequence and is the object known by H through i.[16]

Again, and as we also saw in Chapter Three, 'H ... knows through the sensible intuition in general i' in (K) can be given a weak reading, according to which H knows an object through i but (as far as this assumption by itself goes) no specific claims are made about the nature of the object that H knows, and it is allowed that that object might be, for example, the (i_1, i_2) sequence occurring before H's thought-consciousness or simply i_1 and i_2 taken separately as mental objects of distinct acts of H's thought. Moreover, and as explained in Chapter Three, neither is any claim made here that i_1 and i_2, as they occur sequentially before H's thought-consciousness, put before H's mind (or are) any features, general properties, or aspects (or anything else) at all.

To see the relevance of these two versions of '*H* ... knows through the sensible intuition in general i,' suppose that we adopt the strong reading of (K). Then from (K) and his earlier views Kant can argue directly to (T*i*) (and to H's (T*i*) thought as yielding H's knowledge of a single object through i). After all – as implicitly indicated in the relation-to-an-object discussion – Kant's views about the role of thought in knowledge, about combination as not given, and so on, show that when H knows through i according to the strong reading, H thinks (T*i*). And these views show also that it is H's (T*i*) thought that yields H's knowledge of the single object that is (on that reading) known through i. Moreover, given such Kantian views and the above individuating supposition, this thought yields H a *de re*-like knowledge *of* that object.

However, while such reasoning establishes (T*i*) (and establishes H's (T*i*) thought as yielding H's knowledge of a single object through i), it obviously does not do so in a philosophically satisfactory way. In this reasoning (T*i*) (and H's (T*i*) thought as yielding knowledge) is shown simply by inference from (K) in the special case in which we simply *assume* that H knows a single, individuated object through i with i_1- and i_2-presented features and distinct from i_1 and i_2 taken separately and from the (i_1, i_2) sequence. But in the Deduction Kant needs to demonstrate that when transcendental unity of apperception holds with respect to i, and when H has *any* sort of knowledge through i, then H must think the (T*i*) thought. And Kant must show that this thought yields H knowledge of the sort here just assumed. Reasoning like that above clearly does not establish such results satisfactorily, for in adopting a strong reading of (K) Kant would assume these results or else would assume controversial, basic points that underlie them. And philosophers like Hume will of course urge that there is no ground for making any assumption like the one in the strong reading. Such philosophers will take it to be quite possible that even though unity of apperception holds and *H* knows through *i*, the object (or 'object') that *H* knows is simply i_1 and i_2 taken separately or the (i_1, i_2) sequence.

To establish (T*i*) and his other results satisfactorily, Kant thus cannot argue merely by considering the case in which (K) is given the strong reading. Kant must also consider the case in which *H* is assumed to know through *i*, but as far as this assumption by itself goes, no claims are made about the nature of what *H* knows, and the possibility is left open that what *H* knows is just i_1 and i_2 taken separately or the (i_1, i_2) sequence. Or, in other words, Kant must establish (T*i*) and his other results in the case of the weak reading of (K).

Before turning to Kant's actual arguments for (T*i*) note finally that, when we adopt the weak reading, we evidently regard *i* as an intuition in a wide, not-necessarily-object-representing sense that corresponds to the wide sense of 'representation' remarked in Chapter Two. And when we take Kant ultimately to claim that (T*i*) and his other results hold true on the weak reading, we regard him as claiming that on that reading *i*'s elements must indeed function together for *H* as an intuition in our original, basic sense – that is, as a singular representation of a single, individuated object.

4. B-DEDUCTION § 17 AND KANT'S ATTEMPTS TO PROVE THE UNION OF *i*'s MANIFOLD IN THE CONCEPT OF AN OBJECT

We come now to matters of fundamental importance to the entire argument of the Deduction. (T*i*), as we have just seen, is the claim that, roughly, *H* thinks there to be a single object *x* that has the i_1- and i_2-presented features *F* and *G*. Given the preceding relation-to-an-object discussion, and with one qualification that I will note in Section 5, it is clear that if *H*'s thinking the (T*i*) thought – or, to use our shorthand, *H*'s thinking (T*i*) – yields *H* knowledge of that single object, then that object will be, for Kant, *the* object of *i*, the object known by *H* through *i* in the form that that object takes as it is known by *H*. Hence we can say simply that in § 17 Kant needs to show *both* that *H* thinks the (T*i*) thought *and* that *H*'s thinking that thought yields *H* knowledge of the single object that *H* thinks in that thought.[17]

In order to complete the argument of the Deduction Kant needs to establish both these points. He must show that *H* thinks (T*i*) in order to argue, later in the Deduction, that in thinking that thought, *H* makes a judgment about that object and thereby brings about category application to that object. And he must also show that *H*'s thinking that thought yields knowledge of the single object that *H* thinks. Unless *H*'s thinking (T*i*) yields such knowledge, the fact that *H thinks* (T*i*) and thereby brings about category application to the single object that *H* thinks does not demonstrate that the categories apply to the – or to any – object that *H* *knows* through *i*.

As a careful reading of B-Deduction § 17 shows, Kant himself does not recognize the need for a separate defense of these two points. Rather, he proceeds as though he believes that the first implies the second – that is, that if *H* thinks (T*i*), then *H*'s thinking that thought yields knowledge of the single object that *H* thinks.[18] Nevertheless, we must ourselves sharply separate these points, despite the slight alteration that we thereby produce in the § 17 exposition. In the present section, and ignoring Kant's grounds for the second point, I thus concentrate on his attempts simply to prove that *H* thinks (T*i*). In Section 5 we then turn to the question whether Kant can hope to show that *H*'s thinking that thought yields *H* knowledge, as well as to related questions.

Kant's attempts to prove that *H* thinks (T*i*) – or, equivalently, his attempts to prove (T*i*) – are made puzzling by several factors. At least initially in the Deduction Kant does not acknowledge both the strong and the weak readings of (K). Yet ultimately it becomes clear that he intends to show (T*i*) in the case of the weak as well as of the strong reading. Because of this fact, and because he is well known for claiming, in effect, that the holding of unity of apperception with respect to *i* requires *H* to think the (T*i*) thought, one naturally expects to find Kant at some point arguing to (T*i*) directly and explicitly from a weak reading of (K). However, one does not find, in any great detail, such reasoning. Rather, one discovers arguments in the case of the strong reading from the holding of unity of apperception to (T*i*), along with claims that clearly show Kant to suppose that in the case of the weak reading, too, that holding implies the truth of (T*i*). But Kant does not make it clear exactly how he means to defend those claims, and so the nature and success of his overall argument for (T*i*) remains in doubt.

Kant's § 17 argument for (T*i*) is contained in brief remarks at B137:[19]

Understanding is ... *the faculty of knowledge.* (s_1) This knowledge consists in the determinate relation of given representations to an object. (s_2) An *object*, however, is that in the concept of which the manifold of intuition is *united*. Now (s_3) all unification of representations demands unity of consciousness in the synthesis of them. Consequently (s_4) it is unity of consciousness that alone constitutes the relation of representations to an object, and therefore their objective validity and the fact that they are modes of knowledge; and upon it therefore rests the very possibility of the understanding. [Parenthetical letters inserted.]

As claims (s_1) to (s_3) show, Kant here takes knowledge through *i* to require the union of the manifold of *i* in the concept of an object and so to require *H* to think the (T*i*) thought.[20] However, by (s_3), for *H* to think that thought, unity of apperception must hold with respect to *i*.[21] Hence the holding of unity of apperception with respect to *i* requires (and indeed 'constitutes') *H*'s thinking the (T*i*) thought. (And Kant then leaves it for the reader to note that, given the fact that unity of apperception holds, or has been stipulated by us to hold, with respect to *i*, it follows that *H* actually thinks that thought.)

To make plausible his initial, undefended assertion that *H*'s knowledge through *i* requires *H* to think the (T*i*) thought, Kant clearly must suppose that *H*'s knowledge is to be understood on the strong reading, according to which it is assumed that through *i H* knows a single object with the i_1- and i_2-presented features *F* and *G*. Yet, as the context indicates, in reaching the conclusion in (s_4), he moves on to the central, weak reading of *H*'s knowledge through *i*, according to which no special claims are made about the 'object' known through *i*. And he claims that, on that reading, the holding of unity of apperception with respect to *i* implies *H*'s thinking the (T*i*) thought. (Thus note B138, towards the end, on '*all* knowledge' and '*every* intuition' – my emphases – as well as the B137 claim that, as is implied by the last part of (s_4), in order for the faculty of understanding and so for any knowledge *at all* to be possible, including weak-reading knowledge, unity of consciousness must imply, and indeed must 'constitute,' the (T*i*)-style relation of representations to an object.[22]) A similar course of reasoning can be seen in various A-Deduction texts.[23]

Until Kant makes his concluding inference in (s_4), his claims in the preceding B137 argument are clear and supported by his previous results (or by our Chapter Seven stipulations). After all, if *H* knows through *i* on the strong reading, then *H* thinks (T*i*). (Thus when *H* knows the spruce through *j*, *H* thinks there to be a single object *x* that has the properties and spatial parts of the spruce that elements j_1 to j_4 of Section 2 above present.) And given that unity of apperception *does* hold with respect to *i*,

H of course cannot think (T*i*) *unless* unity of apperception holds, so the holding of unity of apperception is a *necessary* condition for *H*'s thinking (T*i*), as (s_3) claims. But then in the concluding (s_4) Kant shifts to the weak reading of *H*'s knowledge through *i*, and he asserts that on that reading the holding of unity of apperception implies, and so is a *sufficient* condition for, *H*'s thinking (T*i*). (Thus consider the holding of unity of apperception with respect to i_1 and i_2, where the object of *H*'s knowledge through i_1 and i_2 is not assumed to be anything more than those elements themselves, taken separately or in sequence. Kant here asserts that that holding *by itself* implies that *H* thinks that there is a single object that has features that i_1 and i_2 present.) How can Kant go in this way from the strong to the weak reading and, in the process, from the holding of unity of apperception as a necessary condition for *H*'s thinking (T*i*) to the holding of unity of apperception as a sufficient condition for that thinking? In § 17 Kant simply does not make this inference clear, and an air of blatant fallacy hovers over this part of his reasoning.[24]

It may be that Kant *is* arguing here in a flatly fallacious way. But a study of other texts suggests that his reasoning is *not* fallacious but instead rests on the idea that, on the strong reading of *H*'s knowledge, the holding of unity of apperception can be shown to be the source of, and so to imply, *H*'s thinking (T*i*).[25] And he supposes that if that holding is such a source on the strong reading, then he can show that that holding also is such a source on the weak reading, thus confirming the (s_4) conclusion. (That is, suppose that in cases like *H*'s knowledge of the spruce, it can be shown that the holding of unity of apperception with respect to the relevant intuition-elements contains or implies *H*'s thinking there to be a single object that has the features that those elements present. Then Kant's idea is that in cases in which *H* knows through the arbitrary elements i_1 and i_2 but the object known is not assumed to be anything more than i_1 and i_2 themselves, taken separately or in sequence, it can similarly be shown that the holding of unity of apperception with respect to those elements implies *H*'s thinking there to be a single object that has features presented by them.) For the rest of this section I will concentrate on this sort of 'sources' justification for Kant's § 17 inference to the (s_4) conclusion.[26]

A main example of such reasoning occurs in the preliminaries to the official A-Deduction at A106 ff.[27] In effect, that text says that *H*'s thinking (T*i*), as is required for *H*'s knowledge through *i*, is *itself* necessary and yields a necessary, synthetic unity of *i*'s elements in the concept of an object in general. Like all necessity (or like all necessary synthetic unities), this necessary unity – and the necessity of *H*'s thinking (T*i*) – has a unique transcendental source, or 'ground.' As we have seen, however, for Kant transcendental unity of apperception holds necessarily with respect to *i*'s elements. Hence, he concludes, that holding (or, as he says at A107, the 'pure original unchangeable consciousness' that it involves) is the source of, and so implies, the necessity of *H*'s thinking *(Ti)* and the necessary unity that that thinking yields.[28]

As it stands, this A106 ff. reasoning is unsatisfactory, for even if we grant Kant that the holding of unity of apperception is necessary, the A106 ff. reasoning does not show that the necessity of that holding establishes or somehow yields the necessity that he takes to belong to *H*'s thinking (T*i*) and to the unity that that thinking yields. Nor does the reasoning make it clear why the claimed necessity that belongs to *H*'s thinking (T*i*) must have a source or ground distinct from itself. Nor does it show that the holding of unity of apperception, rather than some other necessary thing, is the source of that claimed necessity. Nor does it explicitly distinguish strong and weak readings of *H*'s knowledge.

At least up to a point, these flaws can be remedied by drawing on B-Deduction § 16 points about the 'originality' of unity of apperception that we have ignored up to now, as well as on the views that seem to underlie Kant's A106 ff. claims. Suppose that a necessary unity is indeed yielded by (and, as Kant will hold, only by) *H*'s thinking (T*i*); and suppose that the existence of this necessary unity is implied by the fact that *H* knows through *i*. Then, for Kant, the necessity of this unity will be, in effect, the necessity that, somehow, i_1 and i_2 should occur related together before *H*'s thought-consciousness in such a way that they there form a unity through the concept of an object in general. (They will form such a unity by functioning for *H* to represent, by means of *H*'s use of that concept, one thing.) But such a unity of i_1 and i_2 is not present in them as they are given. Nor can the necessity that i_1 and i_2 form such a unity derive from the fact that *H* does indeed think (T*i*), for that fact does not require that *H* *must* think that thought. Rather, the necessity that i_1 and i_2 form such a unity must derive from some other act of *H*'s thought (thought being, for Kant, what unites distinct mental elements). That other act of thought, because it is the ultimate source of that necessity, must itself be necessary – and must itself establish that necessity – in a way that does not require yet a further source. And because, in establishing the necessity that i_1 and i_2 form such a unity, that act of thought establishes also a relation together of i_1 and i_2 before *H*'s thought-consciousness, that act of thought must somehow take i_1 and i_2 to be related together before itself.

Yet consider the act of thought expressed in a unity-of-apperception claim like the actual-consciousness version of (S). (According to that version, *H* is conscious in thought that the *I think* accompanies all of *i*'s elements taken together.) For Kant, such an act of thought is necessary in roughly the sense that it is necessary that *H* should be conscious, in thought, that the *I think* accompanies i_1 and i_2 taken together. And through this necessary accompaniment of i_1 and i_2 by the one representation *I think* within that act of thought, that act of thought, in a certain necessary way, takes i_1 and i_2 each to be related to the same thing, the one *I think*, and so to be related together before itself.[29]

Moreover, as the *I think* in such a necessary way accompanies i_1 and i_2 before *H*'s thought-consciousness within that act of thought, this necessary accompaniment is original and underived and so requires no further source. That is, this necessary accompaniment is required by the very fact of *H*'s first-person thought through the *I think*, in knowledge, and by the very fact of *H*'s having representations.[30] Furthermore, no other representation of *H*'s necessarily accompanies i_1 and i_2 before *H*'s thought-consciousness, requiring no further source.[31] Hence there is no other representation of *H*'s, beyond the *I think*, whose necessary accompaniment of i_1 and i_2 could itself establish the necessity that i_1 and i_2 should form the above sort of unity (or could itself be the source of the necessary accompaniment of i_1 and i_2 by the *I think*). The unity-of-apperception thought expressed in a claim like the actual-consciousness version of (S) must therefore be the source of the necessity that i_1 and i_2 should form the above sort of unity. Or, put otherwise, the holding of (necessary) unity of apperception with respect to *i* is the source of, and so implies, both the necessity of *H*'s thinking (T*i*) and the necessary unity that that thinking (and only that thinking) yields.

Now we have reached this last result simply for the strong reading of *H*'s knowledge through *i* (on which *H* knows a single object having the i_1- and i_2-presented features and aspects). And our reasoning has involved the claim that, on that reading, *H*'s thinking (T*i*) and the unity that that thinking yields are necessary. However, Kant also needs to reach the last result on the weak reading (on which all that *H* knows through *i* might be i_1 and i_2, taken separately or in sequence). Texts like A106 ff. or B-

Deduction § 17 do not make it clear how he proceeds to that result on the weak reading, however.

Kant may simply reason (perhaps quite implicitly) that because the holding of unity of apperception is the above sort of source on the strong reading of *H*'s knowledge – and because that holding is necessary when *H*'s knowledge is understood in *any way* at all – that holding is such a source also on the weak reading. Such reasoning on its own is unconvincing. (If the holding of unity of apperception is the source of *H*'s thinking (T*i*) in a case like the spruce one, why must that holding remain such a source in the case where it is not assumed that *H* knows anything more than i_1 and i_2 taken separately or in sequence?) But Kant can strengthen this reasoning greatly by recalling that, for him, the *I think* is an *a priori* representation.[32] So, he can urge, the fact that necessary accompaniment by the *I think* is the source of the relevant necessary unity should not depend on facts about *i*'s elements that happen to be peculiar to the strong reading of *H*'s knowledge. Hence given that, for Kant, unity of apperception holds whenever *H* knows through *i*, the result at the end of the next-to-last paragraph will be true also on the weak reading. And so the holding of unity of apperception will be the source of the above necessity on the weak reading as much as on the strong reading.

The line of thought developed in the last several paragraphs for *H*'s thinking (T*i*) is fascinating but unsuccessful. By A106 ff., *H*'s thinking *(Ti)* really can be necessary only in a necessity-of-the-consequent sense. It is necessary that *if H* knows through *i* according to the strong reading and if other Kantian points hold, *then H* thinks (T*i*).[33] However, such a necessity derives from the deductive relations that hold between the antecedent and the consequent of this last necessity claim and requires no special explanation in terms of any transcendental source.[34]

Moreover, further probing shows that in any case the above necessity-of-the-consequent is simply irrelevant to the present Deduction line of thought. Consider the conditional claim that if *H* knows through *i* on the strong reading and if other Kantian points hold, then *H* thinks (T*i*). As just argued, this conditional claim is necessary. But, even given the necessity of this claim, *H* as mere knower need not ever make or even consider this claim in order to know through *i*. Nor need anyone else make this claim in order for *H* to know. So there is no fact here about what *H must* think that requires explanation through the holding of necessary unity of apperception as the proposed source.

What *would* seem to require a transcendental source, or anyway some special explanation, would be a direct, categorical, non-necessity-of-the-consequent necessity that *H* should think (T*i*).[35] But in such a direct, categorical sense *H*'s thinking (T*i*) is *not* necessary. After all, what could conceivably justify such a claim to necessity, other than the *conditional* claim that, necessarily, *H* thinks that thought *if H* knows through *i* on the strong reading? And, in any case, in Chapter Seven we saw that Kant has given no convincing reasons to take the holding of unity of apperception to be necessary in any of the senses there noted. So that holding would not be available to be the source of any direct, categorical necessity that might belong to *H*'s thinking were such a necessity somehow to be established.

The argumentative situation here is especially troubling. If the holding of necessary unity of apperception were to be the source of any direct, categorical necessity that might belong to *H*'s thinking (T*i*), then it could really be that holding only in the sense that, given that various representations are *H*'s, it is directly, categorically necessary that *H* should be conscious in thought that the *I think* accompanies those representations taken together. But it is particularly hard to conceive how Kant could establish such forms of the holding of necessary unity of apperception.

We also should note that even if a direct, categorical necessity for *H*'s thinking (T*i*) were established – and even if the holding of unity of apperception itself were shown to be necessary in the appropriate sense – further argument would still be needed to demonstrate that such a direct, categorical necessity could *not* have its source in something that is given (even if not in i_1 and i_2 themselves merely as they are initially given to the mind). Recent studies suggest, after all, that there may well be direct, categorical, *de re* necessities that obtain in mind-unprocessed nature.[36] And it has not been ruled out that such necessities (existing, for example, in the thought-unprocessed part of the mind) could in some way be responsible for the above direct, categorical necessity. Kant would of course object that such necessities cannot exist in mind-unprocessed nature (or within the thought-unprocessed part of the mind) because they then would be known only *a posteriori*. However, it seems that these sorts of necessities *are* known *a posteriori* (given that they exist at all). Moreover, in any case it is hard to see how a Kantian direct, categorical necessity that *H* should think the thought expressed in (T*i*) could itself be known wholly *a priori*. Roughly, that necessity involves, among other things, categorical claims about the operation of i_1 and i_2, and the knowledge of those claims, like that of the existence and character of i_1 and i_2, is surely *a posteriori*.

Finally, one can question the transition from the holding of necessary unity of apperception as the source, on the strong reading of *H*'s knowledge, of the necessity of *H*'s thinking (T*i*) to that holding as such a source on the weak reading. One can doubt Kant's view of the *I think* as being *a priori*,[37] and, given that doubt, the reasons offered above for the transition collapse.

Not all the problems that I have just raised may seem decisive. However, Kant's failure to show the direct, categorical necessity of *H*'s thinking (T*i*) is by itself conclusive. At this crucial point in the Deduction, his reasoning fails.

5. THE UNION OF THE MANIFOLD OF *i* IN THE CONCEPT OF AN OBJECT AS YIELDING *H* KNOWLEDGE; FURTHER QUESTIONS

Had Kant's above line of reasoning succeeded, he would now have shown that, on the strong reading of *H*'s knowledge through *i*, the holding of unity of apperception with respect to *i* is the source of – and so implies – *H*'s thinking (T*i*). And hence it is also such a source on the weak reading. So, given that unity of apperception holds with respect to *i*, Kant would have demonstrated the first main point he needs to show in B-Deduction § 17 – namely, that *H* thinks the (T*i*) thought. By extending the above reasoning for that point, he might then have argued for the second main point – namely, that *H*'s thinking that thought yields *H* knowledge of the single object that *H* thinks in that thought. However, Kant's difficulty is not merely that this line of reasoning fails. The difficulty is that doubts can be raised about *any* attempts to derive these points from *H*'s knowledge through *i* and the holding of unity of apperception. In short, no convincing argument can establish all that Kant needs to demonstrate in § 17 in order to complete the argument of the Transcendental Deduction.

As one might expect, these doubts center on the weak reading of *H*'s knowledge through *i*. On the strong reading *H* does indeed think the (T*i*) thought and *H*'s thinking that thought yields *H* knowledge of the single object that *H* thinks. Consider now, however, the weak reading of *H*'s knowledge. On that reading, no specific claims are made about the nature of what *H* knows through *i*, and it is allowed that what *H* knows may be simply i_1 and i_2 taken separately or the (i_1, i_2) sequence. Thus even if *H* knows through i, unity of apperception holds with respect to i, and H in fact *thinks* the (Ti) thought, it still needs argument to show that H's thinking that thought yields H *knowledge* (of the relevant single object) through i. Unfortunately, it seems impossible to give any convincing such argument, for, on the weak reading, it is perfectly possible that H's thinking that thought does not yield H knowledge.

It is possible, on the weak reading, that, given the holding of unity of apperception with respect to i, all that H *knows* through i is simply the apperceptive fact that the *I think* accompanies the sequentially occurring i_1 and i_2, H nevertheless still *thinking*, as in (Ti), that there is a single object that has the i_1- and i_2-presented features or aspects F and G, and so on. For example, while knowing that i_1 and i_2 are both accompanied by the *I think* (or both belong to *him himself* or to *her herself*), H might have what turns out to be a nonveridical experience, produced through i and H's thinking that there is a single object that has the i_1- and i_2-presented features or aspects. In this nonveridical experience, H's thought would relate i_1 and i_2 to the (illusory) object of the experience but would not yield H knowledge of that object. Such cases of self-ascribed, illusory experiences surely can occur.

Examples of this sort prove that the second main point noted above need not be true, on the weak reading of H's knowledge through i, even if the first point turns out to be correct and H thinks (Ti). These examples show that there is no convincing argument for the second main point from H's knowledge (read weakly) through i and the holding of unity of apperception with respect to i. But the difficulty here for Kant is even worse than the existence of such examples suggests. On the weak reading, there is no convincing argument of this sort even for the first main point. It is possible for H to have knowledge through i, for unity of apperception to hold with respect to i, and yet for H nevertheless not even to *think* the (Ti) thought.

Consider first Kant's own picture of knowledge. As one can see from the exact form of (Ti) presented in Section 3 and from the discussion there, in order for H to think the precise thought expressed in (Ti), i_1 and i_2, as they occur before H's mind in knowledge in independence of H's activities of thought, must put before H's mind (or be) the selfsame properties or features that H in thought takes to belong to the single object x. However, on the weak reading of H's knowledge, no claim is made that as i_1 and i_2 occur before H's mind in knowledge, they put before H's mind (or are) any features, general properties, or aspects at all. And, in fact, given Kant's views as so far established, it is hard to see how it *follows*, simply from the facts that *H* has knowledge (read weakly) through *i* and that unity of apperception holds with respect to *i*, that i_1 and i_2 *are* the relevant properties or features. (Here and later I sometimes simplify 'put before the mind (or are)' to 'are.') Thus Kant's own views, as so far established, do not seem to rule out the possibility that unity of apperception holds with respect to *i* while *H* yet does *not* think the thought expressed in (T*i*). And so, given those views, it seems that there will be no convincing argument, of the sort required in the Deduction, for *H*'s thinking that thought.

Moreover, Kant needs to show that *H*'s thinking (T*i*) results in a *de re*-like knowledge *of* a single, individuated object that *H* thinks (and knows) through *i*. And, in order to show this fact, he must suppose that there is something special about *i* or about i_1's and i_2's presentations of features or aspects *F* and *G* that makes the thought (T*i*) *de re*-like with respect to the object *x*. On the strong reading of *H*'s knowledge through *i*, on which it is assumed that there is in fact a (single, individuated) object that *H* knows through *i*, such an individuating supposition must turn out to be justified. But suppose that Kant begins only with a weak reading of *H*'s knowledge (on which nothing is initially assumed about what, if anything, i_1 and i_2 are). Then it is hard to see how Kant can show, merely by appeal to the fact that *H* knows through *i* and that unity of apperception holds with respect to *i*, that that sort of individuating supposition is justified. And this problem about justifying that supposition would remain, even if Kant were to find some way of showing that *H* thinks (T*i*).

Consider now a second group of reasons for doubting that *H* must think the (T*i*) thought. To appreciate this group of reasons, suppose that, on the weak reading, Kant really *could* argue convincingly for *H*'s thinking that thought from the fact that *H* knows through *i* and that unity of apperception holds with respect to *i*. Then no matter what i_1 and i_2 put before *H*'s mind and no matter what other knowledge or beliefs *H* has, *H* inevitably *will* think the (T*i*) thought. Yet on the weak reading i_1 and i_2 are not assumed to be anything more than two – perhaps entirely unrelated – intuition-elements through which *H* knows. Couldn't it then be that *H* knows through *i* and unity of apperception holds with respect to *i*, but i_1 and i_2 turn out to be features or aspects that (given other knowledge or beliefs of *H*'s) *H* simply does *not* think – in the way expressed in (T*i*) – to belong to a single object?

For instance, i_1 and i_2 might be features that *H* knows or believes to be incompatible in fact or in logic (say the features of being feline and of being a nitrogen-fixer or of being quadrangular and of being triangular). Or i_1 might be the feature of being round, which *H* knows a real cup before *H* to have, while i_2 might be the feature of being a ringing bell, which *H* knows not to belong to any real object before *H* but only to an illusory object that *H* is experiencing along with the real cup. On any of these examples, *H* in normal circumstances will *not* think that there is a single object that has the features that both i_1 and i_2 present, even though *H* will have at least the knowledge that the *I think* accompanies both of i_1 and i_2. So, there can be no convincing argument, of the sort required in the Deduction, for *H*'s thinking the (T*i*) thought.

The situation for Kant in B-Deduction § 17 thus is very grave. As we have emphasized at the start of Section 3, in order to continue the argument of the B-Deduction, Kant needs by the end of § 17 to have established two points: that *H* thinks the (T*i*) thought and that *H*'s thinking that thought yields *H* knowledge of the single object that *H* thinks. Yet the preceding discussion demonstrates that he can produce no convincing arguments for these points. So it appears that in § 17 an unbridgeable gap opens in the overall argument of the Deduction. I think that this appearance is in fact correct. However, it is worth noting that Kant has some room for maneuver here.

In particular, if Kant were willing to reformulate (T*i*), then he could avoid at least some of the problems raised above. For instance, we might revise (T*i*) to claim that *H* thinks there to be a single, objective reality of which i_1 and i_2 present features, aspects, or objective parts.[38] However, the last example above – of roundness and the ringing bell – shows that this defense will not succeed. Nevertheless, other defenses can be suggested. Thus (i) Kant could defend a modified version of (T*i*) as following from the holding of unity of apperception with respect to *i*; and (ii), by using a weakened (T*i*)-style result, he could argue that *H*'s thinking the thought in that weakened result yields *H* knowledge of the object that *H* thinks.

(i) Kant could argue that if *H* has knowledge through *i* and if unity of apperception holds, then the following modified version of (T*i*) is true: Each of *i*'s elements belongs to *some* group of intuition-elements such that with respect to that group *H* thinks that there is a single object that has the features or aspects presented by the members of that group; and *H* may (but need not) take such a single object to be actual. (i_1 and i_2 might belong to two separate groups of intuition-elements and *H* might think that there is an – actual – object, the cup, that has the i_1-presented feature of being round and the other features presented by the other members of that first group, and another object – not actual – that has the i_2-presented feature of being a ringing bell and the other features presented by the other members of that second group. Or, again, *H* might think that there is one – actual – object that has the i_1-presented feature of being feline and the other features presented by a group of intuition-elements to which i_1 belongs; and another – also actual – object that has the i_2-presented feature of being a nitrogen-fixer and the other features presented by a different group of intuition-elements to which i_2 belongs.)

One can make a strong case for such a modified version of (T*i*), by appeal to the intentionality – or the directedness – of mental phenomena. I believe that every self-ascribed experience belonging to a being like us is in fact intentional in some suitability defined sense. Thoughts are always about things – even if about inner or abstract things; we see patterns of light and dark, color and shade; we hear sounds, taste tastes, smell odors, and so on; and in our experience of them even pains, tingles, itches, and feelings of 'objectless' anxiety involve – perhaps phantom – bodily locations.) So unity-of-apperception-governed experiences through which we know always in fact involve thoughts or some other sort of judgment like the thought expressed in the above modified version of (T*i*).

In addition, this sort of modified-(T*i*) idea would seem to allow Kant to pursue roughly the same goals as would that original B-Deduction § 17 claim. Kant means to use the claim that *H* thinks the (T*i*) thought to argue that *H* judges with respect to the object that *H* thinks and thereby brings about category application to that object. As we have seen, he cannot demonstrate that *H* thinks that thought. And so he cannot use these later Deduction arguments – however convincing they may be on their own – to prove such category application. Suppose, however, that Kant could demonstrate something like the modified-(T*i*) idea. Could he proceed, in parallel with the later Deduction arguments, to show category application to the object that *H* thinks? Yes, for he could now argue similarly from *H*'s thinking the sort of thought expressed in a modified version of (T*i*) to *H*'s judging with respect to the object that *H* thereby thinks and hence to *H*'s bringing about category application to that object.

(ii) Kant also could try to bypass our first objection (based on the possibility of nonveridical experience through *i*) to the result that *H*'s thinking (T*i*) yields *H* knowledge. He could do so by arguing for a weakened (T*i*)-style result, which would also allow him to defend category application to the objects that *H* knows.

In particular, the following weakened (T*i*)-style result might still be true: *H* cannot have knowledge through the unity-of-apperception-governed elements of *i* unless *H* has *some* unity-of-apperception-governed intuition-elements that present features that *H* thinks to belong to a single object; and *H*'s thinking that thought yields *H* knowledge of that object.[39] (For example, *i* might itself be only the imagination-generated product of a dream or delusion, and so through *i H* might know only the apperceptive fact that the *I think* accompanies both of *i*'s elements taken together. But there would still be *some* group of unity-of-apperception-governed intuition-elements of *H*'s, say the spruce-representing elements of *j* of Section 2, through which *H* does know, by means of the relevant thought, a single object.)

This result seems *in fact* true for human beings (all of us surely have some such knowledge-yielding intuition-elements or mental states). And attempts to argue convincingly for this result are not open to the objection above about the possibility that the (T*i*) thought may not itself yield *H* knowledge. Moreover, by appeal to this weakened result, Kant could now argue for category application to the object *H* knows. Given the above weakened result, he could show that *H* has some unity-of-apperception-governed intuition-elements with respect to which *H* thinks a (T*i*)-like thought. And he could show, further, that *H*'s thinking that thought yields *H* knowledge of the object that *H* thinks. Since that thought is a judgment, it would follow that the categories apply to that object.

Kant could thus defend claims akin to the two basic points of B-Deduction § 17 while avoiding various of the objections raised above. It is therefore important to see, briefly, what this reasoning actually would accomplish, given the original goals of the Deduction. From the conclusion in the last paragraph, Kant could generalize to the final claim that if a being like us has any unity-of-apperception-governed intuition through which that being knows, then that being has some unity-of-apperception-governed intuition-elements through which that being knows a category-subsumed object. This claim is weaker than the original Transcendental Deduction conclusion that any unity-of-apperception-governed intuition through which a being like us knows yields that being knowledge of a category-subsumed object. But it still has considerable philosophical interest and its proof would be a significant achievement by Kant.

I see four problems, however. First, it is obviously not enough that points like the modified-(T*i*) idea and the weakened (T*i*)-style result seem *in fact* true for beings like us. In order to achieve anything like the basic purposes of the Deduction, Kant must show that these points are in the relevant ways *necessarily* true for beings like us. Yet, as a Humean philosopher will object, why *must* all our unity-of-apperception-governed (or self-ascribed) mental states be intentional in something like the way sketched above? Even if all such mental states are intentional in some such way, why *must* at least some mental states have connected with themselves thoughts (or judgments) that yield us *knowledge* of objects (as against mere beliefs about objects or even mere conceptions concerning how objects might be)?

Second, in order to establish category application, Kant needs to show that *H*'s thinking a thought of the sort noted in the modified-(T*i*) idea or in the weakened result really does bring about category application to the object that *H* thereby thinks. In addition, both the modified-(T*i*) idea and the weakened result imply that when unity of apperception holds with respect to *i*, it follows that elements of *i* – or some other intuition-elements that *H* has – put before *H*'s mind features or aspects that *H* takes to belong to one or more single objects. And it is hard to see why such a point does follow, for there seems to be no connection between the *I think*-accompaniment of intuition-elements and the fact that those elements put anything before the mind. Of course if Kant really could prove the modified-(T*i*) idea and the weakened result, then even if we could not understand why *H*'s intuition-elements behave in this way, we would still have to agree that they do. But, in the absence of such a proof, the difficulty in understanding why this behavior should occur raises doubts about the possibility of any such proof.[40]

Third, and to note an issue of a sort that recurs in Chapter Nine, it is not enough to show that the object known, on the weakened result, is distinct from the intuition-elements – taken separately or in sequence – through which *H* knows that object. If the weakened result is to have real philosophical interest, that object must be shown to be distinct from *any* intuition-elements or other representations of *H*'s, taken separately or in sequence.[41] Otherwise that result allows the possibility that *all H* knows through *any* of *H*'s mental states is various of *H*'s own intuition-elements or mental states. Such a Humean possibility is precisely the sort that Kant, in the Deduction, hopes to eliminate. But even if it could be shown that the object *H* knows is in fact category-subsumed, why could not that object turn out to be, say, simply an associatively organized sequence of *H*'s representations that is itself category-subsumed and is known by *H* as a part of *H*'s empirical mind? Unless such issues can be dealt with, the weakened result will appear to many philosophers not to be of exceptional significance.

Fourth, the above sort of reasoning requires a considerable revision in the argument of the Deduction as Kant presents that argument. Central to Kant's presentation is the view that the holding of unity of apperception with respect to a given manifold of intuition is the source of the union of that manifold in the concept of an object in general through a knowledge-yielding thought of the sort expressed in (T*i*). According to this view, that union – and hence, ultimately, the holding of unity of apperception itself with respect to that manifold – implies, in turn, category application to the object that is known. This basic Deduction view is not radically altered through the use, in the above reasoning, of the modified-(T*i*) idea and the subsequent inference to category application to the object that is thought in the relevant thought. But the appeal in that reasoning to the weakened result breaks radically with this view.

The break occurs because that appeal abandons the notion that the holding of unity of apperception with respect to intuition *i* is itself the *source* of the application of the categories to the object that *H* *knows* through *i*. On the weakened result, after all, when unity of apperception holds with respect to *i*, the object known may well be known through intuition-elements that are utterly different from *i* itself. And even though the fact that the categories apply to that object is argued to follow from the holding of unity of apperception with respect to those different intuition-elements, neither the holding of unity of apperception with respect to those different elements nor the holding of unity of apperception with respect to *i* will itself then be responsible for the fact that *that* particular object is in fact *known* and is therefore an object of knowledge that falls under the categories. So in appealing to the weakened result the reasoning of the sort envisaged above introduces a fact that would have to be argued on grounds outside Kant's own presentation of the Deduction. Consequently in this respect that reasoning does not defend, but rather simply replaces, Kant's own argument in the Transcendental Deduction.[42]

Reasoning of the sort envisaged above thus may – or may not – succeed, and in any case it does not fully support the original argumentation of the Deduction. Because that argumentation is the subject of our investigation, I will not pursue such reasoning further here. Rather I will hold in abeyance the question whether, through such reasoning or any other reasoning, a Kantian could manage to show anything like category application to objects that we know through unity-of-apperception-governed intuitions or mental states. And in the next two chapters I will focus on the main topic of concern to us in the rest of this book, the remainder of Kant's own argument in the first half of the B-Deduction. As I do so I will assume that in B-Deduction § 17 Kant has in fact demonstrated the two basic § 17 points that we have emphasized throughout this chapter: that *H* thinks the (T*i*) thought and that *H*'s thinking that thought yields *H* knowledge of the single object that *H* thereby thinks. This assumption is of course mistaken, given what our above objections have shown. But for simplicity and accuracy of exposition I make it below. In Chapter Ten we return briefly to questions about the overall success of arguments like the Deduction. Of course one might at this stage ask why we should continue our investigations at all, given that Kant cannot prove his two § 17 points. One answer is scholarly: to follow an influential argument through to its results. Another is philosophical: The remainder of that argument is filled with interesting claims about judgment, objectivity, and the categories.

Before closing the present discussion, I should note a qualification (referred to at the start of Section 4) and make two concluding comments. The qualification arises thus. I said in Section 4 that, given the relation-to-an-object discussion, and with a qualification to be made later, it is clear that if *H*'s thinking the (T*i*) thought yields *H* knowledge of the single object that *H* thinks in that thought, then that single object is *the* object of *i*, *the* object known through *i* in the form that that object takes as it is known by *H*. The qualification ought now to be clear. As I have indicated several times above, *H*'s unity-of-apperception thought that the *I think* accompanies i_1 and i_2 taken together should itself presumably amount to knowledge that *H* attains through *i*. So while the single object that *H* thinks in the (T*i*) thought is indeed what we have been calling '*the* object of *i*, in the form that that object takes as it is represented to *H* by *i*,' i_1 and i_2, as they are thus accompanied by the *I think*, should also themselves count as objects that *H* knows through *i*. (Evidently a similar qualification applies if we consider – as on the above weakened result – *H*'s knowledge through the relevant (T*i*)-like thought.)

This qualification does not affect the results of our above discussion in Sections 4 and 5. But it points to a gap in Kant's overall Deduction attempt to prove category application to all the objects that *H* does or can know by proving category application to the single objects that *H* thinks with respect to sensible intuitions. Given the qualification, intuition-elements like i_1 and i_2, as they occur before *H*'s mind accompanied by the *I think*, *are* objects known through *i*; and yet in this Deduction attempt itself, they are not proved subject to the categories. This sort of gap and the questions that it raises for the Deduction are considered further in Chapter Ten.[43]

Finally, we have noted several times in earlier chapters that Kant has not really proven either his idealism or the position that combination cannot be given. And so it will be useful to comment on the fate of the two basic B-Deduction § 17 points if he were to abandon that idealism or that position. Here the main fact to observe is simple. Briefly, Kant could maintain both of the specific § 17 points even if he were to reject his idealism or his position on combination. After all, suppose that, as § 17 claims, the holding of unity of apperception with respect to *i* does indeed require the two § 17 points that *H* thinks the thought expressed in (T*i*) and that *H*'s thinking that thought yields *H* knowledge (or, for that matter, requires the more complex points stated in the modified-(T*i*) idea and the weakened (T*i*)-style result). Then nothing is implied, in these two points, about the ontological status of the single object that *H* thinks and knows or about whether the combination of *i*'s elements before *H*'s thought-consciousness in the concept of an object in general is or is not given.

Thus, as Kant himself holds, the object that *H* thinks and knows to have the i_1-and i_2-presented features may have existence only insofar as it is thus represented to *H* by i_1 and i_2. And i_1 and i_2's joint functioning to present features of that single object may be due to *H*'s (T*i*) act of thought and may not be given. Equally, however, the object in question may, in the form that it takes as it is thus thought and known by *H*, have existence in itself. And i_1 and i_2's joint functioning to present features of that object may be given to *H* along with those elements (and may then merely be recognized by *H*'s (T*i*) act of thought). (On this latter, non-Kantian situation, the failure of any such object to exist in itself, or the failure of i_1 and i_2, as given, jointly to function in the way indicated, would simply mean that unity of apperception would not hold for *H* with respect to i_1 and i_2.)

Kant himself would of course reject these last sorts of non-Kantian possibilities. But in earlier chapters and above in Section 4 we have already criticized the Kantian views that would underlie such rejections (for example, Kant's views on the direct, categorical necessity of *H*'s thinking the (T*i*) thought). The philosophical interest of such non-Kantian possibilities is this. They show that neither of the most central claims, in the Deduction so far, about unity of apperception and *H*'s knowledge of objects really depends for its truth on either Kant's idealism or his position that combination cannot be given. We will have to note later how far this lack of dependence holds also for other fundamental B-Deduction claims.

6. SUMMARY

Given the holding of unity of apperception with respect to *i*, Kant in B-Deduction § 17 wishes to argue that *H* unites *i*'s manifold in the concept of an object in general. And he wishes to show also that the object in whose concept that manifold is united is known by *H*. This union of *i*'s manifold occurs through what we have called *H*'s act of thought expressed in (T*i*) – roughly, *H*'s thought that there is a single object that has the features that are presented by *i*'s elements. And so Kant in effect wishes to show the two basic B-Deduction § 17 points that *H* thinks the (T*i*) thought and that *H*'s thinking this thought yields *H* knowledge of the single object that *H* thereby thinks. Moreover, not only must he show these two points on the strong reading of the assumption that *H* knows through *i* – the claim that through *i* *H* knows a single, individuated object that has features presented by the elements of *i*. But, also, he must show these points on the weak reading of that assumption – the claim that, while *H* knows through *i*, nothing special is affirmed about the nature of what *H* knows.

We can view Kant as arguing, in B-Deduction § 17 and elsewhere, as follows. On the strong reading of *H*'s knowledge through *i*, *H*'s thinking the (T*i*) thought is necessary, and that necessity must have a transcendental source in (and so must be implied by) the holding of unity of apperception with respect to *i*. Hence it follows that, on the weak reading, the holding of unity of apperception is again the source of, and so implies, *H*'s thinking the (T*i*) thought. If Kant could defend this argument, he could then extend it to show that *H*'s thinking that thought yields *H* knowledge of the single object that *H* thinks.

Unfortunately this last Kantian argument fails, for Kant fails to show that *H*'s thinking the (T*i*) thought is necessary in any sense that requires a transcendental source. Moreover (as the results of Chapter Seven imply)

Kant has not shown that the holding of unity of apperception is itself necessary in such a way that it could be such a transcendental source. Furthermore, Kant does not establish that it would be such a source on the weak reading of *H*'s knowledge if it were such a source on the strong reading.

As we saw, the situation for the § 17 argument is even gloomier than the failure of the above reasoning suggests. As examples show, on the weak reading of *H*'s knowledge unity of apperception may hold with respect to *i* and yet it may not be the case either that *H* thinks the (T*i*) thought or that *H*'s thinking that thought yields *H* knowledge. Hence Kant cannot give any convincing argument for the two basic B-Deduction § 17 points. And so the argument of the Transcendental Deduction, as he presents that argument, cannot succeed.

Nevertheless, Kant may be able to modify his basic § 17 points so as to defend results akin to the original, fundamental Deduction conclusion. But it is not clear whether he really can establish modified versions of these point in such a way as to demonstrate such results. And, in any case, adopting modified versions would require substantial alterations to the Deduction's own argument. Given our focus on that argument, we will thus assume Kant's two § 17 points below in order to follow out his remaining reasoning in the first half of the B-Deduction.

CHAPTER NINE

OBJECTIVE UNITY OF APPERCEPTION AND THE LOGICAL FORMS OF JUDGMENT: B-DEDUCTION § 18 AND § 19

1. INTRODUCTION

Assuming the correctness of B-Deduction § 17, Kant has shown that (T*i*) holds and the manifold of *i* is united in the concept of an object through the (T*i*) thought, thereby yielding *H* knowledge of the single object that *H* thinks. ((T*i*), it will be recalled, is the claim that, roughly, *H* thinks that there is a single object that has the features presented by *i*'s elements.) In the brief § 18 Kant urges that this unity in *i*'s manifold is objective, not subjective. Then in § 19 he argues that because *H*'s thinking the (T*i*) thought produces such an objective, knowledge-yielding unity, (A) that thought is or is part of a knowledge-yielding judgment about the object that *H* thinks and knows through *i*. This judgment has a logical form, roughly a set of relations obtaining among the concepts in the judgment, whose holding is determined by the logical functions of thought in judgment. In § 19 Kant urges also that (B), the logical form of this judgment amounts to or derives from the objective unity of apperception that belongs to the concepts (or further judgments) in the judgment. As he argues in § 20, however, (C) because the logical functions of thought in judgment, through the holding of objective unity of apperception, determine the logical-form relations together of those concepts, the logical functions determine, also, the relations together of the conceptual elements of *i*'s manifold in such a way that the object that *H* thinks and knows through *i* falls under the categories.

Especially in § 19 Kant's reasoning becomes quite elliptical, and above and in the main discussion later I suggest what I think is the most plausible understanding of his course of thought. In the present chapter we consider Kant's § 18 claims about objective unity of apperception, and then we turn to points (A) and (B) of § 19. § 20 and point (C) are discussed in Chapter Ten.

2. OBJECTIVE UNITY OF APPERCEPTION

In § 18 Kant contrasts the objective unity that intuition-elements have through the holding of transcendental unity of apperception with what he calls the subjective unity of consciousness. He describes this contrast not in terms of the being like us, *H*, with *H*'s sequentially (but not necessarily temporally) ordered representations, but in terms of us human beings and our inner-sense awareness of temporally ordered representations; and for convenience I will follow him in this practice.[1] According to Kant, the objective unity of intuition-elements is the unity that they have insofar as they are united in the concept of an object through a thought like that in (T*i*). The subjective unity is a 'determination of inner sense' (B139), a specific arrangement of time-ordered representations before our thought-consciousness that arises through the imaginative association, based on empirical conditions, of different representations. This associatively organized group of representations provides the sequentially ordered manifold that is itself subject to unity of apperception and is objectively combined in a (T*i*)-like thought. But the particular associative organization of these representations is in general peculiar to their possessor and wholly contingent. (Thus one person may associate a given representation with one thing while another person may associate that same type of representation with another thing.) In contrast, the objective union of these same representations in the concept of an object is, Kant implies, the same for all possessors of such representations and is necessary.

Kant's comments in § 18 and his related remarks in § 19 suggest that by the objective unity belonging to the elements of an intuition, as I have just sketched that notion, he in fact means two things, which he does not bother to distinguish sharply from each other.[2] First, and has already been intimated, he means to contrast (a) the 'objective' organization that exists among a group of intuition-elements when (through a (T*i*)-like thought) those elements represent a single object distinct from them taken separately or taken in sequence with (b) the 'subjective' organization that exists among a group of intuition-elements simply when those elements are linked together by association before their possessor's mind. (Of course both types of organization can exist in one group of intuition-elements.) As our earlier discussions indicate, i_1 and i_2 have objective unity of this first sort just when *H* thinks, with respect to them, the (T*i*) thought.

Second, Kant means to note that if a person knows through a given group of unity-of-apperception-governed intuition-elements, then the following is a fact: Those elements have a type of organization that, necessarily, is possessed by any similar group of intuition-elements through which a similar person knows. Kant takes this fact to express the second sort of objective unity. Thus i_1 and i_2 have objective unity of this second sort just in case there is some type of organization such that i_1 and i_2 have that type of organization and it is necessary that, for any intuition-elements similar to i_1 and i_2 that belong to any being like *H*, those intuition-elements have that type of organization also. And Kant means to contrast the obtaining of this sort of objective unity with the point that if, say, I have a group of intuition-elements that happen to be organized associatively before my mind, then it need not be true that if a similar person has a similar group of intuition-elements, then that similar group of intuition-elements will be associatively organized in the same way before that other person's mind.

Ignoring, for the moment, the exact nature of the appeal to necessity made in this second account of objective unity, we can see that both sorts of objective unity will belong to the elements of *i* if – as we are assuming – *H* thinks and knows through the (T*i*) thought. Thus, and first, because we have assumed that *H* thinks and knows through that thought, in order to show that the first form of objective unity belongs to i_1 and i_2, we need only argue that the object that *H* knows is distinct from i_1 and i_2 taken separately and from the (i_1, i_2) sequence. This object is the object that is represented by *i* and that has the i_1 – and i_2-presented features *F* and *G*. Now actually to *prove* that this object is distinct from the items noted would be a complicated task which I cannot pursue here.[3] But the distinctness follows from familiar Kantian considerations about the distinction between the objective and subjective orders. For example, in general i_1 will lack *G* (and *F*) and i_2 will lack *F* (and *G*). But the object of *H*'s knowledge when *H* thinks (T*i*) has both *F* and *G*. Again, the (i_1, i_2) sequence necessarily has two members that sequentially put before *H*'s mind *F* and then *G*. But, in general, the object of *H*'s knowledge when *H* thinks (T*i*) need have no parts or properties that sequentially present *F* and *G*. So that object is distinct both from i_1 and i_2 taken separately and from the (i_1, i_2) sequence. Hence when *H* thinks (T*i*), *i*'s elements are objectively organized, and so form an objective unity, in the first way.[4]

Second, i_1 and i_2 are by assumption unity-of-apperception-governed and yield *H* knowledge. So, as just argued, they have objective unity of the first sort. But then, given B-Deduction § 17, it is true that, necessarily, the holding of unity of apperception with respect to any intuition-elements implies that the possessor of those intuition-elements thinks, with respect to them, a (T*i*)-like thought. Therefore any intuition-elements, belonging to a being like *H*, that are similar to i_1 and i_2 in being subject to unity of apperception and in yielding that being knowledge are united by that being through a (T*i*)-like thought and thus form an objective unity of the first sort. Hence it will be necessary that any intuition-elements similar to i_1 and i_2 in the way just indicated and belonging to a being like *H* will themselves have the type of organization that is yielded by the first sort of objective unity. And so i_1 and i_2 form an objective unity of the second sort.

The first sort of objective unity provides a 'relation-to-a-distinct-object' validity for the intuition-elements that possess it. The second sort provides what Kant calls a universal validity for the intuition-elements that possess it. We could write a great deal, both constructive and critical, about these two sorts of objective unity. Such a discussion is not needed here, however. But because of Kant's emphasis on matters of necessity, we should comment on the role of necessity in the second sort.

As we saw in Chapter Eight, if he could prove (T*i*), Kant would show it to be necessary – in a necessity-of-the-consequent sense – that (T*i*) holds and *H* thinks (T*i*), *given* that *H* knows through *i* and that unity of apperception holds with respect to *i*. And, as we saw, he may well also suppose that *H*'s thinking that thought is itself necessary in a direct, non-necessity-of-the-consequent, categorical sense. The necessity that he associates with the second sort of objective unity is distinct from either of these latter sorts of necessities, however. The latter necessities apply in one way or another to *H*'s thinking (T*i*); but the necessity associated with the second sort of objective unity is the necessity that, *given* that intuition-elements belonging to a being like *H* are similar to i_1 and i_2 in the way indicated earlier, that being should think with respect to those intuition-elements a thought *like* the (T*i*) thought. Moreover, it is clear that – whatever Kant himself may have believed – the necessity associated with the second sort of objective unity is simply a necessity of the consequent.

Of course there are relations between the various necessities that we have just noted, and relations also between Kant's two sorts of objective unity. While we need not try to work out those relations in detail, we should observe that in B-Deduction § 18 and § 19 he seems to imply – and in the *Prolegomena* he asserts – that his two notions of objective unity or objective validity are at least extensionally equivalent.[5]

Now this claim by Kant can be interpreted in two different ways, which are important to note here. (I) The claim may perhaps apply simply to the specific case of i_1 and i_2. In that case, not only does Kant take unity of apperception to hold with respect to all knowledge-yielding intuition-elements, including i_1 and i_2. But, also, and as noted earlier, he supposes himself to have established the position that the holding of unity of apperception with respect to any intuition-elements implies that the possessor of those intuition-elements thinks, with respect to them, a (Ti)-like thought. And he can appeal to that position in arguing for the extensional equivalence, in the specific case of i_1 and i_2, of the two forms of objective unity. (II) As the language of B-Deduction § 18 and § 19 and of the *Prolegomena* strongly suggests, the claim may apply to the case of *any* (unity-of-apperception-governed) intuition-elements. And the claim then asserts that in that case the presence of each form of objective unity implies the presence of the other form, just because of the natures, taken by themselves, of these notions of objective unity, and so in independence of Kant's position that the holding of unity of apperception implies that the possessor of the relevant intuition-elements thinks a (Ti)-like thought.

If Kant's claim is understood in way (I), then the claim to extensional equivalence is clearly correct. After all, in the specific case of i_1 and i_2 we have granted this last position of Kant's. Hence – and as can be seen from our preceding discussion of why i_1 and i_2 form an objective unity of the second (universal-validity) sort – Kant, insofar as he grants that position, is correct to claim that i_1 and i_2's forming an objective unity of the first (relation-to-a-distinct-object) sort implies i_1 and i_2's forming an objective unity of the second sort. Furthermore, he is obviously correct to claim that, granting the above position, i_1 and i_2 do form an objective unity of the first sort. So, and trivially, insofar as he grants that position he will be correct to claim that *if* i_1 and i_2 form an objective unity of the second sort, *then* they form an objective unity of the first sort.[6]

The situation is, however, not as clear when the claim to extensional equivalence is understood in way (II). Half this claim – the implication from the presence of objective unity of the first (relation-to-a-distinct-object) sort to the presence of objective unity of the second (universal-validity) sort – can be defended, at least under certain conditions, when the claim is understood in that way. However, the other half of the claim – the converse implication from the presence of objective unity of the second sort to the presence of objective unity of the first sort – is incorrect when the claim is understood in way (II).

The defensible half of Kant's claim can be supported thus. Understood as on way (II), this half amounts to the following. Suppose, say, it is a fact that two arbitrary intuition-elements *m* and *n* have objective unity of the first sort and so are such that their possessor thinks, with respect to them, a (T*i*)-like thought. Take this fact by itself, and independently of Kant's position that the holding of unity of apperception implies that the possessor of the relevant intuition-elements thinks a (T*i*)-like thought. Then from this fact, so taken, it follows that *m* and *n* have a type of organization that, necessarily, belongs to any similar group of intuition-elements, say *p* and *q*, through which a being like the possessor of *m* and *n* has knowledge. And such a result does obviously (and trivially) follow, at least if we take any such intuition-elements *p* and *q* to be similar to *m* and *n just* insofar as *p* and *q are* unity-of-apperception-governed and themselves yield their possessor knowledge through a (T*i*)-like thought.[7]

As I have said, understood in way (II) the other half of Kant's claim – from universal validity to relation to a distinct object – is incorrect. It is best, however, to postpone this issue until after our discussion of judgment in Sections 4 and 5 below.

The immediate interest of the fact that i_1 and i_2 possess both of the above sorts of objective unity is this. Our knowledge through a given unity-of-apperception-governed intuition must always be of an object that is distinct from the elements of that intuition taken separately or in sequence (and an object that has the various features or aspects that those elements present). And that knowledge must always hold also for any other being like us who knows through similar unity-of-apperception-governed elements and so is in a cognitive situation similar to ours.

3. OBJECTIVE UNITY OF APPERCEPTION AND THE LOGICAL FORMS OF JUDGMENT

Kant calls the objective unity (or unities) that he considers in § 18 the objective unity of apperception in order to indicate that that objective unity has its source ultimately in the holding of unity of apperception with respect to the intuition-elements to which it belongs. As we noted in Section 1, in § 19 he now tries to show two things: (A) that *H*'s objective-unity-yielding (T*i*) thought is or is part of a knowledge-yielding judgment about the object that *H* thinks and knows through *i*, and (B) that the logical form of this judgment amounts to or derives from the objective unity of apperception that belongs to the concepts (or further judgments) occurring in this judgment. (Then in § 20 he will argue for point (C), which concerns the categorial determination of the object of *i* through the logical functions of thought in judgment and which we consider in Chapter Ten.)

Kant's discussions of these points are often highly condensed, and he does not always make every distinction (for example, between the act of judging and the product of that act, the judgment that is thereby produced) on which modern standards of rigor would insist. To avoid confusion, I will introduce clarifications of my own as I proceed. But I will not try to sharpen every claim that Kant makes. And I will leave the detailed working out of some of his claims, as those claims apply to the thought expressed in (T*i*), until Chapter Ten.

I also should note that, in earlier chapters, we have distinguished between Kantian concepts as presenting and as being general properties. We have, in addition, distinguished between intuition-elements as putting before the mind and as being potentially general properties. These matters, and other issues noted in Chapter Ten (for example, the whole question, bypassed in Chapter Three, of the precise relation of intuition-elements to concepts in knowledge), turn out to affect in complex ways our exact, detailed understanding of points (A) and (B) – and of course also of point (C). However, rather than to complicate unduly our present discussion, it is best simply to present in this chapter Kant's basic understanding of (A) and (B) and to note some main questions that it raises. When we focus, in Chapter Ten, on point (C) and the categorial determination of *i*'s object, we can then return to issues like those just mentioned.

In B-Deduction § 19 Kant does not explicitly mention, let alone argue for, point (A) but, instead, proceeds immediately to matters related to (B). So he presumably thinks that (A) is obvious given § 18 and his earlier first-*Critique* discussions of judgment, especially in the A67/B92ff. Metaphysical Deduction of the Categories. And if we omit, for the moment, some points about his views on logic, then it is easy to see, in outline, why he accepts (A).

Kant says in § 19 that a judgment is – or is established through the obtaining of – 'a relation which is objectively valid' (B142). And, by § 18, an objectively valid relation (or an objective-unity relation) in fact holds between i_1 and i_2. If we follow our present policy of bypassing, in this chapter, complications about the precise bearing of intuition-elements on concepts in knowledge, then we can see that insofar as i_1 and i_2 stand in that objectively valid relation they (or what they put before the mind) make up – or are part of – a judgment. And because *H*'s (T*i*) thought yields *H* knowledge of the single object that *H* thinks through *i*, that judgment itself yields *H* knowledge.[8]

To this § 18 and § 19 evidence for taking *H*'s (T*i*) thought to be – or to be part of – a knowledge-yielding judgment, we can add the A132/B171 claim that the faculty of judgment is the faculty of subsuming under rules. Given A106 on rules and concepts, this claim shows that the act of judging will be the act of subsuming an entity under one or more concepts.[9] However, *H*'s thinking the (knowledge-yielding) (T*i*) thought obviously is – or is part of – such a (knowledge-yielding) judgment, for in that thought *H* thinks the entity *x* to be an object and to have the i_1- and i_2-presented features *F* and *G*. And, given Chapter Three on concepts and property-attribution, as well as our Chapter Eight comments on the concept of an object in general, it is clear that *H* regards *x* in these ways just insofar as *H* subsumes *x* under the concept of an object in general and under the concepts of being *F* and of being *G*.

In order to understand the deeper significance, for the Deduction, of (A) and the claim that *H*'s (T*i*) thought is or is part of a knowledge-yielding judgment – and also ultimately to reach point (B) – we must now turn to the Metaphysical Deduction of the Categories. In the Metaphysical Deduction, as we see in Chapter Ten, Kant argues that the categories derive from the logical functions of thought in judgment. In order to do so, he first discusses judgment, concepts, and the logical functions (A67–76/B92–101), and it is on these preliminary comments that I focus here.

In these comments, Kant in effect distinguishes two groups of judgments: simple, basic, categorical judgments (for example, the judgment that the tree is conical) and compound judgments (for example, the judgment that if the tree is conical, then it will shed snow).[10] He takes the basic, categorical judgments to be combinations of concepts, and he takes the compound judgments to be combinations of other, further judgments. He regards the various ways in which a judgment unites together the concepts (or further judgments) that it contains as analogous to the way in which, he holds, a concept unites the representations or objects that fall under that concept. A concept, he says, rests on a function – a function being 'the unity of the act of bringing various representations under one common representation' (A68/B93). By this remark, he means at least roughly that (as seen in Chapter Three) a concept operates as a common mark that is found in – and so that allows us to think mediately of – all the objects that fall under it. And this operation of the concept depends on the fact that the act of considering each of those objects and taking it to fall under the concept is an act that unites *all* those objects by taking them *all* to have contained in themselves, or to fall under, one and the same common mark. But then, and somewhat similarly, Kant takes a judgment to involve *logical* functions or types of unity through which all the concepts (or further judgments) in a certain group are united by being taken all to be organized together, under the *same* logical form, so that they make up *one* judgment possessing that form.

Aside from a few comments below, Kant's further descriptions of the logical functions of thought can be postponed until Chapter Ten. We need only note here that, following Aristotelian – or quasi-Aristotelian – logic (with modifications of his own), Kant takes concepts, as they occur before thought-consciousness in basic, categorical judgments, to be organized in four general types of ways that jointly determine those concepts to occur in a judgment with a specific logical form. Each of these four types of ways – of the quantity, quality, relation, and modality of a judgment – contains three specific logical functions and, roughly, the concepts occurring in any basic, categorical judgment are related together with respect to one logical function from each of the four overall types.[11] In turn, the compound judgments that Kant recognizes are hypothetical- or disjunctive-judgment combinations of basic, categorical judgments.

Thus take the judgment that the tree is conical. This judgment has a logical function of relation, and in this judgment the concepts of being a tree and of being conical occur related together *categorically*, as subject to predicate of the judgment. Furthermore, this judgment has a logical function of quantity, and in the judgment *one* individual thing falling under the subject concept of being a tree is considered. In addition, the judgment has a logical function of quality, and in the judgment the predicate is *affirmed* of that individual thing. Finally, the judgment has a logical function of modality, and in the judgment this affirmation – or the predication of being conical of the individual thing – is taken to be *actual*. Again, take the compound judgment that if the tree is conical, then it will shed snow. This judgment has a logical function of *hypothetical* relation and through that relation it combines, as its antecedent and consequent, the two basic, categorical judgments that the tree is conical and that the tree will shed snow. And similarly, for compound disjunctive judgments.

The fact that concepts in a judgment fall under – or combine through –

the relevant logical functions so as to make up a single judgment is a fact of combination. (Hereafter I often write 'concepts' for 'concepts or further judgments.') So, for Kant, this fact, like all facts of combination, should occur before thought-consciousness. As we see in Section 4, the proper Kantian interpretation of this fact – and of its occurrence before thought-consciousness – raises serious difficulties. For the moment, however, we may avoid issues of interpretation. Following our earlier discussions of combination, we may hold that this fact, as a fact of combination, is produced by and can exist only through an act that thinks the concepts to function in the ways specified by the logical functions in question. In Section 4 we then return to the issues that are raised by this treatment of the fact.

Given such a treatment, because *H*'s (T*i*) thought is or is part of a judgment, the occurrence of that thought before *H*'s mind should involve the following. The concept of an object in general and the concepts of being *F* and of being *G* must occur before a single act of *H*'s thought-consciousness in such a manner that in that act those concepts operate together in various logical-function ways. In so operating, they must make up, for that act, a single thought, which is itself a judgment or part of a judgment. That single thought is the thought that the object *x* has the features *F* and *G* that *i*'s elements present. And for those concepts to operate together in those logical-function ways, that act must think those concepts to operate in those ways.

These last remarks allow us to clarify the point, noted above and in (A), that *H*'s (T*i*) thought is or *is part of* a judgment. A reading of § 19 (and its B142 example of the singular judgment 'it, the body, is heavy') can suggest that when unity of apperception holds with respect to *i*, *all* that follows, in regard to any judgments that *H* may make about the object of *i*, is that *H* thinks the single (T*i*) thought and so makes the single judgment that amounts to that thought. However – and although Kant is not immensely clear about these matters – this last view seemingly should not be exactly what he means to hold. After all, that view creates the problem of demonstrating, in the Deduction, that all the categories other than those associated with the specific (T*i*) judgment apply to *i*'s object. And (as our Chapter Ten discussion shows) Kant has no satisfactory solution to that problem.

It therefore seems that Kant should maintain that, when unity of apperception holds with respect to *i*, *more* follows, in regard to *H*'s judging about *i*'s object, than simply that *H* thinks the thought (and makes the judgment) expressed in (T*i*). Yet from the mere fact of the holding of unity of apperception with respect to the arbitrary intuition *i*, it surely cannot follow that *H* makes any specific judgment about *i*'s object beyond the basic (T*i*) claim that there is an object that has the features that *i*'s elements present. Rather, Kant should claim at least the following: When unity of apperception holds with respect to *i*, *H* makes *some* overall (knowledge-yielding) judgment – involving, but not necessarily restricted to, the (T*i*) judgment – about the object that *H* thinks and knows through *i*. Given that claim, he can then argue that that object falls under the categories associated with that overall judgment.[12]

Thus, for example, while judging that there is an object that has the *i*-presented features *F* and *G*, *H* might also judge that things having *F* and *G* have feature *K*. So *H*'s judgment that there is an object having *F* and *G* would be one conjunct in an overall judgment whose other conjunct is the judgment that all *FG* things have *K*. Or *H*'s judgment that there is an object having *F* and *G* might be one conjunct in an overall judgment whose other conjunct is the judgment that if a thing has *F* and *G*, then something else has *L*. Or *H* might simply judge that there is an object that has *F* and *G*, without *H*'s then making any other judgment about that object.[13]

Problems evidently exist for the idea that, when unity of apperception holds with respect to *i*, *H* makes some (knowledge-yielding) judgment involving, but not necessarily restricted to, the (T*i*) judgment. We will return to those problems below. But, for the present, let us accept the claim, in (A), that *H*'s judgment, as expressed in (T*i*), is or is part of an overall judgment of the sort just indicated.

We may now turn to point (B) – the claim that the logical form of that overall judgment amounts to or derives from the objective unity of apperception that holds with respect to the concepts (or further judgments) occurring in that judgment. (B) follows from Kant's general § 19 position, which I will call (P), that the logical form of *any* judgment consists in [*besteht in*] or derives from the objective unity of apperception – and so from the transcendental unity of apperception – that holds with respect to the concepts occurring in that judgment (B140, § 19 heading).[14] We can reach what seem to be Kant's grounds for (P) by noting that, given our above discussion, a judgment amounts to a combination of concepts occurring before an act of thought-consciousness and related together in various logical-function ways so as to make up, for that act, a single thought having one overall logical form.[15]

As we have held above, such a combination of concepts into a single thought is produced through and can exist only through an act of thought. However – it seems that Kant must mean to argue – the logical functions, as the ways in which concepts are combined together in a single thought, are themselves *a priori* and so necessary, for they determine the *logical* structure of the judgment. So the combination in question and the logical form of judgment that is thereby determined are *a priori* and necessary. And therefore this combination and that logical form must have an *a priori*, necessary source. (Here recall Chapter Eight on the 'sources' reasoning underlying § 17.)

Now unity of apperception is supposed to hold necessarily with respect to all intuition-elements through which we do or can know. So Kant will take it that unity of apperception – and, by reasoning like that in § 18, an objective unity of apperception – holds necessarily with respect to the concepts occurring in a judgment. And it seems he will argue (again by 'sources' reasoning) that the necessary holding of objective unity of apperception with respect to those concepts must be the *a priori*, necessary source of the combination and of the logical form of judgment that is here determined. Hence in that sense the logical form of a judgment consists in or derives from the holding of objective unity of apperception with respect to the concepts occurring in that judgment. Thus we reach Kant's general § 19 position (P). And from this position and point (A), (B) follows. Given, by (A), that *H*'s (T*i*) thought is or is part of a judgment, the logical form of that judgment amounts to or derives from the objective unity of apperception that holds with respect to the concepts in that judgment.

Kant himself simply asserts his general § 19 position (P) – and then, by implication, point (B). He gives no very explicit defense of either view. I think, however, that something like the above reasoning represents the sort of basic defense that he intends of that position and of (B). Now although such reasoning exists, it is impossible to rest content either with it or with claims (P) and (B), for extremely serious problems face both the reasoning and those claims.

First, about the reasoning itself we need say no more than that it is open to various of the sorts of objections raised in Chapter Eight to Kant's § 17 'sources' argumentation.[16] And I see no better prospects for answering these objections now than I do for answering the ones raised earlier. Second, Kant's general § 19 position (P) itself is unsatisfactory. And so also is (B), which follows directly from (P).

To show that (P) is unsatisfactory, suppose that a judger makes – or at least mentally contemplates – some specific judgment *z*. Then (P) guarantees that the logical form of *z* has its source in or derives from the holding of unity of apperception with respect to the concepts in *z*. However, there are very many logically distinct judgments, each with its own distinct logical form (and knowledge-yielding or not) that the judger can make – or at least can contemplate – and that involve just the concepts that occur in *z*.[17] Given (P), unity of apperception holds with respect to those concepts.[18] So, by (P), the holding of unity of apperception with respect to those concepts amounts to or is the source of all the distinct logical forms of all the logically distinct judgments just mentioned.

However, from this result it follows that when the judger makes the specific judgment *z*, the judger also actually makes – or at least actually contemplates – each of these logically distinct judgments, with its distinct logical form.[19] When the judger makes *z*, the concepts in question exist before the judger's thought-consciousness actually subject to objective unity of apperception. And so, given (P), the logical forms of these judgments should then actually inform or structure all those concepts and hence should yield, before the judger's thought-consciousness, all the logically distinct judgments, themselves. Yet it is clear that when the judger makes the specific judgment *z*, the judger need not (and in the usual case the judger will not) make or even contemplate all these further judgments. And so (P) leads to an unacceptable conclusion about what judgments the judger makes or contemplates.

Given the above problems, there may seem to be good reasons to abandon the general § 19 position (P) and no longer to maintain (B), which simply instantiates (P) and faces the same sort of difficulties as does (P). And in this connection we should observe that (P), with its B140 view of the logical form of judgment as literally consisting in (or deriving from) the holding of objective unity of apperception with respect to the relevant concepts, is not the only position about logical form and judgment that Kant may put forward in § 19. As various texts suggest, he may also – or alternatively – propose that the logical form of a judgment does not literally *consist in* the holding of objective unity of apperception. Rather, apperception and its unity somehow contain, and manifest themselves judgmentally in terms of, the logical functions.[20] Whenever objective unity of apperception holds with respect to a given group of concepts, various of the logical functions operate on those concepts in such a way that the concepts come to make up *some* judgment or other having *some* specific logical form. That specific logical form is thus simply one way, out of a number of possible ways, in which the relevant concepts can be organized before thought-consciousness so as to fall under objective unity of apperception. That logical form is not determined to hold with respect to those concepts *merely* through the holding, with respect to them, of such unity.

This new Kantian proposal avoids the 'very many logically distinct judgments' problem. Moreover, by coupling this new proposal with the fact that, as Kant sees it, the holding of unity of apperception implies that *H* thinks (T*i*), we see that this new proposal itself commits him to the claim that when unity of apperception holds with respect to *i*, *H* makes *some* judgment, involving but not necessarily restricted to the (T*i*) judgment, about the object of *i*. And so the new proposal itself supports the idea, in (A), that *H*'s (T*i*) judgment is or is part of a (knowledge-yielding) judgment about that object. And that idea we have already seen to be helpful to the Deduction. Hence we might consider adopting the new proposal, ignoring the § 19 position (P) and its supporting texts, modifying (B) in the light of this new proposal, and proceeding on with the Deduction.

Unfortunately, however, the new proposal leads to difficulties of its own. Most seriously, it claims that the holding of each specific logical form of judgment, or of each specific sort of logical combination of concepts in judgment, does not have its source simply in – and is not implied by – the holding of unity of apperception with respect to those concepts. So, given that Kant takes such specific logical forms to be *a priori* and hence necessary, this proposal in effect abandons his fundamental Deduction idea that the holding of unity of apperception is the source of all necessary combination whatsoever.[21] Now, as it happens, our example of the very many logically distinct judgments that can arise from one group of concepts shows rather conclusively, I think, that that idea cannot be correct.[22] But, given the depth of Kant's adherence to that idea, the new proposal cannot be anything that he himself, working within his own Deduction framework, should wholeheartedly accept. So we should not attribute this proposal, and nothing else, to Kant in the Deduction.[23]

Furthermore, if Kant were to abandon the Deduction's fundamental idea that the holding of unity of apperception is the source of all necessary combination whatsoever, then he would not yet have any good reasons to accept the new proposal instead. Abandoning that idea, he could no longer argue that because the specific logical functions of thought are *a priori* and hence necessary, those functions of thought (and so the logical forms that they determine) have their source in the holding of unity of apperception. But then what grounds could he give for the claim, in the new proposal, that the logical functions are somehow *contained in* and *manifestations of* apperception and its unity? Even if the holding of unity of apperception with respect to given intuition-elements requires that the knower makes a judgment, why does not the logical structure of that judgment turn on factors that are independent of that holding?

The upshot of the preceding discussion thus is that Kant suggests two different accounts of the relation of the logical forms of judgment and unity of apperception. The first agrees with a fundamental (although mistaken) Kantian 'sources' idea about the holding of unity of apperception but leads to serious problems. The second escapes those problems by rejecting that idea but thereby makes itself both un-Kantian, in many ways, and – it seems – unprovable by Kant. (Nor do I see any further, plausible approach to Kant's § 19 views.) This situation is most unhappy for the overall argument of the Deduction as Kant presents it. However, the existence of this situation does not really undermine what we may call the strict argument of B-Deduction § 19 and § 20 for category application to the object of *i*.

After all, Kant's § 19 claims about the logical forms of judgment and the holding of objective unity of apperception, however exactly those claims are read, are intended to establish some relation between that holding and those logical forms. (Presumably those claims are so intended because in § 16 and § 17 he has just emphasized the holding of unity of apperception as a source of necessary combination.) Yet nothing in the strict B-Deduction § 19 and § 20 argument for category application requires that Kant should commit himself to any such relations. All that that strict argument really requires is that he first show that *H* makes a (knowledge-expressing) judgment about the object of *i* (§ 19), next show that that judgment has a logical form determined by the relevant logical functions of thought (also § 19), and, finally, show that the operation of the logical functions leads to category-application to the object of *i*

(§ 20).[24] This specific line of argument claims nothing about the source or underlying origins of the logical functions.

We may therefore proceed as follows in considering the remainder of B-Deduction § 19 and § 20. We will continue to acknowledge Kant's verbal expressions of specific views about objective unity of apperception in claims (B) and (C) (claims which, for simplicity, I have couched in terms of the general § 19 position (P) to which Kant's overall 'sources' idea about unity of apperception commits him), for Kant himself attaches great importance to such views. As we proceed in this way, we can also hold in readiness the new proposal noted above. But, as we develop the actual Deduction argument for category application, we will focus on what I have called the strict B-Deduction § 19 and § 20 argument. And for the most part we will bypass the details of Kant's two accounts of the relation of unity of apperception and the logical forms of judgment.

As we follow this course, we will, in acknowledging the verbal expression of Kant's views in claims (B) and (C), also acknowledge the claim in (A) that *H*'s (T*i*) thought is or *is part of* a knowledge-yielding judgment. As we saw above, this 'is part of' idea can help Kant in his attempts to prove that all the categories, and not just the specific categories associated with the (T*i*) judgment, apply to *i*'s object. And this idea is positively required by the new proposal about the relation of unity of apperception to the logical forms of judgment.

However, while acknowledging this idea, we should not pretend that it does not bring along difficulties. Among other things, in order for Kant to use the idea to show application of all the categories to *i*'s object, he needs to demonstrate that when unity of apperception holds with respect to *i*, *H* actually makes some knowledge-expressing judgment that not only involves but goes beyond the (T*i*) judgment. Yet in Chapter Eight we saw that Kant really cannot prove that when unity of apperception holds with respect to *i*, *H* thinks the thought – and so makes the judgment – expressed in (T*i*). Evidently it will be even more difficult for him to prove that when unity of apperception holds, *H* thinks a thought – and so makes a judgment – that incorporates but goes beyond the (T*i*)-thought and judgment itself.[25] Moreover, we see in Chapter Ten other difficulties related to this idea. So we will not automatically assume that *H* actually makes any knowledge-expressing judgment that not only involves but goes beyond *H*'s (T*i*) thought and judgment. Nor will we assume that the idea that *H* makes such a judgment is unproblematic. Rather, we work with the 'is part of' idea simply because of its usefulness in exposition.

4. QUESTIONS ABOUT THE LOGICAL FUNCTIONS

Before turning to the remainder of B-Deduction § 19 and then in Chapter Ten to § 20 and the final proof of category application to *i*'s object, I should note two last points about the Metaphysical Deduction claims spelled out above. First, these claims are of course couched in terms of Kant's Aristotelian (or quasi-Aristotelian) views on judgment and logic; and Kant's acceptance of such views and his claims about certain of the logical functions (for example, about 'infinite judgments' at A70/B95ff.) have been much criticized. I will not examine such criticisms here (though in Chapter Ten I consider some basic points about the derivation of the categories from the logical functions), for the criticisms are familiar from many other sources and in general are not directly relevant to the points that I will make. I should say immediately, however, that – as we see in detail in Chapter Ten – Kant's use of his form of logic is crucial to the B-Deduction § 20 argument for category-application to *i*'s object. And thus the frequent suggestion that the overall Transcendental Deduction reasoning, as Kant develops it, can easily be severed from the views advanced in the Metaphysical Deduction is mistaken.

Second, serious issues arise when one asks what should be the proper Kantian interpretation of the fact that concepts in a judgment fall under (or combine through) logical functions so as to make up a single judgment. As we noted, this fact is a fact of combination, and we previously took this fact to obtain just when the judger *thinks* that the concepts have – or function in ways specified by – the relevant logical functions. Indeed, not only is this interpretation supported by our earlier account of Kant's general views on combination, but also it agrees with a number of texts.

For example, and as already noted, at A68/B93 Kant describes a logical function as the 'unity of the act [which is an act of judgment] of *bringing* different representations *under* one common representation' (my italics). At A69/B94 he urges that we can reduce or trace back all acts of the understanding – and so all acts of assigning logical functions to concepts – to acts of judgment and so to acts of the subsumption, in thoughts, of entities under concepts. Again, at B112 he holds that in the case of disjunctive judgments – and (I take it) through the understanding's assigning of logical functions – 'the subordinate concepts ... are *thought as* coordinated with ... each other' (my italics). And in various texts he says clearly that in applying the categories – which derive from the logical functions – we bring the relevant concepts or intuitions under the categories and thereby determine that we *think* those concepts as, say, subjects and never as predicates.[26] It is natural to read these texts as implying that in logical-function application itself we think concepts as having logical-function roles in judgments.

Nevertheless, and despite such textual support, our above interpretation faces a glaring philosophical difficulty. Suppose that our bringing of a concept under a logical function does occur through our conscious *thought*, with respect to that concept, that it has the logical function. Then the difficulty is that any such logical-function-application thought surely will itself amount to a Kantian judgment (a taking of an entity, here a concept or further judgment, to have a feature, here a logical-function feature). So that thought will have a logical form that must be determined through the application of logical functions within itself. But, given the above interpretation, that thought will then require a still further logical-function-application thought – and so a still further Kantian judgment – to determine *its* logical form. And so this process will go on, in such a way that we never get to the point of making the judgment that we initially intended to make (or else, impossibly, we simultaneously make an infinite number of logical-function-application judgments).[27]

As far as I know, Kant never considers this specific problem; and his frequent tendency to focus on judgments about objects distinct from our own representations may well hide it from his eyes. But his A132–33/B171–72 infinite-regress argument against the existence, in general logic, of rules for judgment shows that he would take this problem very gravely were it pointed out to him. And it is clear that the problem proves that our previous interpretation is philosophically unsatisfactory, even though it agrees with various texts.

One might, of course, question whether that interpretation is the sole (or even the correct) reading of Kant's view of the fact that concepts fall under logical functions so as to make up a single judgment. In various places (including passages from texts already cited in connection with the previous interpretation) Kant writes in ways that do *not* seem to require logical-function application to occur only through the conscious thought that the concepts involved have the relevant logical functions. For example, he claims that the logical functions are '*forms* for the relation of concepts in thought'; he holds that the logical moments of all judgments (the logical functions of thought) 'are [simply] so many possible *ways* of unifying representations in a consciousness'; and he avers that in the categorical judgment that *the stone is hard*, 'the *stone* is *used* as subject and *hard* as predicate.'[28] However, despite their existence, texts like these, taken by themselves, suggest no clear alternative to our previous interpretation. And it seems quite possible that – as happens with other parts of his work – Kant's treatment of logical-function application is simply to some extent ambiguous.

If Kant is to solve the above problem, then he can do so plausibly only by rejecting the idea that if concepts are to have logical functions, then the judger must consciously think that these concepts have those logical functions. One can imagine various ways of carrying out this rejection. It seems simplest, however, for Kant to hold that when I make a judgment, the concepts involved exist in my mind and have the logical functions required for them to make up that judgment. Nevertheless, although those concepts do make up, in my mind, the judgment, I need not consciously *think* that they have those logical functions and make up that judgment. Rather, their having those logical functions and making up the judgment will be due to activities of my mind of which I am usually not conscious (just as the thoughts that I spontaneously think while writing this book or walking in the woods arise in my mind, with the subject-predicate and other structures that they have, through mental activities of which I am not conscious).

Now, as it stands, this solution to the above problem is actually too simple. Suppose that I judge, and so think, that the tree is conical. Then, in the usual case, I do not have a structurally undifferentiated thought-content before my consciousness. Instead, in some way I consciously think, in this very thought, that the *tree* (that thing or subject) is *conical* (has that property or predicate). For Kant, however, this sort of differentiation in thought-content can arise, it seems, only through consciousness, in thought, of logical-function application. As a simple, philosophically unsophisticated judger, I need not be conscious that the concept (or property) *c* of being a tree and the concept (or property) *d* of being conical have the exact subject-predicate structure described in Kantian logic.[29] But, for my thought-content to be differentiated in the above way, it must at least be true that there is an *R* that is in fact identical to the Kantian subject-predicate structure; that belongs to *c* and *d*; and that is such that in my thought that the tree is conical I think, *of R*, that *c* has *R* to *d*, without my necessarily being able to think with any great exactness about the nature of *R* itself.

These last points may now seem to re-create the above problem, but the preceding solution can be developed further so as to avoid it. The problem may seem to be re-created simply because – it may be objected – if I have to think that the concept *c* of being a tree has the subject-predicate relation *R* to the concept *d* of being conical, then in order for me to think *that* thought, *c*, *d*, and *R* must have logical functions, say *V*, *W*, and *Z*, in order to make up that thought. But then in order for, say, *c* to have logical function *V*, I will have to think, *of* that logical function, that *c* has *it* (it being understood by me in however nontheoretical a way). And so a regress begins again.

This last objection is, however, unconvincing. It is true that if I think the single, whole thought that the tree is conical, then (at least in the usual case) in that thought I in some way think or am aware, in a differentiating fashion, that the tree, a thing or subject, has the property or predicate of being conical. But although my original single thought involves this differentiating thought or awareness, there is no reason to suppose that this differentiating thought is *itself* further differentiated for me. That is, there is no reason to suppose – and in the usual case it is simply not true – that in thinking that (a) the tree, a thing or subject, has the property of being conical, I *also* for example *further* think that (b) the tree is a thing or subject that has the property of having the property of being conical.[30]

In order to allow for the differentiation in the content of my thought that the tree is conical, Kant should therefore hold that, in thinking that thought, I consciously think – in the perhaps quite nontheoretical way indicated – that the concept *c* of being a tree has the subject-predicate relation *R* to the concept *d* of being conical. And then *c*, *d*, and *R* will indeed have the logical functions *V*, *W*, and *Z*. But in thinking that the tree is conical, I need not consciously think that *c*, *d*, and *R* have those logical functions; for *c*, *d*, and *R*, as they exist in my mind, can have *V*, *W*, *Z*, through the operation of mental activities of which I am not conscious, without my being aware that they have those logical functions.

We thus see that we can resolve the above problem as follows. Concepts exist in the mind and have there logical functions in such a way as to make up single judgments. In thinking the thoughts that constitute those judgments, the judger is aware, in thought (in perhaps a quite nontheoretical way), that the concepts have those logical functions. But although the concepts, as they exist in the judger's mind outside his or her thought-awareness of them, have the additional logical functions that underlie this last thought-awareness by the judger, the judger need not be aware, in thought – and, in the usual case, the judger will not be aware, in thought – that those concepts do have those additional logical functions.

This last solution (perhaps with some further refinements) allows us to avoid the above problem. But I do not claim that it is perfectly Kantian. It is neither explicit nor, I think, implicit in the texts. And it has features that Kant cannot – or should not – accept, given his overall position in the first *Critique*.

For one thing, if Kant identifies a concept with a property thought as general in an act of thought (here recall Chapter Three), then I do not see clearly how, as the solution requires, a concept can exist in the mind and have a logical function without the judger's grasping, in that act of thought, the fact that the concept has that logical function. For another thing, Kant in general treats concepts as existing and as functioning judgmentally in the mind only insofar as they occur before acts of thought. And so again it is not clear how he can allow that a concept can exist in the mind and have a logical function, and so can operate in judgments, without conscious thought, by the judger, that the concept has the logical function in question. Rather than accepting that idea, Kant himself seems always to treat concept use and judgment as turning on activities of conscious thought.[31]

Instead of being Kant's own solution, the present solution thus is simply the sort of solution to the above problem that he ought to offer (making the necessary changes in his other views) were the problem to be made explicit. Because the details of that problem do not bear directly on the further argument of the Deduction, I will not pursue this solution here but will simply suppose that we could adopt it if necessary. I will ignore most of the complications introduced by the solution, including especially the claim that there are logical functions of whose applications to concepts the judger is unaware in thought. However, because of its bearing on the further argument of the Deduction, I will appeal below to the view – seen already in the first group of texts cited above and so not an artefact of the solution itself – that, in judging, the judger is aware in thought that the concepts involved have the relevant logical functions.

5. THE COPULA, OBJECTIVE UNITY, AND NECESSARY UNITY

Before plunging deeper into Kant's account of the categories and so moving to claim (C) of B-Deduction § 20 – the claim that, roughly, the categories apply to the object of *i* through the operation of the logical functions in judgment – we should complete our discussion of § 19 by considering the topics remaining in B142. These topics concern the role of the copula in judgment, the relation of the copula and judgment to Kant's § 18 distinction between objective and subjective unity, and some further points about the necessity that Kant attaches to objective unity of apperception.

We need not discuss Kant's account of the role of the copula in judgment at length. Kant begins § 19 by denying the view – of a sort common in his time – that a judgment represents or expresses a relation between two concepts.[32] He notes that this view neither takes into account hypothetical or disjunctive judgments (in which a relation is established between further judgments) nor explains the nature of the relation in question. As already remarked in Section 3, he then asserts that a judgment, in contrast to a merely associative relation of representations, is 'nothing but the way in which different pieces of knowledge [or representations] are brought to objective unity of apperception' (B141). We use the copula 'is' to mean this relation of representations to that unity, in order to distinguish the objective from the subjective unity of those representations.

Kant's language here may suggest that he sees such a role for the copula in every judgment. But it is clear both from the opening § 19 comments about hypothetical and disjunctive judgments and from his other first-*Critique* comments on the copula, as well as from the later B142 'bodies are heavy' example, that in the present remarks he really has in mind only what he takes to be the copula's function, in a categorical judgment, to indicate the relation of the subject-term and predicate-term to unity of apperception.[33] In studying the Transcendental Deduction, however, we are interested not just in categorical judgments but in all types of judgment. So while we will note, immediately below, Kant's other § 19 points about the copula, we need not delve further into his overall views on that subject.

The second of the remaining B142 topics, the relation of the copula and judgment to the § 18 objective-subjective-unity distinction, is linked, in Kant's exposition, to the third topic, the further points about necessity and objective unity that were mentioned above. As in part just noted, Kant claims that the copula is used to distinguish the objective from the subjective unity of representations because – he now adds – the copula indicates the relation of objectively unified representations to 'original apperception and its *necessary unity*' (B142). Through, and only through, the relation of representations to apperception and its unity do those representations come to form a judgment, an objectively valid relation of the representations to an object, as against an imagination-yielded association that has mere subjective validity.

At first glance these last claims look unsurprising given our above discussion of § 19. Thus set aside, for the moment, Kant's claims here about necessity and objective unity, and ignore his views on the copula. Then in these claims Kant asserts that, first, it is precisely through the relation of representations to objective unity of apperception that those representations come to form a judgment and, second, a judgment is an objectively valid relation of representations to an object. And such points are exactly what might be anticipated in the light of what we have already seen § 19 to hold. After all, the first point follows from Kant's view of judgment as involving a logical-function-determined combination of concepts and of such a combination as itself having its source in the holding of objective unity of apperception with respect to those concepts. And the second point simply expresses again the basic § 19 account of judgment that we discussed in Section 3.

Nevertheless, further claims that Kant makes at the end of § 19 go beyond anything that we have so far attributed to him and raise serious problems. At the end of § 19 Kant contrasts the genuine judgment that the body is heavy with what he treats as a claim expressing merely an associative relation of the same representations as occur in that judgment – namely, the claim that 'if I support a body [if I feel an impression of supporting a body], then I feel an impression of weight' (B142). And he asserts that to say that the body is heavy is to say that the two representations (of body and heaviness) 'are combined *in the object*, no matter what the state of the [knowing] subject may be' (B142).

Now in Section 3 we took Kant to regard a judgment as an objectively valid relation of representations to an object that is distinct from those representations themselves, taken separately or in sequence.[34] (Hereafter I often omit 'taken separately or in sequence.') We also took Kant to claim that the holding of unity of apperception with respect to i implies that H thinks the (Ti) thought. And we regarded that thought as being or being part of a judgment that yields knowledge of the single object – distinct from i_1 and i_2 – that is known through that thought. However, on a natural interpretation of the preceding, 'If I support-a-body' example and of the last, 'no matter what' clause in the B142 quotation, at the end of § 19 Kant views a judgment as an objectively valid relation of representations to an object that is distinct not only from those representations themselves but also from *any* representations of the judger's *whatsoever*, taken separately or in sequence.[35] And thus, on the basis of that view, he should claim that when unity of apperception holds with respect to *i*, *H* thinks the (T*i*) thought and that thought is or is part of a judgment that yields *H* knowledge of a single object that is distinct from *any* of *H*'s intuition-elements or other representations whatsoever.

Yet given these last claims trouble now arises swiftly. If Kant takes a judgment to be an objectively valid relation of representations to an object distinct from any intuition-elements whatsoever, then he rules out the possibility of any judgment by *H* that is about merely the (subjective) organization of intuition-elements or other representations in *H*'s mind. But, as many readers have noticed, in his 'If I support a body' example he himself surely gives just such a judgment. And there are obviously many other such judgments, for we all can describe accurately much of the course of our own sequences of representations.

I will not go into detail about how Kant is led to the view of judgment that creates this last difficulty.[36] (I suspect that, underlying his acceptance of the view, is an implicit assumption, from B-Deduction § 17 on, that *H*'s knowledge through *i* – as taken on the strong reading of that knowledge – concerns an object that is distinct from any intuition-elements or other representations of *H*'s.) What is important to note now is simply that it is Kant's need to answer Hume, together with his belief about what § 17 has proved, that leads him to the present view of judgment and this difficulty. After all, in order, in the Deduction, fully to answer Humean claims, Kant must establish that, given the holding of unity of apperception with respect to *i*'s elements, *H* knows something other than merely those or any others of *H*'s representations.[37] But in the Deduction the only means which he can use to establish such a result is his central § 17 claim that the holding of unity of apperception with respect to *i* implies that *H* thinks and knows an object through the (T*i*) thought. So in the Deduction he will want to claim that that object is distinct from any of *H*'s intuition-elements or other representations.[38]

Yet in claiming this last point, Kant now inevitably creates the above view of judgment and the difficulty. Suppose that a judgment always involves a logical form determined through the holding of unity of apperception with respect to the components of that judgment, and suppose that the holding of unity of apperception with respect to any elements or components always implies a (T*i*)-style, knowledge-expressing thought about an object distinct from any of the judger's representations. Then *every* judgment will be an objectively valid relation of representations to an object that is distinct from any intuition-elements of the judger's whatsoever.[39] Of course Kant might conceivably try to escape this result by supposing there could be a sort of judgment that directly inspects, and describes the subjective sequence of, representations that are themselves subject to unity of apperception and are thereby referred to a distinct object of the sort just noted.[40] But such an attempted escape runs afoul of the fact that unity of apperception holds with respect to this judgment just as much as with respect to any other thoughts of ours.

In fact, as long as he holds to his view of what § 17 has shown, Kant's only immediate prospect for escaping from the present difficulty is to abandon the view that the object known through the (T*i*) thought is inevitably distinct from any of *H*'s representations whatsoever. As Kant presents it, the Deduction suggests no way of doing this without undermining his answer to Hume. And thus we reach a serious problem for which the official argument of the Deduction has no answer.

It seems, however, that there are at least two moves open to Kant at this point, if he is willing to modify earlier claims. First – and to remain as close as possible to the actual argument of the Deduction – he could retain his § 17 view that the holding of unity of apperception with respect to *i* implies that *H* thinks the knowledge-yielding (T*i*) thought. But he could deny that it follows simply from that view that the object known is distinct from all of H's representations taken separately or in sequence. Rather, a separate, supplemental argument must – and can – be given to show that while in at least some cases of the holding of unity of apperception the relevant (T*i*)-like thought is about such an object, in other cases that thought may be about the subjective sequence of intuition-elements.

Second, Kant could simply abandon his § 17 attempt (already criticized in Chapter Eight) to show that the holding of unity of apperception inevitably implies the thinking of a knowledge-yielding (T*i*)-like thought. Rather, he could aim at showing that when unity of apperception holds with respect to *i*, the Chapter Eight weakened (T*i*)-style result holds true; and so *H* knows, through a (T*i*)-like thought, *some* object (not necessarily the object of *i*) that is distinct from any of *H*'s representations taken separately or in sequence. Such a conclusion (coupled with revisions in Kant's view of the relation of judgment to unity of apperception) then would allow *H* also to make knowledge-expressing judgments about the subjective sequences of *H*'s intuition-elements.

Of these two possible solutions to the above difficulty, the second is philosophically the more satisfactory, given our criticisms of Kant's § 17 position. But the second also departs considerably from Kant's own views in the Deduction. Given our focus on those views in this book, I therefore propose that, while bearing in mind the second solution, we see Kant as working along the lines of the first solution when he is confronted with the difficulty. We also may return to our original Section 3 discussion of judgment and abandon the specific § 19 account that has led to that difficulty. However, to avoid overlaying our exposition with a mass of reconstruction, I will not try to develop the first solution any further than I have above. I also will postpone until Chapter Ten further comments on the issues arising out of the present discussion, including especially the question of how Kant can show that the object known through *i* – or any other relevant object – is in at least some cases distinct from any of *H*'s intuition-elements or other representations.

The third and last of the remaining topics from B142 concerns the necessity of objective unity of apperception. Above in Section 2 we discussed, as far as we need to in this book, the type of necessity that is involved in the second sort of objective unity that we there isolated. As we did so, we commented very briefly on the relation of that necessity to the sorts of direct, categorical necessities that, in Chapter Seven, we saw Kant to attribute to *H*'s thinking the (T*i*) thought. And in Chapter Seven we noted the correctness of Kant's § 19, B142 assertion that the (claimed) necessity of the holding of unity of apperception and of *H*'s thinking that thought – at least if that necessity is taken in a (UA)-style or in a direct, categorical (N_1)- or (N_2)-style – is compatible with the contingency of the judgment that *H* thereby makes.[41] All that remains for us to observe about necessity in this chapter is a point that we postponed in Section 2 and that Kant now suggests, using his ideas about necessity and judgment, at the end of § 19.[42]

As it was developed in Section 2, that point was what I there said is the incorrect half of Kant's claim about the extensional equivalence of the two forms of objective unity. In particular, the point amounts to his assertion that the presence, in a group of intuition-elements, of objective unity of the second (universal-validity) sort implies the presence, in that group of intuition-elements, of objective unity of the first (relation-to-a-distinct-object) sort, this assertion being understood in way (II) – and so being understood to apply to any group of intuition-elements just because of the natures of these notions of objective unity, taken by themselves. We can now see such an assertion as being made, in terms of Kant's ideas about necessity and judgment, at the close of § 19. Kant there implies that just because the holding of unity of apperception is necessary and yields an objective unity of the second sort (a unity that holds among unity-of-apperception-governed intuition-elements when those intuition-elements have a type of organization that, necessarily, belongs to any similar intuition-elements), the presence of the second sort of objective unity implies the presence of the first sort (a unity that holds among intuition-elements when those intuition-elements represent an object distinct from themselves).

Or, to put this last point in terms of his § 19 ideas about necessity and judgment, at the end of § 19 Kant certainly appears to claim the following. The mere fact of the holding of necessary unity of apperception with respect to any two intuition-elements *m* and *n* produces in *m* and *n* a type of judgmental organization that is such that, necessarily, any intuition-elements similar to *m* and *n* in being subject to unity of apperception and in yielding knowledge will themselves have that same type of judgmental organization. Then Kant implies the basic point now at issue – the claim that, in present terms, just because of the necessity that is here involved, the type of judgmental organization just referred to must involve, specifically, a (T*i*)-like thought to the effect that there is a single object (distinct from *m* and *n* taken separately or in sequence) that has the features that *m* and *n* present. And so that type of judgmental organization must involve the first sort of objective unity.

Kant's underlying reasoning in these claims – reasoning made explicit in *Prolegomena*, § 18 – is that this sort of 'necessary universal validity' of *m*'s and *n*'s judgmental organization can arise only if *m* and *n* (and any intuition-elements similar to *m* and *n*) are in fact involved in a judgment about an object distinct from *m* and *n* (and from the intuition-elements similar to *m* and *n*) taken separately or in sequence. As Kant in effect argues in the last paragraph of *Prolegomena*, § 18, suppose that, necessarily, the judgment in which *m* and *n* are involved (call this 'the *m*, *n* judgment') is such that all other persons' judgments involving intuition-elements similar to *m* and *n* agree with that judgment. Then only if all these judgments – including, in particular, the *m*, *n* judgment – are about the same object distinct from all these intuition-elements can we understand the *necessity* that *all* these judgments should agree with one another.[43]

Kant's basic point at the end of § 19 raises very interesting questions. But all that we need now to note is that this point is incorrect. Suppose that *m* and *n* are indeed knowledge-yielding and subject to unity of apperception and that they are involved in a type of judgment that is such that, necessarily, any intuition-elements similar to them are involved in a judgment of that same type. Then, given just this supposition, it does not follow that the *m*, *n* judgment need be about any object that is distinct from *m* and *n* (let alone about any object that is distinct from any of the knower's intuition-elements whatsoever).

This result does not follow simply because the necessity, here, that any intuition-elements similar to *m* and *n* should be involved in the same type of judgment as are *m* and *n* may turn merely on shared subjective features of all these intuition-elements that, for whatever reasons, each possessor of such intuition-elements must notice. For example, suppose that, by assumption, intuition-elements *p* and *q* are similar to *m* and *n* in being subject to unity of apperception and in yielding their possessor knowledge (on a weak reading that does not require that knowledge to concern an object distinct from them). Then subjection to unity of apperception is itself a type of organization, belonging to *m* and *n*, that the possessor of *m* and *n* recognizes to belong to *m* and *n* in a judgment (namely, the judgment that *m* and *n* belong to *him himself* or to *her herself*). Moreover, it is necessary that any intuition-elements such as *p* and *q* that are similar to *m* and *n* in the way indicated will themselves share that type of judgmental organization. However, given just this last fact by itself and the necessity that it involves, we cannot infer that *m* and *n* are involved in any judgment about a distinct object of the sort that we have been discussing.[44]

I thus see no good grounds for accepting Kant's basic point at the end of § 19 that, in the way explained above, the presence of the second sort of objective unity requires the presence of the first sort. As I have already noted, we could say much more about this point than I have here. However, because it plays no additional role in the strict argument of the B-Deduction, I will consider it no further.

6. SUMMARY

As we observed in Chapter Eight, Kant cannot argue convincingly from the holding of unity of apperception with respect to *i* to the conclusion that (T*i*) is true and *i*'s manifold is united in the concept of an object through the (T*i*) thought, thereby yielding *H* knowledge of the single object that *H* thinks. ((T*i*), we have seen, is the claim that *H* thinks there to be a single object that has the features that *i*'s elements present.) However, assuming the correctness of his B-Deduction § 17 reasoning, Kant takes himself to have demonstrated this conclusion. In B-Deduction § 18 Kant now infers that the unity of *i*'s manifold that is thus produced is objective and not subjective.

Coming next to B-Deduction § 19, Kant introduces the idea of the logical form of a judgment – the set of relations that obtain among the concepts (or further judgments) in the judgment in virtue of the operation of the logical functions of thought in judgment. Given this idea and the results of § 18, he then makes two main points: (A) *H*'s (T*i*) thought is or is part of a knowledge-yielding judgment about the object that *H* thinks and knows through *i*; and (B) the logical form of that judgment amounts to or derives from the objective unity of apperception that belongs to the concepts in that judgment. (A) will be correct because in the (T*i*) thought an object is subsumed under a concept, so that that thought is a judgment about the object that *H* thinks. (B) is inferred from Kant's basic B-Deduction § 19 claim, which I called (P), that the logical form of any judgment consists in the objective unity of apperception that holds with respect to the concepts in that judgment.

There are problems both with Kant's apparent reasoning for (P) and (B) and with (P) and (B) themselves. Yet the strict B-Deduction § 19 and § 20 argument for category application does not require us to adopt (P). Thus while we will hereafter acknowledge, verbally, (P) and (B) because of their importance to Kant's own presentation of the Deduction, we will not ourselves assume (P). (Nor will we assume, although we will keep in mind, an alternative proposal that we noted about the relation of the logical form of judgment to objective unity of apperception.) And while we will acknowledge, also, the claim, in (A), that *H*'s (T*i*) thought is or is part of an overall knowledge-yielding judgment, we will not suppose, without further argument, that *H* has been shown actually to make any knowledge-yielding judgment that involves but goes beyond the thought and judgment expressed in (T*i*).

Following this discussion of (A) and (B), we examined various further points about judgment, objective unity of apperception, and the logical form of a judgment. These included Kant's Aristotelian (or quasi-Aristotelian) position on logic; a regress that arises in connection with his view of logical-function application; his account of the role of the copula in judgment; his § 19 suggestion that a judgment is always about an object distinct from any of the knower's intuition-elements or other representations whatsoever; and, finally, his claim that if intuition-elements have a type of judgmental organization that is necessary and universally valid, then those intuition-elements are united in a (T*i*)-like thought that concerns an object distinct from those intuition-elements.

We are now ready to turn, in Chapter Ten, to the concluding § 20 of the first half of the B-Deduction and its basic claim (C) that links the logical-function structuring of concepts in judgment to the categorial structuring of the object known through the judgment. And in that chapter we will consider, also, the extent to which Kant's argument in the first half of the Deduction has been – or can be made – successful.

CATEGORY APPLICATION TO THE OBJECT OF INTUITION: B-DEDUCTION § 20

1. INTRODUCTION

(T*i*) is the claim that, roughly, *H* thinks that there is a single object that has the features that are presented by *i*'s elements. As we saw, in B-Deduction § 19 Kant argues that *H*'s (T*i*) thought is or is part of a knowledge-yielding judgment about the object that *H* knows through *i*. He holds also that the logical functions of thought, through the holding of objective unity of apperception, determine the logical-form relations together of the concepts (or further judgments) in that judgment. Claim (C) of § 20 then is Kant's claim that, because the logical functions determine the logical-form relations of the concepts in the *judgment*, the logical functions determine, also, the relation together of the conceptual elements of *i*'s manifold in such a way that the *object* that *H* thinks and knows through *i* is category-subsumed. In the present chapter we will see the details of Kant's grounds for (C), and we will note various surprising aspects of these grounds and of his conception of (C). In Sections 2 to 7 we discuss in more detail than we so far have Kant's view of the role of concepts in judgment; his accounts of the logical functions of thought, of the concept of an object in general, and of a category; and his exact understanding of the relation of logical-function application in a judgment to category application to the object known through that judgment. We also will note briefly the interesting relations that hold between Kant's view of category application and Aristotelian views of judgment. In Sections 8 and 9 we then consider various questions about his treatment of the categories that we have postponed from earlier chapters. And we comment also on the ultimate significance of the results of § 20 and of the Transcendental Deduction as a whole.

2. KANT ON CONCEPTS AND THE LOGICAL FUNCTIONS OF THOUGHT IN JUDGMENT

In Chapters Eight and Nine we supposed that Kant has shown that *H* thinks the (T*i*) thought, that thought yielding *H* knowledge of the object that *H* thinks through *i*. As just noted, Kant also holds that that thought is or is part of a knowledge-yielding judgment that *H* makes about that object. Given these points, I will simplify the official Chapter Eight statement of (T*i*) by dropping all but the part of (T*i*) that expresses *H*'s thought. So our Chapter Eight and Nine results now amount to the assertion that, when *H* knows through *i* and unity of apperception holds with respect to *i*, *H* thinks the thought expressed in the simplified (T*i*) claim which I label, as before,[1]

(T*i*) *H* is conscious in thought that there is a single object *x* such that first i_1 puts before *H*'s mind (or is) *F* and *x* has *F* and then i_2 puts before *H*'s mind (or is) *G* and *x* has *G*

or, formally:

H is conscious in thought that $(\exists x)$(x is an object & first i_1 pm (or is) *F* & *x* has *F* & then i_2 pm (or is) *G* & *x* has *G*)

And what follows '*H* is conscious in thought that' in this claim expresses a knowledge-yielding thought of *H*'s that is or is part of a knowledge-yielding judgment about the object of *i*.

Now, as we will see in more detail later in this chapter, a category is a concept of a thing such that that thing is an object in the most general sense of 'object' (that is, that thing is an object in general) and that thing is playing (or elements of it are playing) one of the roles that are specified by the logical functions of thought in judgment. Yet consider the judgment of which *H*'s (T*i*) thought is or is a part. By B-Deduction § 19 the concepts (or further judgments) that occur in that judgment are related together through the application to them of the logical functions that determine the form of that judgment. (Hereafter I often omit 'or further judgments.') Furthermore, that judgment involves the attribution of features *F* and *G* to the object *x* that is known through *i*. And, as already suggested in Chapter Three, these features, taken as general, themselves are identical to (or are presented by) the concepts through which *H* thinks them to belong to the object.

For concreteness, suppose that *R* is one of the logical functions that determine the logical form of the judgment that is here at issue. Then in thinking the overall thought that constitutes that judgment, *H* is conscious in thought that *F* and *G* – or the concepts that present them – are related together through *R*.[2] Hence when *H* knows through *i* and unity of apperception holds with respect to *i*, it is not merely that *H* makes the overall knowledge-yielding judgment about *i*'s object that is (or includes) *H*'s (T*i*) thought. In making that judgment, *H* thinks the following specific knowledge-yielding thought:

(C*i*) *H* is conscious in thought that there is a single object *x* such that first i_1 puts before *H*'s mind (or is) *F* and *x* has *F* and then i_2 puts before *H*'s mind (or is) *G* and *x* has *G* and *F* and *G* are structured by logical function *R*

or, formally:

H is conscious in thought that $(\exists x)[x$ is an object & first i_1 pm (or is) F & x has F & then i_2 pm (or is) G & x has G & $R(F, G)]$

But, in thinking that thought, *H* is thinking that the object *x*, the object known through *i*, is an object that has elements – the features *F* and *G* – that play the logical role that is specified by the logical function *R*. So (as B-Deduction § 20 infers) *i*'s object, in *H*'s knowledge of it, falls under the category associated with *R*. And, as we see in Section 4, *i*'s object is here brought under that category in such a way that claim (C) is correct.[3]

The points made in the last several paragraphs really have only the character of a promissory note about the B-Deduction § 20 treatment of category application, and to redeem this note we need to go more deeply into Kant's views about concepts, intuition-elements, and the logical functions of judgment than we so far have. This project will occupy the remainder of this and the next section of this chapter.

I begin by observing that in the present discussion i_1 and i_2 evidently can be taken to be the members of *i*'s conceptual manifold. So, given the results of Chapter Three, i_1 and i_2 put before the mind or, alternatively, *are*, the potentially general features *F* and *G* that occur *in* the object *x* itself. *H* attributes those features to that object by operating mentally with concepts. Those concepts, we remarked in Chapter Three, Kant sometimes regards as representations, distinct from those features themselves, that present those features (taken as general) to the mind. But, as we saw – and as will prove important below – Kant frequently regards those concepts as representations that simply *are* those features (taken as general).

As we noted in Chapter Three, the situation that I have just described with respect to concepts and intuition-elements leaves open the exact relation that holds between those two sorts of representations. We deferred the discussion of that relation until the present chapter. However, we can see now that there is no real need to specify that relation any further than the facts above imply. As I have just indicated, in § 20 Kant argues from the logical-function structuring of concepts in a judgment to the logical-function structuring of features in the object judged about. In the case in which a concept is simply a feature or property taken as general and an intuition-element is simply a (potentially general) feature, we can infer that the concept itself is just the intuition-element taken as general. And in that case, as I explain below, Kant can straightforwardly argue that the logical-function structuring of the concept in judgment directly implies the logical-function structuring of the feature in the object. In the case in which the concept is the feature or property taken as general but the intuition-element is regarded as putting before the mind (and as not being the same as) that feature, he can give essentially the same argument (turning on the identity of the concept and the feature taken as general) without having to decide the exact relation of concept to intuition-element.

Finally, in the cases in which the concept is regarded as presenting, and as *not* being identical to, the feature taken as general, Kant faces the very serious problem, emphasized below, of explaining why the logical-function structuring of the *concept* (in the *judgment*) has any implications at all for the logical-function structuring of the *feature* (in the *object*) that is presented by this concept. However, this problem will exist whether, in addition, the intuition-element itself is taken to put before the mind the feature in the object or the intuition-element itself is taken simply *to be* that feature. Moreover, it does not seem that the solution to this problem will turn on what exact relation the concept (conceived as presenting, but as not being identical to, the feature) bears to the intuition-element (whether the intuition-element puts before the mind the potentially general feature or the intuition-element *is* that feature). So in these present cases, too, there is no need to decide the exact relation of concept to intuition-element.

My own belief is that Kant, who does not focus on these fine points about concepts and intuition-elements, usually treats concepts simply as properties taken as general. Then, without attending explicitly to the differences between the following two possibilities, he thinks of those properties either (a) as being, in the object, potentially general features that are put before the mind by intuition-elements (as on his appearing theory) or else (b) as being, in the object, potentially general features that are identical to intuition-elements (as on his appearance theory).[4] After all, such a treatment of concepts accords well both with his Deduction view of the logical functions as structuring features in objects by structuring concepts in judgments and with what I suggest below is his Aristotelian view of the operation of concepts in judgments. However, I will not develop this belief further, and it is not necessary to accept it to grant the points that I have just made about concepts and intuition-elements.

Given the preceding discussion, when *H* thinks (T*i*), *H* must have before thought-consciousness a concept c_1 that presents or that simply is feature *F* (taken as general) and a concept c_2 that presents or that simply is feature *G* (taken as general). c_1 and c_2 must occur before *H*'s thought-consciousness organized together through the use of the relevant logical functions of thought (and of the concept of an object in general) in such a way that they yield *H* the thought that the single object *x* of *i* has *F* and *G*. And, as they so occur, c_1 and c_2 must also be part of the overall, knowledge-yielding judgment that *H* makes about the object of *i*. Moreover, except in the first case noted several paragraphs ago, in which a concept turns out to be identical to an intuition-element taken as general, we need draw no conclusions, here or below, about the exact relations of c_1 and c_2 to i_1 and i_2.

3. CONCEPTS IN JUDGMENTS AND FEATURES IN OBJECTS

Given the Deduction's goals, the most obvious problem that faces the above account of the operation of concepts in the overall judgment about *i*'s object is one that we noted in the previous section. Suppose that, whether from uneasiness about the idea of concepts as being features in the mind or for some other reason, we take c_1 and c_2 to present and not simply to be *F* and *G*. Then it is hard to see why the logical-function structuring of c_1 and c_2 in way *R* in the *judgment* at issue should in any way imply or yield the logical-function structuring, in that same way *R*, of *F* and *G* as those features occur in the *object x* of *i*.

After all, if c_1 and c_2 are distinct from *F* and *G*, then there is no direct connection between the logical-function structuring of c_1 and c_2 and the logical-function structuring of *F* and *G*, even given Kant's idealism about the object of knowledge. Moreover, and from a modern standpoint, it may seem that we ought to distinguish (i) the 'syntactic' organization of concepts (or further judgments) into a judgment, through the application to those concepts of the logical functions, from (ii) 'semantic' facts about the relation of that judgment or its elements to the object judged about and its features (or from facts simply about the nature of that object and about any special organization that its features may have).[5] We may then note that syntactic facts (say about the order of conjuncts in a first-order-logic conjunction or about specific case endings in natural languages) often have no bearing on the nature of the objects discussed in claims that embody those facts. So why need there be any connection at all (let alone any direct connection) between the logical-function structuring of c_1 and c_2 in the overall judgment and the existence of any sort of Kantian logical-function structuring of F and G in the object known through that judgment?

Nevertheless, despite the problem here Kant insists on such a connection. We have seen this insistence, in our own terms, already in discussing (C*i*) above, and it emerges in Kant's B-Deduction § 20 argument (to which we return in more detail later) that

> ... that act of understanding by which the manifold of given representations (be they intuitions or concepts) is brought under one apperception, is the logical function of judgment (cf. § 19). All the manifold, therefore, so far as it is given in a single empirical intuition [representing or functioning as a single *object*], is *determined* in respect of one of the logical functions of judgment ... (B143)

It emerges also in Kant's immediately following § 20 comment that 'the *categories* are just these functions of *judgment*, insofar as they are employed in the determination of the manifold of a given *intuition*' (B143; the last two italics are mine). And it is found in many other places, for example in the important B128–29 remarks on the nature of the categories (added at the end of the B-edition version of the A92/B124 ff. 'Transition to the Transcendental Deduction of the Categories') that we discuss later.[6] Given his insistence on the existence of this connection, Kant clearly must solve – or at least avoid – the above problem if his B-Deduction § 20 reasoning is to be at all plausible.

Kant himself ignores this problem. As we see later, he apparently does so because he focuses largely on the view that concepts *are* features (taken as general). On that view, as I have already intimated, the logical-function structuring of concepts in a judgment amounts to or straightforwardly implies the logical-function structuring of the features as they occur in the object as it is known through the judgment. And so on that view Kant simply avoids the problem.

To spell out this last point, when he writes in terms of the view that concepts *are* features, Kant considers the concepts c_1 and c_2 that are logical-function-structured and that yield H the thought that the object has the features F and G. Identifying those concepts with those features, he then in effect also identifies the logical-function organization of those concepts before H's mind – the organization whose occurrence before H's mind is involved in H's *thinking* that thought – with an organization that occurs within the *content*, itself, *of* that thought, the claim or fact *that* the object x of i *has* the features F and G. Because this thought itself yields knowledge of that object, this treatment of the logical-function-structured conceptual *means* whereby H thinks *that* the object has F and G implies that that thought itself is or involves the knowledge-yielding, and so *true*, thought *that* the *object* has the logical-function-structured *features* F and G.[7] So the logical-function structuring of c_1 and c_2 in way R yields the logical-function-structuring of features F and G as they occur in the object of i as that object is known through the judgment.

There are serious philosophical questions about this last resolution of the above problem. And there are serious questions about Kant's whole project of inferring category application from facts about the logical structure of judgment. However, for the present we will ignore such questions and focus on understanding Kant's own development of that project. And here we should note that while the line of thought in the last paragraph resolves the specific above problem, it does not remove all the textually puzzling aspects of his views about the logical-function structuring of concepts in judgments and the logical-function (and categorial) structuring of objects.

In order to present these aspects as clearly as possible, I will from here on, except where otherwise noted, work with the following specific case. That is the case in which concepts c_1 and c_2, as they occur in the overall judgment that H makes about the object of i, are organized through the logical function of subject and predicate (compare A245) in such a way that c_1 functions as the subject term of that judgment and c_2 functions as its predicate term. I also will assume that, through the use of the preceding line of thought, Kant can resolve the above problem, so that the functioning of c_1 as subject term to c_2 as predicate term in the judgment implies the functioning of F as subject to G as predicate in the object x. Thus suppose, as before, that H thinks (Ti) and so thinks that x has F and G. Then in the present case we take the overall judgment that H makes not just to include the judgment that there is an object x that has the i_1-

presented F and the i_2-presented G. Rather, H's overall judgment will include the judgment roughly to the effect that there is an object x, which is a subject thing of type F (F here being presented by i_1), that has the predicate G (G here being presented by i_2). And then that overall judgment will proceed in one of the ways in which it can be further developed (for example, as involving the further claim that all subject things of type F are Gs). This present, subject-predicate case offers Kant what is perhaps his most plausible transition from a logical function (of subject and predicate) to a category (of substance and attribute). And working with this case lets us note significant issues while being as concrete as possible.

In terms of this case, there are three puzzling aspects of Kant's views about the logical-function structuring of concepts in judgments and the logical-function structuring of objects. First (and to remark a point that is actually independent of this specific case), we have already appealed to the fact that Kant frequently identifies concepts with features. In Chapter Three I did not explore possible rationales for this identification. Why (even given his transcendental idealism) would Kant treat the concepts that we form of objects simply as being general properties of those objects as those general properties occur in, and are operated on by, the mind? It is no answer to say that that treatment avoids the specific problem above, for the treatment occurs very frequently in Kant's writings (and in those of various of his contemporaries).[8] And, as I have noted earlier, he ignores that problem.

Second, we should note a specific aspect of the application of the subject-predicate logical function to features like F and G as they occur in the object known. That application is to follow from the application of the logical function to the concepts, in judgment, that present or that simply are those features. Therefore just as c_1 functions as subject term, in the relevant judgment, *to* c_2 as predicate term in that judgment, so F must function in the object as *subject to* G as predicate. However, taken literally this last point means that F *itself* is a subject that *has* G as a predicate or property. And to modern ears such a result sounds flatly confused or mistaken.[9] One wants to argue that the subject of the judgment that H makes about the object x is the object x itself and not the *feature* or *property* F that x has. And the feature G is the predicate or property of that object and not a predicate or property of the *feature* F. (If I say that an object is a tree and is conical, it is the object, and not its feature of being a tree, that is the subject of my judgment. And I attribute the property of being conical not to the *feature* of being a tree – a feature that, understood as an abstract entity, surely has no shape at all – but to the object that has that feature.)

If we take literally and seriously Kant's idea that the subject-predicate logical function is applied to F and G as they occur in the object x, we thus seem forced to conclude that Kant identifies – or else assumes an unexplained relation to hold between – (a) the feature F that is attributed *to* the object x by the subject term of the relevant judgment and (b) the object x, itself, that is the true subject of the judgment.[10] And, in doing so, he takes the predicate G to apply to F. The view of F, x, and G that is embodied in this conclusion will seem unacceptable to modern readers. Yet careful attention to the texts shows that Kant accepts points that commit him to it.

Thus in the early "False Subtlety of the Four Syllogistic Figures," § 1, he writes (and there is no indication that he later withdraws this point) that the thing itself about which a judgment is made is the subject of that judgment.[11] And in Reflexion 3921 he urges that

> Through understanding we know in bodies not the real subject, but rather the predicates of extension, solidity, rest, motion, etc… . predicates without a subject and without a final subject cannot be thought, *the unchanging* [or permanent] *predicates then are called together the subject.*[12]

Given these two quotations in conjunction, Kant certainly seems to identify the subject of the judgment – the object or thing about which the judgment is made, and the object to which various properties are attributed in that judgment – with certain 'unchanging' predicates of that object. Or else he takes there to be some tight and unexplained relation between that object and those predicates.

Again, in the important *Metaphysical Foundations of Natural Science* footnote in which he first presents the considerations about judgment and the categories that he later incorporates into the B-Deduction, Kant evidently identifies concepts with the features (like those of being a stone or of being hard) that they present. And, using this identification, he says that

> … in the categorical judgment *the stone is hard*, the *stone* is employed as subject and *hard* as predicate, so that it remains permissible for the understanding to interchange the logical function of the concepts and say something hard is a stone.[13]

However, suppose now that I make such a judgment as a knowledge claim about an object, the stone, that is known through this judgment. Then, as Kant immediately continues in the above quotation,

> ... when I represent to myself *in the object* as *determined* that the *stone* in every possible determination of an object and not of the mere concept must be thought only as subject and the hardness only as predicate, the same logical functions now become pure concepts of the understanding for cognizing objects, namely, *substance* and *accident*.[14]

Ignoring points about the categories (or 'pure concepts') to which we will return in a later section, we see that here Kant seems clearly to be claiming that when I know an object through the judgment that the stone is hard, that object involves the two features of being a stone and of being hard in combination. In this combination, the feature of being a stone plays the role of subject, and the feature of being hard plays the role of predicate (or accident) to that subject.[15]

Given Kant's comments, in the last *Metaphysical Foundations* text quoted above, about *determining* the feature of [being a] *stone* to be thought only as subject and never as predicate, there is at least a structural parallel between what he says about that feature and his Reflexion 3921 talk of the 'unchanging [or permanent] predicates' as themselves being 'called together the subject.' Of course without further premises (from the Schematism) one cannot argue that what is thought only as being a subject must be something that is unchanging in the object (or the converse).[16] Nevertheless, despite this difference between the *Metaphysical Foundations* view and the Reflexion 3921 view of a feature's being treated as a subject, it is still true, according to both texts, that a feature or property *becomes* or *functions as* a subject (and, correlatively, other features become predicates) simply by taking on a special role, namely, the role of being thought only as a subject or of being unchanging.

Moreover, this same structural parallel between the *Metaphysical Foundations* view and the Reflexion 3921 view can be seen also to hold between those views and the A399–400 position cited already in Chapter Three, a position that has the ring of Reflexion 392l. As we saw, at A399–400 Kant writes that

> If I am to declare a thing to be substance in the appearance, predicates of its intuition must first be given to me, and I must be able to distinguish *in these* [predicates] the permanent from the transitory and the substratum (the thing itself) from what is merely inherent in it. (My italics)

Not only does this text indicate, as I urged in Chapter Three, that the elements of the manifold of intuition include matters for concepts (here, predicates). But, also, it strongly suggests that we take some of these predicates *to be* or to function as the permanent subject (or the substance, insofar as the pure concept of subject is schematized through the notion of permanence in time) and others of these predicates *to be* the transitory accidents that are merely inherent in (and so *belong to*) that permanent subject.[17] We take these predicates in these ways insofar as – in structural parallel with the *Metaphysical Foundations* footnote and Reflexion 3921 – we take them to have (or we 'distinguish in them') certain roles: here at A399–400, the roles of being permanent and a substratum and of being transitory and merely inherent in a substratum. Furthermore, as I have just indicated, at A399–400 Kant once more treats a feature as a subject to which other features function as properties (or genuine predicates). And once more he either identifies that subject-feature with the object itself or else assumes that the subject-feature has some special, unexplained relation to the object.

We will return later to these issues about features determined to function only as subjects. Meanwhile, note, further, the "False Subtlety of the Four Syllogistic Figures" § 2 statement that 'a mark of a mark is a mark of the thing itself (*nota notae est etiam nota rei ipsius*).'[18] And observe also Kant's *Logik*, § 63: 'What belongs to [*zukommt*] the mark of a thing belongs also to the thing itself.'[19] These texts certainly make it sound as though predicating the mark or feature *being mortal* of the mark *being human* of Socrates is predicating the mark *being mortal* of Socrates himself, just as if *being human* were *itself* the (or a) subject of the predicate *being mortal* – and just as if *being mortal*, by belonging to *being human*, therefore belongs to Socrates. (The exact relation of *being human* to the object Socrates would then remain unexplained, as far as these texts go.)

Finally, the Metaphysical Deduction of the Categories introduces a judgmental relation that structurally parallels the predication-relation that, according to the above mark-of-a-mark texts, holds between the predicate feature and the subject feature in the object. At A68–69/B93 Kant asserts that

> In every [subject-predicate] judgment there is a concept which holds of many representations [*für viele gilt*] and among them of a given representation that is immediately related to an object [*auf den Gegenstand unmittelbar gezogen wird*]. Thus in the judgment: *all bodies are divisible*, the concept of the divisible applies to

[*bezieht sich auf*] various other concepts, but is here applied in particular to the concept of body, and this concept again to certain appearances [the intuitions or empirical objects of our knowledge] that present themselves to us.[20]

In the last sentence Kant clearly takes the predicate concept in a judgment to apply to the subject concept, just as in the above mark-of-a-mark texts he has taken the predicate feature in the object to apply to the subject feature. Note also that (as its association with traditional logic would lead one to predict) the general idea of a mark of a mark as being a mark of a thing is found outside Kant's work.[21] And thus the puzzle about features of objects as being subjects of other features as predicates is not peculiar to that work itself.

The third puzzling aspect that we should note concerning Kant's views about the logical-function structuring of concepts in judgments and the logical-function structuring of objects bears on our above, tentative distinction between the syntactic organization of concepts in a judgment and semantic facts about the relation of that judgment (or its contained concepts) to the object judged about. On the one hand, modern readers expect such a distinction, and various of Kant's texts may seem to encourage the drawing of it. Yet, on the other hand, study shows that Kant actually draws no sharp such distinction, at least in the form that modern readers expect. And this fact may seem disturbing.

The usual modern idea is that a sentence is constructed out of linguistic elements in accordance with various syntactic formation rules.[22] This construction proceeds in a way that, standardly conceived, is independent of the way in which meanings – and, in general, relations to objects, facts, and features in the world – are assigned, semantically, to those linguistic elements and to that sentence. Moreover, except in special cases like that of sentences that describe other sentences, this construction – and the syntactic organization that it yields – is taken to apply just to sentences and other linguistic entities and not to the things in the world that the sentences concern.

Applying this idea to the case of a Kantian judgment, the modern reader is likely to focus on the view that concepts present features from which they are themselves distinct (the view of concepts that, as noted in Chapter Three, is most plausible to such a reader). Such a reader is then likely to take the logical functions to order concepts, as so viewed, into judgments according to the relevant syntactic formation rules (understood as in the last paragraph). This reader will see the mind as relating those concepts, and the judgments that they make up, to objects and features in the world through the (semantic) use of the concept of an object in general. Using that concept, the mind thinks there to be objects that have the features that the concepts present. And this reader will also take those objects and features not themselves to have syntactic organizations but to possess their own characteristic nonsyntactical organization together.[23]

As I noted, this modern application of syntactic and semantic considerations to Kantian judgments may also seem supported by various texts. Thus at the beginning of the Transcendental Logic Kant contrasts general logic, which

> abstracts from all content of knowledge, that is, from all relation of knowledge to the object, and considers only the logical form in the relation of any knowledge to any other knowledge ... (A55/B79; note also A52/B76, A54/B78, and A56/B80)

with transcendental logic, which

> should contain solely the rules of the pure thought of an object (A55/B80)

– that is, the rules that concern the use of the categories, the pure concepts of an object in general. Later in the first *Critique* he considers the 'merely logical employment of the understanding' (B128). He says that in this employment, and in the case of a subject-predicate, categorical judgment like the judgment that all bodies are divisible,

> it remains undetermined to which of the two concepts the function of the subject, and to which the function of predicate, is to be assigned. For we can also say: Something divisible is a body. (B128)

However, when through the use of the concept of an object in general the concept of body is brought under the category of substance,

> it is thereby determined that its empirical intuition in experience [that is, it seems, the feature, in experience, that the concept is or presents] must always be considered as subject and never as mere predicate. (B129)

(Compare the *Metaphysical Foundations* footnote quoted above.[24])

To a modern reader these two groups of first-*Critique* texts may appear to show Kant distinguishing, in at least roughly one modern way, between matters connected with the syntax or logical form of a judgment and matters connected with the semantical relations that hold between judgments (and their contained concepts) and objects and features in the world.[25] Thus general logic and the logical employment of the understanding would be concerned with inferential relations (between judgments) that, it seems, turn on syntactic structure or logical forms. (Note the first quotation above.) But through the use of the concept of an object in general (and the categories that realize or specify that concept) judgments are related semantically in certain determinate ways to features in objects.

Nevertheless, and as intimated above, these modern impressions are mistaken. Kant does not in fact distinguish, in anything like the above way, between the syntax and semantics of judgment. This point is shown already by the B-Deduction § 19 claim that the logical form of a judgment derives from the objective unity of apperception that belongs to the concepts in the judgment. On the present sort of modern conception, that logical form should be a matter solely of the judgment's syntax and so should have nothing to do with the reference of the concepts in the judgment to objects. Yet, for Kant, holding of objective unity of apperception with respect to those concepts itself implies the occurrence of the concept-of-an-object-in-general thought, by the judger, that relates these concepts to the object judged about. (Through the (T*i*) thought the judger thinks there to be an object that has the features the concepts present or are.) So the same holding of unity of apperception that *is* or yields the supposedly purely syntactic logical form of the judgment also establishes what the present sort of modern reader will think is the semantic relation of the concepts in the judgment to the object judged about. Moreover, that holding is the only way in which the logical form of the judgment is established; and it is also the only way in which the relation of the concepts in the judgment to the object is established.

These facts show that, for Kant, the logical form (as determined by the logical functions) does not inhere in those concepts in independence of their relation, through the use of the concept of an object in general, to the object judged about. (For Kant, the specific logical form cannot exist in the concepts without that relation's also obtaining; and that relation cannot obtain without at least *some* logical form's existing in the concepts.[26]) Hence he will hardly think of the logical form as a purely syntactic organization that is present in the concepts in independence of the semantic relations of the concepts to the object.

The basic point that Kant does not introduce, in a way that modern readers might expect, the idea that a judgment has a purely syntactic organization is also strongly suggested by a fact noted earlier in this section.[27] That is the fact that he frequently identifies concepts with features taken as general and then identifies the logical-function organization that belongs to the concepts in a judgment with the organization that features have together in the object that is judged about. To the extent that he makes these identifications, he of course thinks of the logical-function organization (the only plausible Kantian candidate for a syntactic organization) as belonging to objects and features in the world as well as to the judgments that concern those objects and features. But, as noted above, the usual modern reader will not regard things in the world (except when those things are themselves linguistic entities or judgments) as having a syntactic organization.

Again, if Kant means to draw the sort of syntax-semantics distinction for judgment that was sketched above, then he should conceive of the logical-function organization of concepts in a judgment as determinately fixed in that judgment in independence of the relations of those concepts to the object judged about, just as in the sentence 'the tree is conical,' the term 'the tree' functions grammatically as subject in independence of whatever object is semantically assigned to be its designation. However, as the B128–29 and the *Metaphysical Foundations* quotations indicate, Kant writes that, through the mere logical employment of the understanding, the logical function of a particular concept (say, the concept of body) is *not* yet fully determined (say to be just the subject of judgment). That full determination comes precisely through the use of the concept of an object in general to think an object as having the features (taken to function as subject) that the particular concept presents.[28] But then, given this fact, it seems clear that through the logical employment of the understanding the mind, in operating with concepts and the logical functions, is not establishing a syntax, of the usual modern sort, for a judgment. And since, within Kant's framework, the only thing that could plausibly be taken to establish such a syntax is the logical employment of the understanding (and the logical-function organization that that employment yields), it therefore looks once more as though Kant is not drawing the above sort of modern syntax-semantics distinction for judgment.

We are therefore left with the puzzle of how Kant can understand a judgment's logical form and relation to objects, if not according to the ideas of syntax and semantics that I have sketched. And this puzzle may seem all the greater simply because his B128–29 comments on the logical employment of the understanding and the use of the concept of an object in judgment, and his similar comments elsewhere, obviously seem in some ways like those ideas.

4. KANT ON THE CATEGORIES (I)

We have now seen three puzzling aspects of Kant's views about the logical-function structuring of concepts in judgments and the logical-function structuring of objects: his frequent identification of concepts with features; his treatment of certain features of an object (and not just the object itself) as functioning as the subject to which other features function as predicates; and the puzzle, just noted, about how he understands a judgment's logical form and relation to objects, given that he does not accept a standard sort of modern syntax-semantics distinction. These puzzles can be resolved, I think. (Here see Section 6 below.) But to do so we need in this and in the next section to turn to the details of Kant's own view of the categories. As we study those details, we will also acquire further evidence that Kant does not accept standard modern ideas about syntax and semantics. And we will gain the information that we require to understand the B-Deduction § 20 argument for category application.

In Section 2 I suggested that a category is a concept of a thing such that that thing is an object in the most general sense of 'object' (that is, that thing is an object in general) and that thing is playing (or elements of it are playing) one of the various roles that are specified by the logical functions of thought in judgment. In order to motivate this description of a category, I have found it useful to classify Kant's characterizations of the categories into four main groups: characterizations of the categories (a) as being concepts of an object in general (or simply as being concepts of objects); (b) as being the logical functions of thought themselves; (c) as being or as involving both (a) and (b) (or as being representations described in ways neutral between (a) and (b)); and (d) as being representations containing the necessary unity of the synthesis of the manifold of any intuitions through which knowledge is had.[29] It will be helpful to have examples of each of these groups.

Under (a) one finds Kant's B113 view of the categories 'as *a priori* concepts of objects'; his A93/B126 description of a category as 'a *concept* of an object as being given [through intuition], that is to say, as appearing'; and his A254/B309 comment, in the discussion of phenomena and noumena, that the categories 'think objects in general.' Other representative (a)-style passages include A111 (through the categories 'we think objects in general for appearances'), A129–30 ('a formal *a priori* knowledge of all objects, so far as they are thought (categories)'), and B146 ('the concept, through which an object in general is thought (the category)').

I have found about the same number of comments falling under (b) as under (a). Among them are some to which Kant gives special prominence.

Thus, as we have noted in Section 3, at B143 of B-Deduction § 20 the categories are described as 'just these functions of judgment, insofar as they are employed in determination of the manifold of a given intuition.'[30] In *Prolegomena*, § 39, Kant says that 'in themselves [the categories] are nothing but logical functions, and as such constitute not the slightest concept of an object in itself.'[31] Notice also A147/B187 ('The categories, ... without schemata, are merely functions of the understanding for concepts and represent no object'); B148 ('as concepts of objects [the categories] are then empty ... they are mere forms of thought'), and a similar remark at B150; A253 ('the category is a mere function of thought'); B288 ('the categories ... in themselves ... are merely *forms of thought* for the making of knowledge from given intuitions'); and B431 ('I should understand by these concepts the merely logical functions'). Perhaps the same view of categories is embedded in the pre-1781 Reflexion 4638 – 'The determined logical function of a representation in general is the pure concept of the understanding.'[32] But because this fragment also mentions 'concepts which should express the modes of thinking an object in general,' I prefer to classify it under (c).

Clearly among the large number of examples under (c) – which in fact seems the largest of our four groups – are the central B128 definition quoted in Section 3 (categories 'are concepts of an object in general, by means of which the intuition of an object is regarded as determined in respect of one of the logical functions of judgment'); the A245 comment that the categories are 'so many modes of thinking an object for possible intuitions ... . [They] are nothing but representations of things in general, so far as the manifold of their intuition must be thought through one or other of these logical functions'; the A247/B304 claim that the pure category 'expresses only the thought of an object in general, according to different modes' (these modes being, I take it, the logical functions of thought); and the B159 description of the categories 'as *a priori* modes of knowledge of an object in general' (to assume the same reading of 'modes' as in the two preceding texts).

Outside the first *Critique* one finds in group (c) the *Prolegomena*, § 21, description of the categories as 'falling exactly parallel to' the moments of the understanding in judgment and as 'being nothing more than concepts of intuitions in general which are determined in themselves as judgments, necessarily and with universal validity, in respect of one or other of these moments.'[33] Again, in the *Metaphysical Foundations* footnote quoted in part above in Section 3, Kant describes the categories both as 'determinations of our consciousness borrowed from the logical functions of judgments in general,' which determinations 'are nothing but mere forms of judgment insofar as these forms are applied to intuitions' and as 'derived' from 'the formal operations of the understanding in judgments, from which [operations the categories] also differ in nothing except that in the concept of the understanding, an object is thought as determined in regard to one or the other function of judgment.'[34]

Reflexion 5932 contains similar (c)-style definitions – for instance, a category is 'the *concept of an object in general*, so far as *it is determined in regard to a logical function of judgment a priori* in itself (that one must through this function think the combination of the manifold in its representation).'[35] Moreover, besides such (c)-style passages, I am for present purposes classifying under (c) texts that are neutral between (a) and (b), as for example the A57/B81 description of what are obviously the categories as 'concepts which relate *a priori* to objects ... solely as acts of pure thought' and A79/B105 on pure concepts which we are entitled to regard 'as applying *a priori* to objects.'

In my last group (d) I count such comments as B151 on 'the synthesis which is thought in the mere category in respect of the manifold of an intuition in general, and which is entitled combination through the understanding'; A119 on the categories as containing 'the necessary unity of the pure synthesis of imagination in respect of all possible appearances'; A125 on the categories as 'grounds of the recognition [of course through synthesis] of the manifold'; A138/B177 on the category as containing 'pure synthetic unity of the manifold in general'; and A220/B167 on pure concepts as, like other concepts, containing a synthesis, the synthesis of the pure concepts being 'an *a priori* condition upon which experience in general in its formal aspect rests.'

By Kant's parenthetical remark in the Reflexion 5932 quotation given in the next-but-last paragraph, the categorial combination of a manifold of intuition involves the determination of that manifold according to a logical function. So I also count as falling under (d) those interesting texts in which Kant says not that the categories *are* the logical functions but that they *contain* the logical functions. Here we have, for example, A239/B298: 'In the absence of such object,' Kant writes – an object to which a concept (and here he is clearly discussing among other concepts the categories) may be applied – '[the concept] has no meaning and is completely lacking in content, though it may still contain the logical function which is required for making a concept out of any data that may be presented.' Note also A242–43/B300–301, which describes the categories of substance and of causality as still containing, respectively, the subject-predicate logical function and the ground-and-consequence logical function even if one omits from these categories their schemata. Similarly, but more generally, A245 describes a category as containing 'nothing but the logical function for bringing the manifold under a concept' when the category's schema is omitted.

Studying the above groups of texts, we find both examples of loose phrasing and genuine philosophical questions. It is hard to suppose, for instance, that Kant really means *Prolegomena*, § 21, which, as we saw under (c), describes the categories as falling 'exactly parallel to' the logical functions, to contradict *Prolegomena*, § 39, which, as we saw under (b), takes the categories not to parallel but in themselves *to be* the logical functions. So one or both of these texts surely must embody a hasty or loose expression of his views. Again, and while one may think one can reconcile, at least roughly, various texts in (c) – and in (d) – with those in (a) or in (b), Kant's descriptions of the categories in (a) seem simply to exclude the descriptions of the categories in (b). How can concepts of an object in general, which apply to objects, be the same as logical functions, which serve to order concepts in judgments? Moreover, the views in (a) and (b) cannot be reconciled by supposing that one of these sorts of views is later than the other, for both sorts of views are found in the B-Deduction itself as well as in other A- and B-texts and elsewhere.

Unless we are to see Kant as thoroughly confused, we must regard some of the above characterizations merely as loose or hasty formulations of his true views and others as verbally different presentations of one fundamental account of the categories. And in this connection I think we should take Kant's views in (c) – and particularly in central texts like B128, *Prolegomena*, § 21, and the *Metaphysical Foundations* footnote – as definitive. On the (c)-style understanding that I propose of a Kantian category, the categories are not *identical to* the logical functions (just as Kant in effect says at *Prolegomena*, § 21, and in the *Metaphysical Foundations* footnote). But they involve the logical functions. More specifically, they involve a phenomenon that one can see Kant as describing in the (c) texts in two different ways.[36] First, at B128 he describes this phenomenon as the determination of an empirical intuition in respect of a logical function by means of the use of the concept of an object in general. Second, and as in effect in the *Metaphysical Foundations* footnote, he describes this phenomenon as the determination of an object by means of the application of the logical functions (or forms of judgment) themselves to the intuition of the object.

A (c)-style understanding of the categories which identifies this phenomenon and relates the characterizations in (a) to those in (c) and (d) – although it eliminates the characterizations in (b) as loose or inaccurate expressions of Kant's thought – is the following. Categories are not identical to the logical functions. But – to speak roughly – they contain these functions embedded in a property in their intension, as is suggested by the later remarks under (d). They also contain in their intension – again to speak roughly – the property of being an object, as is suggested by the remarks under (a). These properties are in fact merged in that intension in a way that guarantees that when a thing is subsumed under the category, that thing is taken to be an object in the most general sense of 'object' (that is, that thing is taken to be an object in general) and that thing is taken – or various of its elements are taken – to play one of the logical roles that are specified by the logical functions. Or, in short, and to repeat the account of a category that I suggested earlier, a category is a concept of a thing such that that thing is an object in the most general sense of 'object' (that is, an object in general) and that thing is playing – or various of its elements are playing – a logical role specified by one of the logical functions.

Given this account of a category, a category evidently is – as Kant claims in the (a) texts – a concept of an object in general. (Or, as I have said in earlier chapters, a category is a realization, by means of the particular logical function that the category involves, of the overall concept of an object in general.) Moreover, and as the first (B128) description under (c) has it, the empirical intuition of a thing falling under a category is determined with respect to a logical function by means of the application, to that empirical intuition, of the above sort of concept of an object in general, or category. And, as the second (A245) description under (c) has it, a thing falling under a category is itself determined – or elements of it are determined – through the application of the logical function that the category involves to that thing (or to the features presented by the elements of its intuition).

Furthermore, the above account of a category shows that this application of the logical function proceeds, in the way suggested earlier in connection with claim (C*i*) of Section 2, by the judger's subsuming various elements of that thing – and so various features of that thing that are presented by elements of the manifold of intuition – under the logical function. ((C*i*) was the claim that *H* thinks there to be a single object that has the features *i*'s elements present, those features being structured by logical function *R*.) Therefore category application to that thing will intellectually combine these elements of the manifold (as also can be seen from (C*i*)). And so a category may further be described, as in (d), as containing the necessary unity of the synthesis of the manifold of intuition.

Finally, on the present account we can understand Kant's remarks in (b) as being loose formulations of his (c)- and (d)-view that the categories contain the logical functions and that category application to an object therefore involves the application (in the sense indicated above and in claims like (C*i*)) of the logical functions to that object. (Thus observe the *Metaphysical Foundations* quotation in (c): The categories are derived from the formal operations of the understanding in judgments and differ from those operations 'in nothing *except that* in the concept of the understanding, an object is thought as determined in regard to one or the other function of judgment'; my italics.)

Because of its ability to organize and explain Kant's various characterizations of a category (and for further reasons that emerge below), I propose to adopt the above account of a category. This account may be expressed, for further reference, as the claim that

(CT) A category is a concept of an *x* such that *x* is an object in general and there is a logical function of thought *L* such that *x* or features of *x* play the logical role specified by *L*

Given this (CT) account of a category, and if we accept the Kant's identification of concepts with features taken as general, it will indeed be true, as argued already in Section 2, that the object of *i* falls under a category. Assume, as before, that *H*'s (T*i*) thought is or is part of a knowledge-yielding judgment about the object of *i*; and let *R* be one of the logical functions that determine that judgment's logical form. Then *R* will govern the concepts c_1 and c_2 that *are* the features *F* and *G* of the entity *x* that, within *H*'s thought, *H* thinks to be an object and to have those features.[37] So *H* will think the knowledge-yielding thought that there is an object x such that x has the i_1 and i_2-presented F and G and R structures F and G. That object x of i will therefore fall under a category as characterized above – namely, under the category whose application to x requires that x or features of x should be governed by R and hence should play the logical role R specifies.

Moreover, claim (C) of B-Deduction § 20 will be correct: Because the logical functions determine the logical-form relations of the concepts in the judgment about i's object, the logical functions determine, also, the relations together of the conceptual elements of i's manifold in such a way that the object that H thinks and knows through i, as that object is known through this judgment, is category-subsumed.

5. KANT ON THE CATEGORIES (II): FURTHER DEVELOPMENT

With the establishment of category application to the object of i and the demonstration of claim (C), we reach the main goals of B-Deduction § 20. But much remains unexplained – for example, the three puzzling aspects that we noted in Section 3 about the logical-function structuring of concepts in judgments and of features in objects. Kant's views are complex, and we will not complete our explanation of their details until the end of Section 6. But to begin the explanation, I now want to revert, for the sake of concreteness, to the case discussed in Section 3. That is the case in which concepts c_1 and c_2, as they occur in the overall, knowledge-yielding judgment that H makes about i's object, are organized through the subject-predicate logical function, and features F and G thereby themselves function in the object in a subject-predicate fashion. In that case, H thinks the following version of the knowledge-yielding (C*i*) thought given in Section 2:

(SP*i*) H is conscious in thought that there is a single object x such that first i_1 puts before H's mind (or is) F and x has F and then i_2 puts before H's mind (or is) G and x has G and F is subject to G as predicate

or, formally:

H is conscious in thought that $(\exists x)$[x is an object & first i_1 pm (or is) F & x has F & then i_2 pm (or is) G & x has G & F is S to G as P]

In this case, and for reasons of a sort that we have just seen, the object x, as it is thought and known by H, falls under the category associated with the subject-predicate logical function – namely, under the category of substance (or the pure category of 'inherence and subsistence,' A80/B106). In other words, as x is thought and known by H, x is a thing such that (i) that thing is an object in general (an object in the most general sense of 'object') and (ii) that thing plays the subject-predicate logical role that is specified by the subject-predicate logical function (or features of that thing – here F and G – play that role).

Moreover, it is not just that x falls under the category of substance in the sense that x is an object in general and *features of* x play the subject-predicate logical role. As we saw in Section 3, Kant will take F, insofar as it plays the role of subject to G as predicate, actually to *have* G as predicate. Furthermore, either (a) Kant identifies F, as the subject of G, with the object x itself or else (b) Kant takes some unexplained relation to hold between F, as the subject, and that object x.[38] And in either case (a) or case (b), the result is that the object x has the feature G that is predicated of F. However, given that x thus has the subject feature F and is, further, either identified with that subject feature or else taken to have an unexplained relation to it, Kant also takes x *itself* to be a *subject* of G as predicate. Hence, and as various texts show, the object x itself falls under the category of substance not just in the sense indicated above but also in the further, straightforward sense that x is a thing such that that thing is an object in general and that thing is *itself* a subject that has various features as its predicates.[39]

We can now begin to understand the Kantian claims about general logic and the logical employment of the understanding that we introduced in Section 3 in connection with Kant's rejection of modern ideas about syntax and semantics. As we saw, Kant takes the logical form to inhere in concepts in a judgment in a way that cannot exist independently of the relation of those concepts to the object judged about. He regards the logical-function organization of concepts in a judgment as belonging to objects and features in the world as well as to the judgment that is related to those objects and features. And he holds that the full determination of the logical function of a concept in a judgment is brought about not just through the mere logical employment of the understanding but through the use of the concept of an object in general to think an object as having the features that the concept presents.

Given these points and our Section 4 discussion of category application, Kant should not suppose, in a standard modern fashion, that there are purely syntactic relations that inhere in concepts in a judgment in independence of the relation of those concepts to the object judged about. Thus take the judgment, by *H*, that is or that involves the (SP*i*) thought. In this judgment, Kant will identify the features *F* and *G* with the concepts c_1 and c_2 that present those features. And then he will not suppose that *H* takes the subject-predicate relation to belong to those concepts in independence of *H*'s thinking there to be a single object *x* to which those concepts (identified with those features) belong.

Rather, he will suppose that there is one activity that has two parts or aspects that cannot exist independently of each other: namely, the part that consists in the ordering of concepts c_1 and c_2 into the relevant logical form of judgment through the operation of the logical functions, including the subject-predicate function; and the part that consists in the relating of c_1 and c_2 to the object *x* (and to the features *F* and *G* as belonging to *x*) through the use of the concept of an object in general.[40] And it is only through an abstractive exercise of the mind that we can consider one part of this activity – say that of the subject-predicate ordering of c_1 and c_2 in the judgment – and not attend to the other part.[41]

Such an abstractive exercise of the mind is, I think, what Kant notes in the A55/B79 Transcendental Logic passage quoted in Section 3. He there says that general logic 'abstracts from all content of knowledge, that is, from all relation of knowledge to the object, and considers only the logical form in the relation of any knowledge to any other knowledge.' In other words, general logic ignores the relation of concepts like c_1 and c_2 to the object judged about – a relation that occurs in any judgment – and focuses just on the logical-function-established logical form of those concepts in that judgment.

Moreover, that logical form – and each of the specific logical functions that belong to individual concepts in judgment – is not fully determined in independence of the use of the concept of an object in general to think an object as having the features that the concepts in the judgment present. Kant does not make this idea of full determination of the logical function through the use of the concept of an object in general completely clear.[42] But he seems to claim that in general logic (which ignores that use) we consider simply the various *possible* ways in which concepts can be determinately organized, through the logical functions, into judgments of different logical forms, and the various inferential relations that obtain among those judgments. Thus note his talk, in the above quotation, of considering 'only the logical form in the relation of any knowledge to any other knowledge.' And the same point appears to occur in Kant's B128 position that, as far as the merely logical employment of the understanding goes, and in the case of the categorical, subject-predicate judgment that all bodies are divisible, 'it remains undetermined' which concept plays the (fully determined) role of subject and which concept the (fully determined) role of predicate: 'For we can also say: Something divisible is a body.'[43]

This last discussion explains, at least to an extent, many of Kant's views about general logic and the logical employment of the understanding, and it reinforces the point that he does not describe judgment in ways that fit standard modern ideas about syntax and semantics. But of course it does not explain all aspects of his views on judgment and the categories, including the three puzzling aspects that we noted in Section 3. To make that explanation (to which I come in Section 6 below) as useful as possible – and also to bring out further the nonmodern side of Kant's views – I want to note one last group of points about his treatment of category application.

To see this group of points, observe that in B-Deduction § 20 Kant remarks that the 'act of understanding by which the manifold of given representations ... is brought under *one* apperception, is the logical function of judgment' (B143, my italics). And connect this remark with three more texts: First, with Reflexion 5930:

> the objective unity of the consciousness of the manifold of representations is the connecting of the same [manifold] either with one and the same concept, e.g., All men (in a word, a universally valid combination of concepts in a consciousness), and then the unity is called logical; or this logical unity of consciousness is regarded as determined in *the* concept of a thing and constitutes its concept: that is the synthetic or transcendental unity of consciousness.[44]

Second, with the *Logik*, § 17, definition of a judgment in general:

> a judgment is the representation of the unity of consciousness of diverse representations or the representation of the relations of the same, so far as they constitute *a* concept[45]

And, third, with a passage from Kant's early interpreter Mellin, discussing the A266/B322 definition of form and matter of judgment: In the judgment that the horse is fast, Mellin says,

> horse and fast are here this matter. Both these concepts are to be combined with one another in one judgment, that means ... [that] through this combination of both concepts in judgment the relation of the same [concepts] to one another ought so to be represented to me that they constitute in my representation now only *a* concept, of the horse as a fast animal. This relation or determination of the manner in which diverse representations, here two concepts, as such, belong to *One* consciousness, or now constitute only *a* concept, is the *form* of judgment.[46]

Such texts suggest that Kant held the following view of category application. Suppose that in knowing through *i*, *H* thinks the subject-predicate (SP*i*) thought.[47] For definiteness, suppose, also, that this thought amounts to the specific judgment, by *H*, that this *F* is *G* (this tree is conical, this body is divisible, this horse is fast). Ignore, for simplicity, the quantity, quality, and modality of this judgment. Then when *H* thinks this thought and judges that this *F* is *G*, the concepts c_1 and c_2 – identified by Kant with the features *F* and *G* that they present – occur before a single act of *H*'s apperceptive thought-consciousness. (Here observe the B143 quote above.) Moreover, as c_1 and c_2 so occur, they are organized together through the application, to them, of the subject-predicate logical function. In addition, insofar as they are identified with *F* and *G* they are thought by *H* to belong to the object *x*. And now the above passages suggest that this thought-consciousness by *H* of c_1 and c_2 as being organized together in this way takes the form of a thought-consciousness, by *H*, of a *single concept* (roughly, the category of substance, the category of inherence and subsistence) as *informing* the *logical matter* that c_1 and c_2 constitute.

We, as modern readers, have been strongly influenced by Fregean and subsequent logic and philosophy of logic. For us, when we read Kant, the overall mental object of *H*'s thought-consciousness, when *H* makes the judgment that this *F* is *G*, is likely to be the propositionally expressed *fact* of the *relation* that the concepts c_1 and c_2 bear to each other (and to the object *x* that is judged about) through *H*'s use of the relevant logical functions and of the concept of an object in general. For Kant, however, the overall mental object of *H*'s thought-consciousness when *H* makes the above judgment is apparently rather different. That overall mental object is, roughly, a *single concept* (category) of an *x* such that *x* is an object in general and *x* has features that stand to one another as subject to predicate, with that single concept informing the logical matter that concepts c_1

and c_2 constitute in such a way as to yield *H* the (SP*i*) awareness. That is to say, it yields *H* the awareness of an *x* such that *x* is an object and *x* has *F* and *G* and *F* functions as subject to *G* as predicate.

Or, to be a bit more exact than I so far have (and to bring in, besides the subject-predicate logical function, the other relevant logical functions), observe that the judgment that this *F* is *G* is singular (it is about this particular *F*), assertoric (it asserts that this *F* is actually *G*), subject-predicate, and affirmative (it affirms that this *F is G*). So the overall mental object of *H*'s thought-consciousness is approximately this. It is the concept of a single, actual *x* such that *x* is an object in general and *x* has a feature that functions as subject to another, predicate feature that *x* is affirmed to have, with that single concept informing the logical matter that concepts c_1 and c_2 constitute in such a way as to yield *H* an overall awareness of the expected sort. That is to say, it will yield *H* an awareness of a single, actual *x* such that *x* is an object and *x* has the feature *F*, which functions as subject to the feature *G* that *x* is affirmed to have.[48]

To be concrete, suppose that *H* makes Mellin's judgment that the horse (represented to *H* in intuition) is fast. Then, as modern readers, we will likely suppose that when *H* makes this judgment, *H* has before thought-consciousness the *fact* that the concepts of being a horse and of being fast fall under the relational, subject-predicate logical function.[49] Given the above texts, however, it seems that, for Kant, when *H* judges that the horse is fast, *H* has before thought-consciousness the category of substance informing the concepts of being a horse and of being fast in the way indicated above. The result of this informing is that *H* has before thought-consciousness, roughly, the concept of an *x* such that *x* is an object and *x*, thought through the concept of being a horse to be a horse, functions as a subject that has the predicate feature that is thought through the concept of being fast. (Here I rely on the fact that when the feature of being a horse is taken by *H* to function as subject to the feature of being fast as predicate, *H* then takes *x* itself, which has the feature of being a horse, to function as a subject that has as its predicate the feature of being fast.)

For Kant, the mental object of *H*'s thought-consciousness in the present case is therefore not anything like a proposition, fact, or Fregean thought, considered as a complex of concepts standing in relation. Rather, that mental object is roughly the *single concept* which, by *Logik*, § 17, the concepts of being a horse and of being fast then constitute as Mellin puts it, the concept of a horse as being fast. Moreover (and to follow

Kant's identification of concepts with the features that they present), this single concept, taken with the intuition-presented conceptual matters of being a horse and of being fast that it here informs, *is* the object of knowledge in which the concept of an object in general is made determinate. For Kant, there thus does not seem to be, ultimately, the sharp distinction that modern readers would likely maintain between the *concept* of a horse as being fast (or the *horse itself*, having its feature of being fast) and the *fact* or *proposition that* the horse *is* fast.[50]

To conclude the present group of points, we are now in a position to understand Kant's claim, in a famous but famously obscure sentence at A79/B104 of the Metaphysical Deduction, that

> the same [logical] function which gives unity to the various representations *in a judgment* also gives unity to the mere synthesis of various representations *in an intuition*; and this unity, in its most general expression, we entitle the pure concept of the understanding.

To interpret this claim, return to the above case in which, in knowing through i, H makes the subject-predicate judgment that this F is G; and ignore, in this judgment, all the logical functions and associated categories except the subject-predicate function and the category of substance. Assume also the common Kantian identification of concepts c_1 and c_2 with the features F and G that they present. And recall that (as on the appearing theory) those features are put before the mind by – or else (as on the appearance theory) those features are identical to – the intuition-elements i_1 and i_2. Then we can see from the above discussion that when, in making this judgment, H through an act of mind applies the category of substance and so brings it about that that category informs the concepts c_1 and c_2 in the way indicated above, H's act of mind does two things. It makes c_1 and c_2 occur in *one* judgment for H (before H's thought-consciousness). And, also, it makes i_1 and i_2 function as *one* synthesized intuition i for H (one synthesized intuition i that represents or is the determinate phenomenal object x having feature F functioning as subject to feature G as predicate).

Moreover, and as follows from our earlier discussions, in this act of mind the same logical function that H is conscious of as applying to (and so as uniting) c_1 and c_2 as they occur in the *judgment* is a logical function that H also is conscious of as applying to (and so as uniting) the intuition-element-presented features F and G as they occur in the *object* x about which the judgment is made. Hence suppose that we ignore the unifying role that we have already seen in Chapter Eight to attach, for Kant, simply to the use of the concept of an object in general in synthesis of the manifold of *i*. And suppose, also, that we focus just on the unifying role that these last comments show to belong to the above logical function. (Or suppose that, identifying category – and so the pure concept of an object in general – with logical function,[51] we identify these two unifying roles.) Then we can understand why, as A79/B104 claims, the same subject-predicate logical function that gives unity to representations in the *judgment* that this *F* is *G* also gives unity to the synthesis of various representations in the *intuition i*.[52] And we can see why Kant calls this unity the pure concept of the understanding (here, the category of substance).

6. KANT ON THE CATEGORIES (III): ARISTOTELIAN EXPLANATIONS

The picture that we have now derived of category application and judgment is this. In a knowledge-expressing judgment by *H*, concepts are organized together through the application to them of the logical functions. Those concepts are simultaneously (in the order of logic) related to objects and object-features through *H*'s thought, by means of the concept of an object in general, that there is an object *x* that has the features that the concepts present (or are). Kant, indeed, commonly identifies the concepts with those features, taken as general; and he takes the features, as they occur in the object *x* that the judgment is about, to be organized together through the same logical functions that organize the concepts in the judgment. It follows that the object satisfies the condition of being a thing such that that thing is an object and that thing plays (or various of its features play) the logical roles specified by those logical functions. And so it follows that the object falls under the categories associated with those logical functions.

More specifically, when concepts are organized so as to function as subject and predicate in a judgment, the features that those concepts *are* are then organized so as to function as subject and predicate in the object judged about. This fact means that the one feature (say that of body) functions as the subject that *has* the other feature (say that of divisibility) as its predicate. Kant then either identifies this subject-feature with the object itself or else takes there to be some unexplained relation that holds between the subject-feature and the object. In either case the result is that the object itself functions as a subject (the object is a body) that has the predicate feature as its predicate (the body has the property of being divisible).

One can raise various questions about this account of the categories and judgment, and there are still other characteristics of Kant's account (having to do with the role of necessity in the determination of the manifold of intuition) that we have not yet discussed in detail. But before we turn to these matters, we should ask why Kant adopts just the above treatment of the categories. In particular, why does he accept the three puzzling aspects that we noted in Section 3 as belonging to his views about the logical structuring of concepts in judgments and of features in objects? Why, without arguing the matter, does he proceed as though it is natural (i) to identify concepts with features taken as general, (ii) to treat features of objects as subjects that have other features as predicates and (iii) to give no standard, modern sort of treatment of a judgment's syntax and semantics? Furthermore, and to add a point that we noted above in Section 5, (iv) why is he not uneasy about treating a judgment in such a way that there seems to be no sharp distinction between the concept of an object as having features (or the object itself, having those features) and the fact or proposition that the object has those features?

The fine details of Kant's account of the categories surely have many sources. And Kant's deepest and most fundamental ideas – for example, his central view of the logical organization of concepts in judgment as requiring a comparable organization of features in objects – are, as far as I know, wholly original to him. But the basic answer to the particular questions above is, I think, that in presenting his fundamental ideas about category application Kant relies, for various details of those ideas, on a version of the traditional Aristotelian theory of logic and judgment. And he accepts this version without argument and without ever really considering that it could have any genuine alternatives.[53]

A careful development of the exact relations of Kant's views to Aristotelian views on logic and judgment (and to the Aristotelian or quasi-Aristotelian views of his contemporaries) would require much space and scholarly delving. Also in important ways Kant is not an Aristotelian. For example, his representationalism, which comes to him through earlier Cartesian philosophy (broadly conceived, as in this book, to include the work of both Descartes and empiricist representationalists like Locke), is not, as far as I can see, matched by anything in Aristotle.[54] Nor is his idealism. But in many ways Kant seems clearly to rely on Aristotle's views. And in this connection the facts that we need to note in order to defend the preceding answer to our questions can be set out in a simple, straightforward way. Thus while Kant thought that he could make local improvements to the logic of his time (as he argues in his early "False Subtlety of the Four Syllogistic Figures"), his basically unhesitating acceptance of Aristotelian logic itself is well-displayed in famous sentences at Bviii of the Preface to the second edition of the first *Critique*:

> That logic has already, from the earliest times, proceeded upon this sure path [of a genuine science] is evidenced by the fact that since Aristotle it has not required to retrace a single step, unless, indeed, we care to count as improvements the removal of certain needless subtleties or the clearer exposition of its recognized teachings ... It is remarkable also that to the present day this logic has not been able to advance a single step, and is thus to all appearance a closed and completed body of doctrine.

And various salient points about Aristotle's position can be suggested to underlie the puzzling aspects of Kant's views that we have questioned above.

To restrict attention just to the bare bones of Aristotle's claims – and wholly to ignore many subtleties and complications – there are two main points that we should note here. First, Aristotle attributes a certain subject-predicate, or substance-attribute, ontological structure to the individual, particular objects that we know – those objects (like individual sticks, stones, tables, and human beings) that he calls primary substances. Second, Aristotle supposes that when we know these individual, particular objects, that ontological structure is reflected in the mind, in the structure of the judgments that we make about these objects.

It will be helpful to be more specific about these points than I just have. First, Aristotle regards an individual, particular object, or a primary substance, as a 'this such,' a group of forms or general features (for example, the forms of being human and of being seated) inhering in the matter that belongs to that primary substance. To speak roughly, Aristotle takes these forms themselves to be organized in a subject-predicate (or in a substance-attribute) way as they inhere in that matter. Each primary substance has a form that is that primary substance's essential nature. Roughly, as it inheres in the matter of the object, that essential nature (which is a so-called Aristotelian secondary substance) functions as a subject that has predicated of itself the other forms that inhere in the matter.[55] Thus, for example, the primary substance Socrates has as its essential nature the form of being human (or of being a man). As that form inheres in the matter of Socrates, it functions as a subject that has predicated of itself the various other forms or features – for instance, the form of being seated or of being wise – that inhere in that matter. Moreover, Aristotle holds that insofar as the form of being human is such a subject inhering in that matter, the primary substance Socrates can himself be said to be human (to be such a subject) and to have the forms, like the form of being seated, that are predicated of the form of being human.[56]

To come to the second point about Aristotle's views, suppose that we know the primary substance Socrates, say through the judgment that this human being (the object that we see before us) is seated. Then through sense perception our mind 'receives' or takes on various of the forms belonging to that primary substance, in the present case the forms of being human and of being seated.[57] According to Aristotle, these forms are received in – and so they occur in – the mind without that primary substance's matter itself occurring in the mind. As they occur in the mind, these forms are described by Aristotle variously as 'affections of the soul,' as 'likenesses' of actual things, and as 'images.'[58] Moreover, he supposes that, as they so occur in the mind, the forms are combined by the intellect.[59] When they are so combined (and apparently insofar as they are contemplated mentally[60]), the forms then constitute the unitary thought or judgment that this human being is seated. This judgment is true just in case the combination that the forms have in the mind corresponds to the combination that they have in reality in Socrates himself.[61]

Aristotle does not say explicitly, it seems, that the combination that the forms have in the mind in a true judgment is or involves exactly the same sort of subject-predicate combination that those forms have in reality in Socrates. But since, for Aristotle, in a true judgment the same forms occur in reality and in the mind, it is natural to take his theory to be committed to such a result (or at least to the existence of some sort of strict analogy or structural parallel between the type of combination that the forms have in the mind and the type of combination that they have in Socrates).[62] Moreover, it seems that in taking a judgment (or thought) to amount to a unitary combination of forms in the mind, Aristotle may not always sharply distinguish two things that we modern readers would want distinguished: namely, (a) an object or a form as *having* a certain form (Socrates or the form of being human as having the form of being seated) and (b) the fact or proposition *that* that object or form has that form (the fact that Socrates or the form of being human has the form of being seated).[63]

I do not claim that Aristotle's views of judgment and logic are completely clear or without philosophical problems. (Thus while very interesting ideas are suggested by his position that forms occur, without their matter, in the mind, this position, when taken literally, is very puzzling.[64]) Moreover, as I noted above, there are many differences between Aristotle and Kant. Nevertheless, in various Aristotelian claims we surely have points that are accepted by Kant apparently without conscious argument and then are reflected in the puzzling aspects that we noted above about his views.

This last point is, I think, clear – or at least very plausible – in the case of Kant's treatment of concepts as identical to the features that they present. Kant surely takes this treatment over from Aristotle or from contemporary Aristotelian theories and then proceeds, without qualms, in its terms. (And one would expect Kant's alternative treatment of concepts as presenting features from which those concepts are distinct to result from his representationalist, Cartesian background.)

Again, the last point also is clear – or at least very plausible – when one turns to Kant's otherwise highly puzzling treatment of features, in objects, as being subjects to which other features function as predicates. Such a treatment is just what one would expect if Kant thinks of features in objects as like Aristotelian forms, one of these features being the essential nature of the object – the subject or substance feature – to which the other features then belong as predicates.[65] And given the Aristotelian background, we also can understand why, in developing this treatment, Kant appears either to identify or else to assume an unexplained relation to hold between the subject-feature in the object and the object itself. The explanation will be that, accepting Aristotelian theory, Kant does not literally take the object (say, this stone) *to be* one and the same thing as its subject-feature (say, the feature of being a body). Rather, and without supposing that he needs to give any explanation for doing so, he proceeds as though the subject-feature is the essential nature of the object (for him, the permanent substance of the object) which itself belongs to the object and to which the other features of the object belong as predicates.[66]

The point that I have just made about the presence of Aristotelian claims in Kant's views also casts light on the fact that Kant gives no standard, modern sort of treatment of the syntax and semantics of a judgment. It does not appear that Aristotle's view of judgment itself admits a clear such distinction in any obvious way.[67] Thus for Aristotle the forms that are combined to make up thoughts or judgments are forms that from the start are conceived of as belonging to objects (rather than being conceived of, for example, as uninterpreted abstract entities which must be related, semantically, to objects). When they are combined in the mind, these forms thus do not have a purely syntactic combination that exists in them in independence of their relations to objects and features in the world. Indeed, and as I noted, Aristotle's theory seems committed to the view that the combination that forms have in the mind is the same as – or is in some way strictly analogous or structurally parallel to – the combination of those forms in reality.

Now the Aristotelian view of forms as combined in the mind into unitary thoughts or judgments is of course not worked out in exactly the same way as is Kant's view of the logical-function structuring of concepts (or features) in judgments. (Thus Kant gives a detailed account, in a way that Aristotle does not, of twelve logical functions that govern the combination of concepts in judgments. And various of Kant's logical functions, like those that operate in hypothetical, disjunctive, and infinite judgments, either correspond to nothing in Aristotelian logic or else correspond to nothing that Aristotle would regard as a way of combining forms in objects.) Nevertheless, there are obvious structural parallels between Aristotle's treatment of forms as combined in the mind into unitary thoughts and Kant's treatment of concepts as combined, through the logical functions, into judgments. The fact that (as it seems) Aristotle's view of thought and judgment does not admit of a clear syntax-semantics distinction then is itself paralleled by the similar fact about Kant's treatment of judgment. In addition, we have already seen strong reasons to suppose that Kant's treatment, insofar as it concerns the combination of concepts and features in judgments and objects, is based on Aristotelian ideas. It is therefore not surprising (however philosophically disconcerting modern readers may find it) that Kant's treatment does not show a real awareness of standard modern ideas about syntax and semantics. (We have already seen in Section 5 that Kant's account of general logic and of the logical employment of the understanding – however modern it may perhaps initially appear – does not proceed in terms of standard modern ideas.)

Finally, my above point about the presence of Aristotelian claims in Kant's views suggests at least one reason why Kant would not be uneasy about treating a judgment in such a way that no sharp distinction exists between the concept of an object as having features (or the object itself, having those features) and the fact that the object has those features. Aristotle of course accepts nothing like the idea that leads to this treatment of judgment – namely, the idea of a judgment, as it occurs before thought-consciousness, as an informing of concepts by the various categories that are specified by all the logical functions operating in the judgment. But, as we noted, within his own theory Aristotle himself may not sharply distinguish between an object (or a form) as having a certain form and the fact or proposition that that object has that form. To the extent that Kant's treatment of judgment is analogous to Aristotle's view of judgment and partly derives from it, Kant would not be uneasy about the fact that his own treatment takes there to be no sharp distinction of the sort first noted above, particularly if he accepts this area of Aristotle's theory as a matter of course and has his attention fixed elsewhere.

The puzzling aspects that we have noted in Kant's treatment of judgment thus seem to have a basis in Aristotle's views on judgment and logic. Kant accepts a version of these views and does not consider that there could be any genuine alternative to them.

In concluding the present discussion, I should note that if we step back from the details of Kant's and Aristotle's theories, we can, in an instructive manner, see Kant's overall treatment of judgment as in a certain sense generalizing and reversing Aristotle's account. To speak broadly, Aristotle begins (in the order of logic) with the ontological structure of individual, particular, mind-independent sensible objects.[68] These objects are roughly subject-predicate-organized forms inhering in the matter of those forms. Given this account of objects, Aristotle then moves to the case of judgment by indicating how those forms and their organization in the object are reflected in the mind and are organized into unitary judgments in such a way as to yield knowledge of the objects.

In contrast to Aristotle, Kant begins with the logical structure of judgment. Working with the idea of such a structure, he generalizes – or widens – the specific Aristotelian view of a subject-predicate logical organization of forms in thought and judgment, so that he recognizes explicitly (in a way that Aristotle does not) at least some non-subject-predicate judgments – in particular, hypothetical and disjunctive judgments.[69] Proceeding in this way, Kant arrives at twelve logical functions of judgment or ways that concepts (or further judgments) can be logically structured in judgment. Given this account of judgment, he then moves (reversing Aristotle's order) to the case of the individual, particular, sensible objects that we judge about and know. He argues that the logical functions, because of the way that they operate in judgment and in the synthesis of the manifold of intuition, have to be reflected in the subject-predicate and eleven other logical-function structures of those objects.

The result of Kant's way of proceeding is thus a view of objects as category-structured that differs significantly from Aristotle's view of objects and categories. Aristotle himself of course introduces a list of categories in the course of his discussion. For Aristotle, categories are, roughly, concepts – like those of substance, quantity, quality, relation, place, date, and so on – that specify the different basic ways that reality (as made up of sensible objects and other entities related to sensible objects) exists.[70] And for Aristotle the nature of these categories (and his own list of categories) is not, it seems, determined by the logical structure of the judgments that we can make about sensible objects in reality, although it may bear indirect relations to that logical structure. For Kant, as we have seen, categories are the various logical-function-involving concepts that specify the different basic ontological structures that belong to sensible objects. And for Kant, in contrast to Aristotle, the nature of a category (and Kant's specific list of categories) is directly determined by the logical structures of the various kinds of judgments that we can make about those objects.

7. EVALUATIONS. THE NECESSITY OF CATEGORY APPLICATION

The preceding discussion has shown in some detail what a Kantian category is, how Kant derives the categories from the logical functions of judgment, why he takes the object of sensible intuition in general *i* to be category-structured, and why his account of judgment and category application takes the specific and sometimes puzzling form that it does. But in explaining that account, we have not considered how far it works. In particular, neither in this nor in the previous chapter have we asked, first, whether Kant's accounts of judgment, of the specific categories, and of the logical form of judgment are correct or even plausible. Second, we have not asked whether one can correctly infer, in a Kantian fashion, philosophically significant conclusions about the categorial structuring of the objects of our knowledge from the fact that the judgments that we make about those objects have the types of logical structure that they do.

The first question does not itself require a great deal of discussion here. Kant's version of Aristotelian logic has turned out to be inadequate.[71] Moreover, many of his own Metaphysical Deduction derivations of individual categories from the logical functions that he accepts are highly unconvincing. (For example, observe his notorious derivation of the category of community from the function or form of a disjunctive judgment, or his attempt to connect the logical functions of quality with the categories of Reality, Negation, and Limitation.) We need not rehearse the details of these problems now or note what might be said to try to mitigate them. But we should observe immediately that, as the discussion in this chapter and in the previous chapter has shown, in Kant's own Transcendental Deduction argument ideas from the Metaphysical Deduction and from his version of Aristotelian logic play a very important role. So, and contrary to what I think is the opinion of at least some Kant scholars, the Transcendental Deduction as Kant presents it cannot succeed in proving anything like *Kantian* (and so logical-function-determined) category application if these ideas fail. However, given the problems just mentioned, these ideas do fail. So, whatever the prospects for defending reasoning in the spirit of the Transcendental Deduction, the actual Transcendental Deduction as Kant presents it cannot succeed in proving the application of any concept that he himself would call a category.

These last comments bring us to the second question above. If he is allowed to use them, conceivably Kant's Metaphysical Deduction and Aristotelian-logic ideas may allow him successfully to infer philosophically significant conclusions about category application from facts about the logical structure of judgments. (The success of this specifically Kantian inference is a second-question issue to which we return shortly.) But, as just noted, those ideas are inadequate. So, when we focus on the second question, it is natural to ask at once whether such an inference can succeed in the case in which we consider, instead of Kant's own ideas about logical structure, the sorts of logical structure that are introduced in modern logics.

This case raises large issues. But I do not think that it provides a great deal of hope that such an inference can succeed. As readers know, there are a variety of modern logics, including, besides standard first-order predicate logic, a number of alternative logics (and various extensions and subtle modifications of first-order logic). Moreover, all these logics can be formulated in a number of different ways. First-order logic itself (with its extensions) has a good call to be considered the basic logic by which we usually reason. But in the case of that logic, I do not think that one can draw inferences successfully from the fact that judgments – or, more accurately, sentences – have specific first-order structures to any philosophically very interesting conclusions about the objects to which those sentences (and their contained constants, variables, predicates, and quantifiers) are semantically related. (Thus the mere fact that a sentence is a first-order universal quantification imposes no restrictions on the nature of the objects in the domain of discourse.[72])

I am less clear about the case of alternative logics. To speak loosely, some such logics do not seem to support such inferences. But others may support – or at least may make natural – certain general ways (say non-realist or anti-realist) of viewing the objects to which sentences formulated in terms of those logics are related. Nevertheless, and aside from the fact that such general ways do not seem to yield anything like a Kantian categorial structuring of objects, these latter sorts of alternative logics are not the basic logic – the modern equivalent of Aristotelian logic – in terms of which we usually reason. And so the existence (if it is a fact) of such general ways would not seem to lend great support to the Kantian idea that, appealing to such a basic logic, we can infer philosophically interesting conclusions about the ontological structure of objects from facts merely about the logical structure of our judgments about those objects. Of course some future investigator might nevertheless introduce a new standard idea of logical structure that would allow such conclusions to be inferred. But at the moment I do not think that the prospects are good for this sort of inference.[73]

Given the inadequacies of Kant's version of Aristotelian logic and the inability of current ideas from modern logic to support such inferences, the specific Transcendental Deduction project of deducing a categorial structure of objects from features of the pure logical structure of judgment must be accounted unsuccessful. This fact does not mean that attempts, both by Kant and by later philosophers, to carry out such a project have not been deeply illuminating. But it does mean that, if they are to stand any chance of success, arguments for a categorial structuring of objects must appeal not just to the purely logical features of sentences (or thoughts and judgments) but also to general facts about us as language users (and as thinkers) and about our relations to the situations in which we speak, think, and attempt to know. The development of such arguments is, indeed, a familiar recent way of attempting to defend something in the spirit of Kant's conclusions about the categories. And, whether or not such arguments can themselves succeed (none that I have seen fully convinces me), the challenge of constructing and evaluating them is still open.

Even if sound and interesting arguments of the sort just mentioned can be given, their production is not a goal of this book. To evaluate the success of the Deduction as Kant presents it – and in fairness to Kant's own intellectual situation – we need, however, to return to a point noted above at the start of our discussion of the second question. We must ask (as our final inquiry into that question) whether, *if* one accepts Kant's own Metaphysical Deduction and Aristotelian-logic ideas, one can at least *then* infer philosophically significant results about category application from facts about the logical structure of judgment. If one can, then the Deduction from the point of view of Kant's own historical situation would be, in this respect, a brilliant local success, destroyed, unfortunately, by the subsequent failure (which he could not anticipate) of his starting assumptions.

This final question returns us to the specific B-Deduction § 20 reasoning about category application to the object of *i*. The situation in regard to this question is as follows. Given our Section 3 reasoning (and accepting (T*i*)), Kant establishes a logical-function organization of the elements of *i* and of the elements of any other intuitions involved in the overall judgment of which *H*'s (T*i*) thought is or is a part.[74] (In that thought, recall, *H* takes there to be a single object *x* that has the *i*-element-presented features *F* and *G*.) Kant also shows that the object of *i* falls under the concept of being an object that is – or whose features are – logical-function-structured. And so he shows that *i*'s object falls under a category. Indeed, this logical-function structuring involves all the logical functions that determine the logical form of the overall judgment in question, and so he demonstrates, at least, that that object falls under all the categories associated with those logical functions.

As we have imagined in the (SP*i*) case above, this logical-function structuring may be simply a subject-predicate structuring of *i*'s own elements. (In that case, *H* thinks *F* and *G*, as they belong to *x*, to function as subject and predicate.) But if the overall judgment is hypothetical, this structuring may also be a hypothetical, if-then or ground-consequence structuring that implies that, say, if *i*'s object has the i_1-presented feature *F* (which it does), then some other intuition-represented object has some feature *Z*. Or if the overall judgment is disjunctive, this structuring may also be a disjunctive structuring that implies that, say, either *i*'s object has the i_1-presented *F* or some other object has some feature *W*. And of course this structuring also involves logical functions of quality and of modality.[75]

However, even without working out the further details of this sort of category application, which can be done by considering various ways in which *H*'s (T*i*) thought can be part of the overall judgment in question, we can see that it is subject to two objections that cast serious doubt on the philosophical significance of what (given our Section 3 reasoning and accepting (T*i*)) Kant has shown. First – and to allow ourselves once more to criticize the Metaphysical Deduction – it seems clear that various of these logical-function structurings of objects do not lead to anything like the specific cases of category application that Kant takes them to. He has, indeed, a plausible transition (although see comments below) from structuring by means of the subject-predicate logical function to the application of the Kantian category of substance. (It was for this reason, along with the Aristotelian antecedents of that logical function, that I have focused on the case of that logical function.)

However – and to ignore all the obscurities about Kant's categories of quality – it simply is not true that use of a hypothetical judgment automatically commits one to any sort of genuine cause-effect relation between a substance (or state of a substance) referred to in the antecedent of that judgment and some substance (or state of a substance) referred to in the consequent – or even to any other genuine cause-effect or Kantian pure ground-consequent relation. (Thus 'if 5 is divisible by 8 then 5 is divisible by 4' carries no commitment to any causal relation between substances. Or, consider the claim – clearly involving reference to phenomenal objects and events, and introducing time specifications – that 'if Mary became six feet tall just before Bill became three feet tall, then Mary at that time became over twice as tall as Bill.' This claim turns on a semantic or logical implication and does not commit one to any causal relation whatsoever between what is referred to in its antecedent and what is referred to in its consequent. And in claims like 'if there's smoke, there's fire' or 'if Smith is in New York, then he's visiting his grandmother,' the situation introduced by the antecedent is certainly not standardly meant as any sort of *ratio essendi* for the situation introduced by the consequent.[76]) Nor is it at all clear how, through the use of a disjunctive judgment (and, by A73/B99 and A75/B100, this is *exclusive* disjunction!), one is automatically committed to a relation of causal community between substances referred to in the disjuncts of that judgment.

Second, even if this last objection is accepted, one might argue that Kant at least shows category application in a general, logical-function sense whether or not he succeeds in showing the application of the twelve specific categories listed at A80/B106. (For example, even if Kant does not show the application of the specific category of cause and effect, one might argue that he at least shows the application of what we could call the logical antecedent-consequent category, the category that applies just when the having of a feature by one object functions as logical antecedent to the having of some feature by some object as a logical consequent.) However – and here we reach the second objection that I have mentioned – the philosophical significance of this sort of logical-function category application seems open to doubt. Suppose that it is shown that, say, the having of *F* by *i*'s object functions as an antecedent to the having of *Z* by some other object as consequent. Then this fact (and so the application of the general 'logical antecedent-consequent' category) hardly seems ontologically exciting. After all, in some other judgment the having of *F* by *i*'s object may function as consequent to some other object's having of some feature as antecedent. (And similarly, and even more obviously, for the case of a disjunctive judgment.) The fact that a feature (or the having of a feature) can be both an antecedent and a consequent (or a disjunct and a disjunct) – or can be, depending on the judgment, modally possible, actual, or necessary – hardly seems a result that an anti-Humean critical philosopher should strive to demonstrate.

Moreover, the same point holds for our favored example of subject-predicate judgments. Suppose that the feature of body, in the judgment that this body is divisible (or, say, the feature *F*, in the (SP*i*) judgment), functions as subject to the feature of divisibility as predicate. Nevertheless, in some further judgment the feature of body will function as predicate to some other feature as subject. Thus in the judgment that this divisible thing is a body, which follows logically from the judgment that this body is divisible, the feature of body functions in that way. (In fact, the feature of body here functions as predicate to the very feature to which it previously functioned as subject.) These points hardly show that the feature of body (or the object *x* itself, in the case of the (SP*i*) judgment, as having the feature *F*) is a substance in any philosophically interesting sense. For example, they do not show that that feature (or the object *x* itself) is a substance in the traditional sense – among others – that that feature is a subject that cannot be a (mere) predicate. And there is not a great deal of philosophical interest in the idea that a feature (or object) can be *both* a subject *and* a predicate. (Moreover, and to revert to the first objection, this present point shows also that even the subject-predicate logical function does not really yield the actual category of substance, in the traditional sense just indicated.) But then, given this objection and the first one, the philosophical significance of Kant's B-Deduction § 20 reasoning even from the point of view of his own historical situation is open to serious doubt.[77]

Although I think that it is convincing, this second objection is of a sort which Kant considers. He would believe himself to have an adequate response in terms of the necessity of category application. Mention of this necessity – and of Kant's response – brings us to the last topic that we will consider about the specific details of that application.

As we saw in Chapter Eight, Kant takes a claim like (T*i*), in which the concept of an object in general is applied in synthesizing the manifold of *i*, to involve necessity. It involves necessity in the necessity-of-the-consequent sense that it is necessary that *if H* knows through *i* (and if other Kantian points hold), *then H* thinks the (T*i*) thought. Or else it involves necessity in the direct, categorical sense that it is necessary that *H* should think that thought.[78] It follows that, for Kant, a category-applying claim like (SP*i*) involves the same sort of necessity. Specifically, and in the case of the subject-predicate (SP*i*), it will be true that

> It is necessary that [if *H* knows through *i* (and so on), then *H* is conscious in thought that there is an *x* such that (*x* is an object and first i_1 puts before *H*'s mind (or is) *F* and *x* has *F* and then i_2 puts before *H*'s mind (or is) *G* and *x* has *G* and *F* is subject to *G* as predicate)]

Or else it will be true that

> It is necessary that [*H* is conscious in thought that there is an *x* such that (*x* is an object and first i_1 puts before *H*'s mind (or is) *F* and *x* has *F* and the i_2 puts before *H*'s mind (or is) *G* and *x* has *G* and *F* is subject to *G* as predicate)]

Now of course, given the discussion in Chapter Eight, Kant cannot establish either of these necessity claims. But the issue here is not the adequacy of Kant's grounds for the claims but what he can do with them assuming that he has established them. And what he can do, in connection with the second objection, is clear. Suppose that either claim is true. Then when *H* thinks the (SP*i*) thought and so applies to the object *x* of *i* the Kantian category of substance, *H* does not think that *x*'s feature *F* just *happens* to function as subject to *x*'s feature *G* as predicate. Rather, and in one of the two above ways, it is *necessary* that, as *H* thinks the (SP*i*) thought, *H* should think that *F* is subject to *G* as predicate. And so, it may seem, Kant has answered the second objection. In knowing through *i* and in thinking *x*'s feature *F* to be a subject, *H* has necessarily to think that feature to be a subject; and hence (it may seem) *H* cannot also think of that feature as (merely) a predicate.

Evidently the same sort of argument will show that if, in knowing through *i*, *H* thinks *x*'s having of *F* to function as an antecedent to the having of *Z* by some other object as a consequent, then *H has* to think of *x*'s having of *F* in this way. And similarly for the case of disjunctive judgments and of the other logical functions. So (at least given Kant's Metaphysical Deduction and Aristotelian-logic ideas) the application of the Kantian categories, as derived from the logical functions, *is* philosophically interesting. These are applications which, in one of the two above senses, it is *necessary* for *H* to think to occur, and which – it may seem – *H* cannot think to occur (merely) otherwise.

Kant gives just this last sort of reasoning, or something very close to it, in the B128-29 text quoted in Sections 3 and 5. When we make, for example, the subject-predicate judgment that all bodies are divisible, then, Kant says,

> as regards the merely logical employment of the understanding, it remains undetermined to which of the two concepts the function of subject, and to which the function of predicate, is to be assigned. For we can also say: Something divisible is a body.

However, he holds,

> when the concept of body is brought under the concept of substance, it is thereby determined that its empirical intuition in experience *must* always be considered as subject and never as mere predicate. Similarly with all the other categories. [My italics]

Furthermore, he makes related comments in the *Metaphysical Foundation* footnote quoted in Section 3, in the "Metaphysik Volckmann" text quoted in Chapter Three, and elsewhere.[79]

Kant's B128–29 claims seem to combine reasoning of the sort that I have just suggested with the view of the logical employment of the understanding that we developed in Section 5. In that logical employment, as we saw, we abstract from the relation of concepts (in a judgment) to objects that is established by means of the concept of an object in general. And we consider just the (not yet fully determined) logical functions of the concepts in that judgment and the various inferential relations that obtain between that judgment and other judgments.

Now the idea that we consider such inferential relations seems to underlie Kant's B129 point that when we judge that all bodies are divisible, we can also say or judge that something divisible is a body. (Note the similar example in the *Metaphysical Foundations* footnote. Kant will take the judgment that something divisible is a body to follow by an immediate inference from the judgment that all bodies are divisible.[80]) So his overall B128–29 claim seems to run as follows. Merely to take the concept of body to function as subject in a categorical judgment, as we may in the logical employment of the understanding, is not yet to make the logical role of that concept in our knowledge the categorical role of a subject that can never be a (mere) predicate. Besides taking the concept of body as a subject, we are also committed, because of the immediate inference in question, to taking that concept as a predicate. However, suppose that we take the judgment that all bodies are divisible to express knowledge of objects. In that case, we relate the concept or feature of body to objects through our use of the concept of an object in general to think objects as having that concept or feature as subject. As we do so, 'it is thereby determined that [the] empirical intuition [of the concept of body – namely, the feature of body] *must* always be considered as subject and never as mere predicate' (italics mine).

The above reasoning by Kant is ingenious and shows again the deep importance to him of the necessity of category application. Unfortunately, however, this reasoning does not succeed, and so our second objection remains unanswered. There are various problems with this reasoning; two stand out.

First, and as recalled above, Kant has not established (and it does not seem that he can establish) the necessity that is supposed to be involved in a claim like (T*i*) or (ST*i*). So he is not in a position to assert either the necessity-of-the-consequent necessity claim or the direct, categorical necessity claim that underlies the above reasoning.

Second, even if such necessity claims were true, the above reasoning would be unsuccessful. Consider again the judgment that all bodies are divisible. From this judgment there follows, given Kant's logic, the further judgment that something divisible is a body. Hence if I know the truth of the judgment that all bodies are divisible and know even a modicum of logic, I know also the truth of the judgment that something divisible is a body. However, and by reasoning exactly like that above, when I know the truth of the latter judgment, it is necessary (in one of the two above ways) that I think that the feature or concept of being divisible functions as subject to which the feature or concept of body functions as predicate. So it is now a fact *both* that I must think the feature or concept of body as subject *and* that I must think that feature as predicate. (And this fact holds even though I cannot, given what I must think, think of the feature of body as merely a predicate.) Moreover, this sort of fact holds even when we ignore inferential relations between judgments. In the totality of my empirical knowledge or experience – here compare texts like A110 – there will very likely be judgments like the judgment that the hazy thing over there is a body as well as the judgment that all bodies are divisible. So once more I will have to think the concept or feature of body both as subject and as predicate. Therefore Kant's appeal to the necessity of category application does not allow us to assign a special subject-role to the concept of body but not to the concept of divisibility. Consequently we are returned to the second objection.

The second objection, like the first objection, thus stands. Hence even given his Metaphysical Deduction and Aristotelian-logic ideas, Kant cannot infer philosophically significant results about category application from facts merely about the logical structure (in Kant's sense) of judgment.

The present part of the Transcendental Deduction therefore cannot be defended as a success even within its own historical framework, a success unfortunately vitiated by the inadequacy of the preceding ideas. Nor, for reasons suggested earlier, can this part of the Transcendental Deduction be defended simply by replacing those ideas with ideas from modern logic. Rather, and as indicated above, the interest of this part of the Deduction is more indirect. It lies in the question, whose ramifications are still being explored today, of whether from general facts about us as language users and thinkers – and about our relations to the situations in which we speak, think, and attempt to know – we can draw important conclusions concerning the structure of objects and the world.

8. FINAL ISSUES

The discussion in this chapter and in the preceding chapters has left us with six issues that we should consider before we turn to our ultimate conclusions about the significance of Kant's attempt to deduce the categories of the understanding: first, the problem, raised in Chapter Nine, of how Kant can show the application, to *i*'s object, of *all* the categories; second, the question, observed at the end of Chapter Eight, of how he can prove category application to the subjective sequence of i_1 and i_2, given that i_1 and i_2 (as known by *H* to be subject to unity of apperception) are themselves objects of *H*'s knowledge; third, an issue, connected to a matter raised in Chapter One, about category application to inferred objects; fourth, the problem, noted in Chapters Eight and Nine, of how Kant can show that the object known through *i* – or any other relevant object – is distinct from all of *H*'s intuition-elements or other representations, taken separately or in sequence; fifth, the question of how category application proceeds in the case of the spatial manifold of intuition (and how quality and intensive magnitude enter into category application); and, sixth, the issue of how Kant's idealism and his treatment of combination affect the Deduction's view of the logical functions and the categories. (In connection with the sixth issue I will recall, as well, Kant's problematic treatment of things in themselves.) I will consider these issues in order below.

As indicated in Chapter Nine, the first issue is simple. By showing merely that *H* thinks and knows the object of *i* through the (T*i*) thought (if he could show that result), Kant at best shows the application to that object of just the categories associated with that thought. So the problem arises of showing the Deduction's claimed result that all the categories apply to *i*'s object.[81] Now if Kant could show that in knowing through *i*, *H* makes an overall judgment that involves but goes beyond the (T*i*) thought, then he could at least show the application of the further categories that are associated with that overall judgment. So he would come closer than before to showing the result just noted.

However, there are serious difficulties lurking here. For one thing, and as we noted in Chapter Nine, Kant has not proven that *H* makes any such overall judgment (let alone the specific (T*i*) judgment). Nor has he argued that if *H* makes such a judgment, that judgment expresses knowledge of *i*'s object and so yields the application of the relevant categories to *i*'s object as that object is *known* by *H*. For another thing, even if it could be shown that *H* makes such an overall, knowledge-yielding judgment, it still would not be shown that that judgment involves *all* the logical functions and so yields the application of *all* the categories to *i*'s object.

As our earlier discussions of intensionality indicate, it does not follow that if *H* thinks (and judges) that *p* and if *p* implies *q*, then *H* thinks (and judges) that *q*. So we cannot escape the present problem by arguing that the thought-content in (T*i*) implies thought-contents involving all of the other logical functions (even trivial thought-contents like the claim that if the object *x* is *F*, then the object *x* is *F*). Nor do I see any other way of resolving this problem within the confines of the Deduction itself, for I see no plausible argument that the holding of unity of apperception with respect to *i* requires *H* to judge in a knowledge-yielding way that brings in all the categories.[82] Thus the Deduction does not succeed in showing more than that at least some of the categories identified in the Metaphysical Deduction apply to *i*'s object. (Of course if they were successful the proofs of the individual schematized categories in the Principles would allow Kant eventually to escape the present problem. But those proofs are not at issue here.)

Like the first issue, the second issue raises a problem that I do not think can be solved within the confines of the Deduction itself. As we urged in Chapter Eight, intuition-elements (or other mental states or representations) like i_1 and i_2 are surely entities that, as they occur before *H*'s thought-consciousness, are *known* by *H* to be accompanied by the *I think*. So, given Kant's claim to show, in the Deduction, category application to all the objects that a being like us does or can know, Kant should be able to show, in the Deduction, category application to i_1 and i_2.

The Deduction proceeds, however, by demonstrating that a necessary condition of unity of apperception's holding with respect to a given intuition is that the categories should apply to the object of that intuition, that object being distinct from the elements of that intuition taken separately or in sequence. This argument clearly does not allow us to show category application to that intuition itself or to its elements. (Indeed, given Kant's extreme position on judgment at the end of B-Deduction § 19 – a position we saw in Chapter Nine that Kant is not forced to adopt – all judgments and so all knowledge and category application concern objects distinct from any of *H*'s representations whatsoever, taken separately or in sequence.) Nor do I see, in the Deduction itself, any way of showing that i_1 and i_2 must be objects known through some other unity-of-apperception-governed intuitions and so must be category subsumed.[83] Hence any attempt to show category application to subjective sequences of our own representations must go outside the Deduction. At this point one would think of appealing to the Analytic of Principles (for example, to the First and Second Analogies and to some of the considerations in the B-edition Refutation of Idealism). But here we pass beyond the confines of the Transcendental Deduction. (And we also reach issues about self-knowledge that, as far as I can tell, Kant never completely integrates with his Transcendental Analytic claims about the scope and limits of category application.[84])

The third issue arises because, as follows from points made in Chapter

One and later, the Deduction proof of category application and our own discussion have focused on the case in which unity of apperception holds with respect to *i*. In that case it is argued that *H* must know, in a *de re*-like fashion, a single, individuated, category-subsumed object. Suppose, however, that *H* simply knows, by inference, the *de dicto* claim that all bodies are divisible or that there is something that is a body and is divisible, without knowing, in a *de re*-like fashion, any individual, particular body. Then if the Deduction is to show category application to all the objects that we do or can know, it must be possible for Kant to prove, on the basis of the Deduction's argument, that the categories apply to the objects that make true these last sorts of *de dicto* claims, even though *H* does not know these objects as individual, particular things.

Inferred objects (and *de dicto*-like judgments) raise subtle questions for Kant's account of knowledge. But (as is suggested in part by A225–26/B272–73 remarks in the Postulates) it seems that Kant's basic approach to the third issue would run as follows.[85] As we noted in Chapter One, in the Inaugural Dissertation Kant accepts the view that our having knowledge requires at least the possibility of our – or of some being's – recognizing the object known as the particular, individuated thing that it is by intuiting that object.[86] Now Kant can be seen, in the first *Critique*, to accept a somewhat similar view.[87] Indeed, in the first *Critique* Kant implicitly holds that any object known by us (whether as an individual, particular thing or simply as making true some general, *de dicto*-like claim that we know) can in principle be intuited by us or by some being and, as intuited, can be known through a *de re*-like judgment. However, Kant also argues that we have no grounds to suppose that any being *besides* us has any sort of sensible intuition – or indeed has any intellectual intuition – that allows that being to intuit the object known by us as that object exists in itself (or as that object otherwise exists).[88] Hence any object known by us can in principle be intuited by us and, as intuited, can be known through a *de re*-like judgment. But through that judgment such an object is brought under categories. Moreover, Kant could argue, further, that if that object is the above sort of inferred object, then that object falls under the categories associated with the original *de dicto*-like judgment (that, say, all bodies are divisible). After all, given that this judgment is made by *H* and is true, *H* will take its logical functions to govern the synthesis of the manifold of the intuition that *H* can in principle have of that object. So Kant will suppose that he has a straightforward response to the third issue.

I think, however, that this response is less straightforward than it may look. I have two main worries. First, it is not clear why every object that we know must be intuitable *by us* (as against, say, simply linked lawfully, in accordance with the Analogies of Experience, to objects that we do intuit).[89] Second, there is a problem of intensionality for the further argument given above to show that the object in question falls under the categories associated with the original *de dicto*-like judgment. Even if I think – and know – that all bodies are divisible, and even if I later think (about a thing that I then intuit) that this thing is a body, it does *not* follow, in logic, that I also then think that this thing is divisible. And so it does not follow that I apply to this thing the category of substance associated with my thought that all bodies are divisible. These two worries seem to me genuine, and they show that the question of category application to inferred objects (and category application through *de dicto*-like judgments) needs more attention than Kant gives it in the Deduction. However, because the Deduction's main line of thought – its argument connecting unity of apperception to judgment and category application – is independent of these worries, I will not pursue them here.

I should, however, perhaps note that the response that we have just been discussing to the third issue itself gives rise to further issues about the scope and limits of category application and about the basic sort of Kantian principle that any object known by us can be intuited by us. Thus – and to broach the scope-and-limits issues – one might ask why, if Kant allows category application to phenomenal objects known through *de dicto*-like judgments, he should not also allow that through the pure categories we might know, in a *de dicto*-like fashion, substantive truths about things as they exist in themselves. One might argue that the most that the Transcendental Aesthetic by itself shows is that all our *de re*-like sensible knowledge is of phenomenal objects. Moreover, we of course lack intellectual intuitions through which we could know, in a *de re*-like fashion, things as they exist in themselves. But it would still seem to be possible, as far as Kant's reasoning so far has gone, that through our thought and understanding operating by themselves (but still governed by the categories) we should know objects in a *de dicto*-like fashion as they exist in themselves.[90]

After all, in the Inaugural Dissertation Kant takes us to have a 'symbolic' and *de dicto* like metaphysical knowledge, through intuition-unaided 'universal concepts in the abstract,' of objects existing in themselves.[91] Furthermore, the Deduction argument for category application to the object known through intuition *i* shows that the categories function, in the case of such intuitive knowledge, simply as rules for the synthesis of intuitions and so apply merely to phenomenal objects. Nevertheless, that argument by itself does not rule out the possibility of *de dicto*-like judgments expressing knowledge of objects as objects exist in themselves. And, as logical functions operate in the usual way in such judgments, the categories associated with those logical functions might themselves apply to those objects. There might obtain, with respect to the properties belonging to those objects, a mind-independent, logical-function-like categorial structuring that parallels the logical-function structuring of the concepts that occur in the judgments.

In the first *Critique* Kant does not explicitly consider the possibility of such category application to objects known *de dicto* through thought as they exist in themselves. But he clearly would take himself to have eliminated this possibility. He would argue that our having knowledge requires at least the possibility of our – or of some being's – recognizing the object known as the particular, individuated thing that it is by intuiting that object. But we have no grounds to suppose that any being besides us actually can intuit the object known as it exists in itself. Given this fact and the fact that, as the Transcendental Aesthetic shows, our own intuitions are all sensible and represent objects merely as they appear, none of our own *de dicto*-like knowledge through thought can concern objects as they exist in themselves. And so – arriving at the scope and limits of the categories that he announces in B-Deduction § 22 and § 23 – Kant would conclude that by means of the categories we can know only phenomenal objects.

The preceding first-*Critique* reasoning raises subtle questions that turn on the view that our having knowledge requires at least the possiblity of our – or of some being's – recognizing the object known by intuiting that object. One might ask whether we should accept such a view at all. Couldn't we conceivably have grounds for knowing *that there are* certain things that no being at all ever could, even in principle, recognize in a *de re*-like fashion through intuition? Moreover, even if we ignore such worries and accept the above view, questions arise about the strength of the object-recognizability that it requires.

Thus in the somewhat similar Inaugural Dissertation view, noted above, about the need for such recognizability, Kant seems to require only the logical possiblity that some being should recognize in intuition the object that we know.[92] And, as I have indicated, in the Inaugural Dissertation he supposes that we actually have *a priori* metaphysical knowledge, of a *de dicto*-like sort, that applies to objects existing in themselves – objects that, as they so exist, we cannot intuit. However, the effect of his first-*Critique* reasoning is to defend the basic claim that if we know an object in any manner, then we must have specific, concrete grounds for supposing that there is or could be an intuition, by some being, through which that object is recognized in a *de re*-like fashion as the particular, individuated thing it is. (And then, as I have noted, Kant argues further that such grounds are lacking in the case in which it is supposed that we know objects existing in themselves; and hence there is no *real* possibility, in the Kantian sense, of there being any such objects of our knowledge.[93])

Discussion of this basic, first-*Critique* claim about the need for the above sort of grounds in complicated by the fact that Kant has not really proved his Transcendental Aesthetic position that through our intuitions we can know objects only as they appear. It might be held, however, that if Kant had proved that position, then he would be right to urge that we lack the above sort of grounds in the case of objects existing in themselves. Nevertheless, more thought would still be required before we could decide whether the basic, first-*Critique* claim that we need specific, concrete grounds for supposing that the relevant objects are recognizable in the proper way is preferable to the earlier Inaugural Dissertation view of such recognizability in terms of mere logical possibility. (Thus, for example, without further discussion of issues about evidence, cognitive significance, and various sorts of Kantian and non-Kantian realism, it is not clear that our having knowledge of *de dicto* truths does in general require anything more than the mere logical possibility of some being's recognizing the relevant objects, in a *de re*-like fashion, by intuiting them – if indeed even that possibility is required.) However, interesting as they are, the problems here are again independent of the Deduction's main line of thought, and I will not examine them further in this book.

As we saw in Chapter Nine, the fourth issue is important because if the Deduction is successfully to eliminate the Humean possibility that all we may know is the subjective sequence of our own representations, Kant must show that the object that H knows through i is distinct from all of H's intuition-elements or other representations, taken separately or in sequence. (Hereafter I omit 'taken separately or in sequence.') Or, to be more accurate, recall, from Chapter Eight, that Kant cannot show (T*i*) to follow from the holding of unity of apperception with respect to *i*. Rather – and in a way that goes beyond the actual argument of the Deduction itself – he should attempt to demonstrate the Chapter Eight, weakened (T*i*)-style result that takes the holding of unity of apperception to imply that *H* has some unity-of-apperception-governed intuition-elements through which *H* thinks and knows in a (T*i*)-like thought. And then to eliminate the above Humean possibility Kant must show that the object that *H* knows through that thought is distinct from all of *H*'s intuition-elements or other representations.[94]

The question whether Kant can show the relevant such object to be distinct from all of *H*'s representations is a key – and, I think, an unresolved – issue raised by the Deduction. As far as I can see, we certainly *do* have intuition-elements – or mental states – through which we know objects distinct from all our own representations or mental states. But *must* we have such intuition-elements? Kant might perhaps argue that the fact that the objects of our knowledge are categorially structured through judgments, while our (mere) representations are not, shows that those objects must always be distinct from our representations. However, this argument would be unconvincing. Kant has not really shown the categories, at least as they are listed at A80/B106, to apply to the objects of our knowledge. And even if the A80/B106 categories are shown to apply to such objects, we have already seen reasons, in connection with the second issue above, for Kant to take subjective sequences of our representations to be among the things that we know and so themselves to be subject to categories.

The question thus remains whether we ourselves, or beings like us, could not know only subjective sequences of our own representations or mental states. As far as I can tell, arguments for giving a Kantian, negative answer to this question will have to go beyond the actual claims of the Transcendental Deduction itself. I do not think it is obvious whether such arguments are available. One might, for example, argue (as in the B274 ff. B-Refutation of Idealism) that I cannot take myself to exist and to have a specific, time-ordered sequence of representations unless at least some of those representations give me knowledge of the existence of one or more (outer) objects distinct from all my representations. Or one might appeal to facts about the way in which we come to be able to self-ascribe mental states in the situations in which we find ourselves. On the basis of these facts, one might argue that at least some of those mental states must yield us knowledge of objects distinct from all our mental states. But I am not yet convinced that any such arguments will succeed.

Thus perhaps our normal, actual representations or mental states are not rich enough or stable enough to be all that we (with our capacity to take ourselves to exist and to have time-ordered sequences of representations) can know. And perhaps given the way in which we do come to be able to self-ascribe mental states, those mental states cannot be all that we know. But it is not clear that we could not in certain circumstances have sufficiently rich and stable representations to allow those representations to be the sole objects of our knowledge. Nor is it clear that we, or beings essentially like us, could not come to be able to ascribe mental states in ways – however peculiar or unlikely – that would allow the sequences, in our minds, of those mental states to constitute all that we know.

Of course the failure of twentieth-century sense-datum theories may well suggest that something like Kant's basic position about the possibility of knowing only subjective sequences of representations is correct. And I would like to be able to accept this suggestion. But this failure does not show, specifically, that the holding of unity of apperception with respect to our representations itself implies that we cannot know only such sequences of representations. Nor are suggestions, however attractive, proofs. These matters seem to me, at the moment, open questions. And thus, at this central point in our discussion of the issues that the Transcendental Deduction raises, we have to see Kant as, once more, broaching fundamental topics whose ramifications have not yet been fully explored.[95]

About the fifth issue – the question of how category application proceeds in the case of the spatial manifold of intuition – we can be quite brief. This issue does not itself affect the argument of the first half of the B-Deduction, in which, as we have seen, Kant focuses on the case of the sensible intuition in general i and abstracts from all points about the specific sensible mode (say spatial or temporal) in which i is given to H. But this issue needs comment if only to relate our Chapter Three remarks on the spatial manifold of intuition to our above account of the categories. In connection with this issue I also will say something very briefly about synthesis under the categories of quality.

Kant's own discussions of space (and of the synthesis of objects in space) in the Transcendental Aesthetic, Analytic, and Dialectic are complex and frequently obscure. Moreover, Kant never makes it completely clear how the conceptual synthesis of the manifold of spatial parts proceeds. But if we read various first-*Critique* texts bearing on these matters in the light of our Chapter Three remarks and of our above account of the categories, then we get a fairly clear idea of what that synthesis – and its application of the categories – must involve. The texts I have in mind are those like A102 on drawing a line in thought; A105, on thinking 'a triangle as an object, in that we are conscious of the combination of three straight lines according to a rule by which such an intuition can always be represented' (and the later A105 comment that 'the concept of this unity [of rule] is the representation of the object = X, which I think through the predicates, above mentioned [the three straight lines], of a triangle'); A106, on the way in which the concept of body 'serves as a rule in our knowledge of outer appearances' by necessitating 'the representation of extension, and therewith representations of impenetrability, shape, etc.'; B137–38, on drawing a line; and B162 and A190/B235 ff. on apprehending a house.

In conjunction with our remarks in Chapter Three and our above account of the categories, such texts suggest that category application proceeds as follows in the case of the spatial manifold.[96] A human empirical outer intuition, say the spruce-representing intuition j of earlier chapters, has elements, say j_3 and j_4, that put before the mind spatial parts of the object. In synthesizing this intuition, the knower makes a judgment, say the judgment that the tree is conical. In making that judgment, the judger thinks there to be an object x that is a tree and that has the spatial parts that the elements j_3 and j_4 present. (Here note the A105 comment quoted parenthetically above, on the 'representation of the object = X, which I think through the predicates ... of a triangle,' those 'predicates' evidently being the three straight lines that are the sides of the triangle.[97]) That judgment has a quantity – in the case just given, the quantity of singularity. And so, as the judger thinks there to be such an object x, the judger thinks there to be a single object x that is a tree and is conical (as against thinking, in particular fashion, that there is some object that is a tree and is conical or, in universal fashion, that all objects are trees and are conical).

Thus the judger uses a category of quantity – in the present case, the concept of a single object x that has homogeneous spatial parts – to synthesize j_3 and j_4. As the judger uses that category, the judger also brings to bear one or more shape concepts (like the concept of being conical or the concept of being triangular) that, one can say, are 'specified' by the spatial parts presented by j_3 and j_4 and apply to the single spatial object that those parts, when combined, constitute.[98] And of course as the judger does these things the judger also applies categories of quality, relation, and modality.

I have noted earlier the operations, in synthesis, of the relational and modal categories, as far as we need to consider those operations in this book. About the very obscure questions surrounding the application of the categories of quality, we need observe only two basic points. First, it is difficult to make out in any exact, plausible way the connection of the affirmative, negative, and infinite logical functions of quality (A70/B95 ff.) with the idea that in each appearance 'the real that is the object of sensation has intensive magnitude, that is, a degree' (B207).[99] Second, this degree (by A168/B208, a degree of the object's influence on the sense through which it is apprehended) parallels a degree in the strength of the sensation through which the object is apprehended. The intensive magnitude and degree belonging to the object should not, however, be conceived to emerge from a synthesis of a separate 'intensive-quality manifold' that is given along with the spatial and conceptual manifolds. Rather, and very roughly, because all sensations themselves have intensive magnitudes, as the synthesis of sensations under a quantitative category yields a spatial object, that object is yielded in such a way as to have also a reality with an intensive magnitude of a specific degree.[100]

The above account of category application in the case of the spatial manifold could be sharpened. But I take it to be on the right lines as far as it goes. And, along with the texts that I have cited, it indicates that Kant's views on that manifold accord with the basic account of category application that we have presented earlier in this chapter.

The last, and sixth, issue that we need to discuss concerns the ways that Kant's idealism and his position that combination cannot be given affect the Transcendental Deduction view of the logical functions and of the categories. As I noted, in connection with this issue I will also comment again on Kant's problematic treatment of things in themselves.

As we saw in Chapter Eight, even if he were to reject his idealism or his position that combination cannot be given, Kant still could hold the two specific B-Deduction § 17 points that *H* thinks the (T*i*) thought and that that thought yields *H* knowledge of *i*'s object. (Or, as one can see, he could hold the two related points that can be stated in terms of the weakened (T*i*)-style result that we sketched in Chapter Eight.) Moreover, Kant's earlier § 16 position on the holding of unity of apperception is evidently independent of both his idealism and his position on combination. Kant himself would of course want to reject such claims to independence. But, as we saw in Chapter Eight and earlier, these fundamental § 16 and § 17 conclusions of the B-Deduction argument can be maintained whether one takes the object known through *i* to be mind-dependent or not and whether one takes the combination of features in that object to be mind-created or simply given.

The question thus arises whether this same sort of independence exists also in the case of Kant's fundamental § 19 and § 20 conclusions about the operation of the logical functions in judgment and in the categorial structuring of the objects judged about. (Here I assume for the sake of discussion that these conclusions, which we have challenged above, are correct.) Kant will of course answer this last question negatively. As we have seen, he supposes that the mind acts to combine concepts into judgments in accordance with the logical functions. In the case of judgments expressing knowledge, he takes the mind thereby also to combine, in the mind-dependent object known, the features that those concepts are (or present), in accordance with the logical functions. And he takes the mind, in so proceeding, to effect category-application to the object known. One might ask, however, whether we are forced to accept this negative answer rather than a positive answer that separates Kant's fundamental § 19 and § 20 conclusions from his idealism and his position on combination.

As far as Kant's specific conclusions about the operation of the logical functions in judgment go, his negative answer is correct or very plausible. At least if one assumes, with Kant, the traditional idea that concepts occur in the mind in judgment, it does not seem plausible to claim that the combinations, in logical-function ways, that those concepts have as they thus occur are somehow given to the mind. (Kant's idealism about the objects of knowledge is of course not itself directly relevant to his specific claims about the operations of the logical functions in judgment.) As Kant insists – and as non-Kantian philosophers like Aristotle would surely agree – the organization of concepts in the mind in thoughts and judgments is the product of mental activity and is not (or is certainly not always) simply impressed on the mind from outside.[101]

Kant's negative answer seems harder to defend, however, in the case of his conclusions about the categorial structuring of the objects that are judged about and known. Although the following idea is highly non-Kantian (and would indeed bring us back to Aristotle), it seems possible to hold that the objects of our knowledge are mind-independent things that have features that are combined in the sorts of subject-predicate and other ways that Kant himself takes to be due to the mind's logical functions of thought. (And then one could hold in quasi-Aristotelian fashion that, when the objects are known, these combinations of features are paralleled by the same sort of combinations of concepts – this time determined through activities of the mind – in judgments.) To Kant this idea of objects as being mind-independent things that have, in themselves, what he would take to be necessary, categorial structures would be anathema. But if one rejects his idealism and if one accepts the possibility that such necessary structures might exist independently of the activities of the mind, then this idea must be taken seriously by anyone who sees the objects of our knowledge as having the sort of categorial structuring that is defended (though in terms of the mind's activities of synthesis) in B-Deduction § 20. Thus if one is willing to adopt certain other non-Kantian ideas, Kant's fundamental § 20 conclusions about the categorial structuring of the objects of knowledge, like his § 17 views and his § 16 claim about unity of apperception, are independent both of his idealism and of his view that combination cannot be given. More could be said at this point – and no doubt will be said by defenders of Kant. But I think that this independence is in fact the case.

The comments in the last paragraphs bring us finally to Kant's treatment of things in themselves. As we have noted in Chapter Two, this treatment is a source of unending difficulties. It seems impossible for Kant consistently to deny that we can know objects as they exist in themselves while at the same time he presents a picture of our knowledge – and a picture that, by A11–12/B25 ff., is itself supposed to embody knowledge – that contains detailed claims about the operations of our mind as it exists in itself and about the affection of our mind by objects as they so exist. Yet the presence, in this picture, of such claims is crucial to this picture as Kant presents it. And so he is caught in an impossible situation.

More suggestions have been made – and can be made – about this problem than I can hope to do justice to here. Briefly, it would seem that, given his overall picture of knowledge, Kant's own most promising way out of this situation would be as follows. As various commentators have suggested, he should argue that while we cannot know objects as they exist in themselves, within his theoretical philosophy we must nevertheless *think* or postulate such objects in order to meet theoretical demands of our reason or understanding.[102] However, while this traditional means of escape certainly relieves some of the pressures, it is, I think, itself badly flawed.

Thus Kant's own arguments for the necessity of thinking things in themselves that we cannot know are not, in my opinion, successful.[103] And, even if they were in general successful, these arguments certainly do not demonstrate or even attempt to demonstrate the specific necessity of our thinking of the mind in itself (and of objects as affecting the mind in itself) in the detailed way that Kant proposes in the Transcendental Aesthetic and Analytic. (What, indeed, could show that we have to endorse just this specific view of the mind and of objects – a specific view that, in principle, we can never know? Surely not the need to explain features of what we do know, for in principle there are an infinite number of distinct logically possible noumenal grounds of – or explanations for – such features. And, if we can know nothing of any of these grounds, then it is hard to see what legitimate reason we can have for preferring one to another.) Moreover, in any case Kant's arguments will not allow him to claim to *know* the specific facts about the operation of our mind in itself (and about objects as affecting our mind) that underlie his overall picture of knowledge.

As I suggested in Chapter Two, my own view is that, when all is said and done, Kant has put himself into an impossible situation in making the claims that he does about things in themselves. To say this is not to deny that his claims about such things are developed with a characteristically Kantian depth and subtlety. Nor is it to deny that one can conceive of ways of refining or modifying his picture of knowledge – or of various related pictures of knowledge – that may escape this situation. Rather, it is to say that, as many philosophers have emphasized, Kant simply cannot in consistency present, as knowledge, an account of knowledge that makes claims that imply the impossibility of that account itself's being known.

Unfortunate as this fact is for Kant's overall situation, it hardly undercuts the enormous, permanent interest of his basic ideas. And, as Kant scholars, we must simply record this fact and add it to the list of his achievements and difficulties. But, in closing, I should emphasize again that Kant's fundamental B-Deduction § 16, § 17, § 19, and § 20 views on unity of apperception, unity of the manifold in the concept of an object in general, the logical functions, and the categories can be argued to be independent of his idealism and his position on combination. Given this fact, it is possible to defend at least parts of those views (or related views, as for example the weakened (T*i*)-style result of Chapter Eight) while allowing that the objects of our knowledge are things existing in themselves. This possibility of course itself raises various questions. And it is desirable for scholars to continue exploring ways to escape the above problems about things in themselves while preserving Kant's idealism and as much else of his overall picture of knowledge as possible. But I think that such a possibility (which other recent Kant scholars have noted) provides the most satisfactory way of escaping such problems while defending – to the extent that this can be done – the sorts of fundamental B-Deduction views that I have noted above.

9. CONCLUSIONS. THE OVERALL INTEREST AND SUCCESS OF THE FIRST HALF OF THE B-DEDUCTION

In the B-Deduction (in its first half) Kant presents a basic argument that runs as follows. (I) Unity of apperception holds with respect to the sensible intuition in general *i*. (II) The holding of unity of apperception with respect to *i* implies that the being like us, *H*, thinks the (T*i*) thought that unifies the manifold of *i* in the concept of an object. (Recall again that in that thought *H* thinks there to be a single object that has the features that *i*'s elements present.) (III) That thought yields *H* knowledge of *i*'s object. (IV) The (T*i*) thought itself is or is a part of a knowledge-yielding judgment about the object of *i*. (V) Because of the way that the logical functions of thought structure the concepts that occur in that judgment, those same logical functions also structure the features of *i*'s object that the elements of *i* put before the mind or are. And, as a result of this structuring (given also points (II), (III), and (IV)), *i*'s object, as it is known by *H*, falls under the categories of the understanding. Kant will then generalize this last result to reach the general conclusion, in the first half of the B-Deduction, that the object of any sensible intuition in general through which a being like us knows falls under the categories. And in the second half of the B-Deduction Kant will use this conclusion to argue, finally, that the categories apply, specifically, to the objects of any human, empirical sensible intuitions through which we do or can know.

This five-step argument *is* (with the general, first-half conclusion just noted, and the final argument in the second half) *the* argument of the B-edition Transcendental Deduction of the Categories. (For the A-edition argument, which ignores the role of judgment and the logical functions in

(V) and does not focus first on the case of the sensible intuition in general *i* in the way that the B-Deduction does, recall the remarks in Chapter Four.) As we have seen above and in the preceding chapters, each of the five main steps of this argument either is not established by Kant or else can be shown to be mistaken. Thus the basic argument of the B-Deduction fails. (And, for similar reasons, so does the argument of the A-Deduction.)

The failure of the Transcendental Deduction undermines the central argument of Kant's theoretical philosophy, and it thereby leaves Kant without any proof of the objective validity of the categories. Because this failure rests on difficulties in (or on mistaken inferences about) the fundamental Kantian ideas that are involved in steps (I) to (V) above, it carries with it many serious problems for those ideas. Neither the Transcendental Deduction nor Kant's views on unity of apperception, on unity of the manifold in the concept of an object, and on the logical functions in their roles in judgment and in category application can be maintained in the form in which Kant presents them. Nor can many other Kantian points that we have noted above, for example various of Kant's claims about necessity.

Readers of Kant should not try to disguise these facts from themselves. Nor should they try to defend the Transcendental Deduction itself, or the sorts of Kantian views that I have just noted, by observing that arguments in some ways akin to the Deduction, or views in some ways like those just noted, may perhaps still succeed. For example, we have noted in Chapter Eight that a weakened version of Kant's claim that *H* thinks and knows through the (T*i*) thought may still hold true – namely, the claim that when unity of apperception holds with respect to *i*, *H* has some unity-of-apperception-governed intuition through which *H* thinks a knowledge-yielding thought like the (T*i*) thought. And in the present chapter we observed that by appeal to facts not just about logical structure but also about us as thought and language users and about our situation in the world, one might conceivably be able to argue that the objects that we know have to be structured in some sort of categorial way. These points are deeply interesting, and they originate in ideas of Kant's. But they are not Kant's own claims or arguments in the Deduction, and their correctness would not vindicate the Deduction itself.

The career of the Transcendental Deduction is the career of many other great philosophical ideas. A theory is presented; it is defended by its supporters as correct in all its major details; doubts creep in; the theory, in the form in which it is presented, is seen to be incorrect in all or many of those details. It is by this time realized, however, that that theory has proposed ideas that so impinge on the rest of philosophy that little or no significant progress can be made without coming to terms with them. And the theory functions itself as a source of fundamental suggestions for approaching those ideas.

The Transcendental Deduction has reached this last stage of theory development. It itself does not succeed. But the arguments and ideas that it suggests – including the points noted in the next-to-last paragraph and a multitude of others – go as deep, and ramify as widely, as any other arguments and ideas in philosophy. Kant, as the author of the Transcendental Deduction, should be honored not for the correctness of his exact views – beautifully squeezed (with an occasional rough edge or bulge!) into the recesses of the architectonic – but for the depth of his insights and the immense fertility of his views. One can no longer unthinkingly hold, after the Deduction, that, in Aristotelian fashion, our judgments give us knowledge of category-subsumed objects whose structure is reflected by, but is independent of, the structure of those judgments. Nor can one unthinkingly hold, in Cartesian fashion, that at least some of our mental representations give us knowledge of the existence and character of mind-independent objects. One also cannot unthinkingly hold to traditional views that we have a special knowledge of our self as it exists in itself or that our self-knowledge is independent of our knowledge of objects. Nor can one unthinkingly accept the radical skepticism implied by parts of Hume's work or hold to a myriad other points that Kant has challenged or reconceived.

Our concern in this book has been with the Transcendental Deduction (and, specifically, with the first half of the B-Deduction) as Kant presents it. As Kant presents it, the Transcendental Deduction fails; and much of our discussion has been a detailed charting of that failure. But only an argument of the greatest reach and depth could, I think, sustain anything like that detailed examination. And the fact that Kant has not reached the Deduction's goal hardly detracts from the interest of his progress toward it.

10. SUMMARY

In B-Deduction § 17 Kant argues that H thinks the (Ti) thought, the thought that there is a single object that has the features F and G that i's elements present. In § 20 he now defends what we have called claim (C). In (C) Kant notes that the logical functions of thought in judgment, through the holding of unity of apperception, determine the logical-form relations together of the concepts occurring in the judgment that is or involves *H*'s (T*i*) thought. And he holds that, because the logical functions determine those logical-form relations, the logical functions determine, also, the relations together of the conceptual elements of *i*'s manifold in such a way that the object that *H* thinks and knows through *i* is category-subsumed.

Given the results that Kant has now reached, his reasoning for (C) is straightforward. By B-Deduction § 19, *H*'s (T*i*) thought is or is part of a knowledge-yielding judgment about the object of *i*. That judgment involves the attribution, through the use of the concept of an object in general, of the features *F* and *G* to that object. Kant takes those feature to be (or to be presented by) the relevant concepts c_1 and c_2 in *H*'s thought (T*i*). And he takes those concepts, and so *F* and *G*, to be related together through the application to them of the logical functions that determine the form of that judgment. Hence when *H* knows through the judgment that involves *H*'s (T*i*) thought, *H* judges that there is a single thing *x* such that *x* is an object, *x* has the features *F* and *G* that are put before the mind by intuition-elements i_1 and i_2, and *F* and *G* are structured by the relevant logical functions. For Kant, however, a category is a concept of a thing such that that thing is an object in the most general sense of 'object' (that thing is an object in general) and that thing is playing – or elements of that thing are playing – one of the roles specified by the logical functions of thought. Yet in judging in the way that I have just indicated, *H* is thinking that *x* is such an object. So in that judgment *H* brings that object under the categories associated with the relevant logical functions.

As we observed, the most obvious problem for this § 20 argument for category application is this. Why should the logical-function structuring of the concepts c_1 and c_2 in this last knowledge-yielding *judgment* imply an identical logical-function structuring of the features, in the *object*, that those concepts present? Kant himself ignores this problem, largely because of his identification of concepts with features taken as general. But even given this way of avoiding the problem, there are three further puzzling aspects of his views. First, and in a way perplexing to the modern reader, Kant does frequently identify concepts with features taken as general. Second, and in the case of the subject-predicate logical function, he treats the feature, in the object, that functions as the subject as being itself a feature that *has* the feature, in the object, that functions as the predicate. Third, he does not seem to accept a standard, modern distinction between the syntax and the semantics of judgment.

Faced with these puzzling aspects, we turned to the further details of Kant's claims about the categories. By examining these claims, we came to ratify the account of a category that was stated in the next-but-last paragraph. Adopting this account, we saw that, for Kant (and in contrast to what would be the case for standard modern views of syntax and semantics), the assignment of a logical-function structuring to concepts in a judgment and the relating of those concepts to the object judged about are two inseparable parts of the overall act of judgment. We saw also that the three puzzling aspects of Kant's views seem to derive from his acceptance, without explicit argument, of Aristotelian views on logic and judgment. Thus for Aristotle the individual, particular objects of our knowledge have a subject-predicate (or a substance-attribute) ontological structure. And when we know those objects, that ontological structure is reflected in the mind in unitary judgments that the mind makes about those objects. We argued that, when they are spelled out in detail, such Aristotelian views can be seen to account for the puzzling aspects of Kant's views, although in many fundamental respects – for example, in regard to his position that the object known is a mind-dependent thing whose structure is determined by the logical structure of judgment – his views differ greatly from Aristotle's.

Given our above understanding of Kant's account of the logical functions and the categories, we next asked how far this account succeeds. The answer to this question was, unfortunately, negative. First, and as is well-known, Kant's own version of Aristotelian logic has turned out to be inadequate in comparison with modern logic, and his derivations of the individual categories from the logical functions that he recognizes are unconvincing. Second, it seems very unlikely that, in a Kantian fashion, one can successfully infer facts about the categorial structuring of objects from facts merely about the logical structures of the judgments that we make about those objects.

The Transcendental Deduction of the Categories is, I think, the deepest and the most far-reaching argument in philosophy. The result of our discussion in this chapter and in the preceding chapters is that the Transcendental Deduction fails. This failure must not be denied. Nor can it be evaded by noting the multitude of fundamental ideas about self-awareness and self-knowledge, objectivity, idealism, and the logical and categorial structure of judgment that the Deduction introduces. Nor is the Transcendental Deduction itself vindicated by noting the possibility (if it exists) of arguments to conclusions in some ways like the Deduction's conclusion by reasoning in some ways like the Deduction's reasoning. However, the existence of such ideas and the possiblity of such reasoning are one measure of what Kant achieves in the Deduction. One may feel that it would have been better had the Transcendental Deduction not failed. However, failing, the Deduction still has, in its influence and depth, a unique success. And the outcome of reasoning should not in any case be the subject of philosophical regret.

NOTES

NOTES TO CHAPTER ONE

[1] For these familiar points and the introductory matters below, see the A- and B-Prefaces, A1–2, B1–6, B19–24, A84–94/B116–26, B127–29, the Transcendental Doctrine of Method, especially Chapter I (The Discipline of Pure Reason), and Chapter Four below.

[2] Note, for example, B72, B138–39, and B145–46.

[3] For this characterization of intuition, see Howell (1973) and its text references; Hintikka (1965) and (1967), reprinted in Hintikka (1974, 126–34 and 160–83), and also Hintikka (1969) and (1972); Parsons (1969), reprinted with a new Postscript in Parsons (1983, 110–49); and Thompson (1972). Relevant texts are A19/B33; A24–25/B39; A31–32/B47; B136 note; A319–20/B376–77; A713 = B741; *Prolegomena*, § 8; Kant to J. S. Beck, July 3, 1792 (Ak. 11, 347–48); *Logik*, § 1 (Ak. 9, 91). As the papers just cited bring out, the interpretation of some details of Kant's view of intuition is controversial. The approach of this book to the Transcendental Deduction is meant to be independent of the points of controversy.

[4] For the points about concepts here and below, see B39–40; B133–34 note; A137 = B176; A320/B376–77; *Logik*, § VIII.C and §§ 1–8 (Ak. 9, 58 and 91–96). See, further, the papers cited in note 3, and also Chapter Three. Parsons observes that the idea noted here – of a concept as a representation mediately related to the relevant objects through the relation that it bears to the general property that those objects possess – appears circular. '[F]or what could mediation by marks or characteristics [in my present terms, by the general property in question] be but some *predicative*, in Kantian terminology conceptual, content of the representation?' (Postscript to Parsons, 1969, in Parsons, 1983, 144). The question of this circle requires further study (see Parsons' footnote 44). My own view is that (a) Kant may well not be clear about how far, in talking of the mediacy of a concept, he is partially defining or is, rather, simply characterizing a concept; (b) the mediacy talk is in any case Kant's and represents his own understanding of concepts; and (c) the apparent present circle would not exist if Kant is proceeding thus. He is not trying, in his mediacy talk, to define or explain the specific predicative element in concepts or properties via that element itself. Rather, he is trying simply to indicate how, given the notion of a property or predicative element as already understood, we can regard certain entities in the mind, which he calls concepts, as representing objects by standing in relation to predicative elements (properties or marks) that occur in the objects represented. Any Kantian attempts to define or explain the predicative or property notion itself (to the extent that Kant makes such attempts) should then be sought in his conceptualistic approach to general properties (here see Chapter Three, especially note 7), where a

circle may indeed lurk. (However, because nothing important in this book turns on the success of Kant's conceptualism, I will not pursue that circle further here.)

5 See A51–52/B75–76, B68, B72, B138–39, and B145.

6 Inaugural Dissertation (Ak. 2, 385–419), § 25, and also § 1 and § 10. As we observe in Chapter Ten, in the first *Critique* Kant continues to hold something like the present Inaugural Dissertation view, with its requirement of the possibility of some being's intuiting the object known. But he argues in effect that we have no grounds for taking this possibility of intuiting the object to be realized by any beings other than ourselves. And he concludes by rejecting the existence of this possibility, as far as it concerns the intuition of the object by beings other than ourselves. Hence, given the present view, Kant denies that we have any knowledge of objects that we ourselves cannot recognize in sensible intuition. However, in the Inaugural Dissertation (for example, at § 25), he accepts the existence of this possibility, as far as it concerns intuition of the object by beings other than ourselves. He holds that we actually have a metaphysical, 'symbolic' knowledge of general truths concerning objects that we ourselves cannot recognize in sensible intuition. (See, for instance, §§ 8 to 10.)

7 I abstract here and in the next paragraphs from differences among these philosophers about the precise nature of the inner consciousness involved and also from specific Kantian matters like the inner-sense/apperception distinction.

8 Recall the *Third Meditation* and the third paragraph of *Replies to Objections I.*

9 *Essay* IV.4.8 and 11.3, etc.; note also Locke's talk at II.8.23 of our having, through primary qualities, 'an idea of the thing as it is in itself.'

10 Kant's many first-*Critique* remarks about the distinction between objects as they appear to us through sensible intuition and objects as they exist in themselves support the present understanding of 'object as it exists in itself.' (See Bxviii–xix note, Bxx, Bxxvi–xxviii, A26–30/B42–45, A32–36/B49–53, A36/B53 ff., A41–49/B59–66, B66–72, A92/B124 ff., A114, A128–29, B156, B157–58, B164–65, A181/B223, A190/B235, and the Phenomena and Noumena Chapter.) See also below. The distinction could be refined or reconstructed in various ways, but we need not attempt that task in detail here. A simple, clear reconstruction is in terms of the possible-worlds (or, strictly, perceptual-worlds) semantics for perceptual claims introduced by Hintikka (1962 and later works). The actual world of that semantics will be the world or realm of objects as they exist in themselves; the phenomenal world, or world of objects as they appear via sensible intuition, will then be very roughly the result of packing together, into one world, all the worlds perceptually alternative to the actual world. See Chapter Two, note 30, and the references there, and also Chapter Five, note 6. There is of course a rich, voluminous, and controversy-laden literature about objects existing in themselves. See, for example, Kemp Smith (1962), Adickes (1924), Paton (1936), Strawson (1966), Prauss (1974), Walsh (1975), Allison (1983), and Guyer (1987).

11 I ignore Cartesian complications (which could be dealt with) about God's mind and the conservation of objects.

12 B40–41, A31–32/B47, and B48–49; also A39–41/B56–58 and A46–49/B64–66. The Metaphysical Expositions of Space and Time at A22–25/B37–39, B39–40, and A30–32/B46–48 are fundamental to understanding the Aesthetic itself; but they are not, in my view, as plausible as the Transcendental Expositions. My account below of

those Expositions (of which the argument about space is clearly the more satisfactory) fleshes out Kant's actual claims while, I think, respecting his basic intentions. When I speak of plausibility I of course mean plausibility relative to Kant's view of mathematics.

[13] For transcendental idealism, empirical realism, and other points mentioned in this paragraph, see A26–28/B42–44, A34–36/B51–53, A37–39/B53–56, A41–43/B59–60, B274, and A369 ff.

[14] See A320/B376–77, A34/B50, A50/B74, and A197/B242.

[15] See texts like A34/B50 and A98–99.

[16] Note texts like A20/B34 and A23/B38, and comments in Chapters Three and Ten.

[17] Or, if one accepts arguments by Kripke and others for the existence of what in Kantian terms would appear to be (or to be akin to) synthetic (and *de re*) necessary truths known *a posteriori*, then it seems that some such truths might describe a structure of (physical) space or time belonging to objects as they exist in themselves. Here and below I enter a dispute going back to Kant's time. See Kemp Smith (1962, 112–14); Vaihinger (1922, vol. 2, 134–51 and 290–326); and, recently, Allison (1983, Chapter 5) (defending Kant) and Guyer (1987, Chapters 15 and 16) (criticizing Kant and Allison). My own views were reached independently of these last two discussions.

[18] Allison (1983, 113), argues that because, for Kant, space as we know it *a priori* is mind-dependent but space conceived of as belonging to things in themselves would not be mind-dependent, there could be no sort of significant similarity between these two spaces. Allison's reasoning is not entirely clear but seems based on the false assumption that if x depends on a but y does not, then x and y cannot be significantly similar. (Table t may depend for its identity on the wood w from which it is made, while a qualitatively identical but distinct table u does not depend on w; gesture g may depend for its identity on its maker m, while a similar gesture h does not depend on m. These examples cannot be rebutted by arguing that in the case of mind-dependence, in contrast to the logical situation in the table or the gesture case, mind is unlike things as they are in themselves. We cannot know the nature of things in themselves; and at A358–59, A379–80, and B427–28 Kant suggests that minds in themselves may be – or may be like – objects of outer sense as they are in themselves.)

[19] Compare the idea in note 17. But observe that because of the possibility, indicated in the next sentence below, of an *a priori* knowledge of such supposed synthetic necessary truths, the present suggestion is importantly different from that idea.

[20] Thus the present suggestion of such synthetic necessary truths would seem to escape Kant's chief B168 objection to the somewhat similar, but nonevolutionary, idea of a 'preformation-system of pure reason.' (And given its evolutionary underpinnings – if they were really to exist – the present suggestion would also escape Kant's B167 objection that there would be no limit to the number of such 'predetermined dispositions to future judgments [judgments expressing *a priori* knowledge, by minds, of the relevant synthetic necessary truths]' that such a 'preformation-system' idea would have to assume.) I do not myself take this present suggestion to express a truth. But it indicates a possibility that a Kantian, accepting the existence of *a priori* knowledge of synthetic necessary mathematical truths about space and time, should take seriously. Guyer, whose views on Kant on space and time are generally

consonant with mine, assumes that were there really synthetic (and *de re* – 'absolute') necessary truths pertaining to space and known by us, then it would be plausible to regard space as a mind-dependent, mind-imposed feature of objects as objects appear to us (Guyer, 1987, 364). If I am right about the present suggestion, Guyer's assumption is wrong.

21 A bit more precisely, it is the idea of the ontological independence of the object from any cognitive or other mental states (except possibly for the intellectual intuitions of a God – or except for itself, if the object were self-representational) through which the object is or could be grasped. This idea allows for the possibility that there are some entities, like God (see B71, B306, and B308–309), that have existence in themselves but that cannot appear to us in sense experience. Ontological independence here means that the existence or occurrence of the (in-itself) object does not entail the existence or occurrence of a mind or mental state distinct from that object. Further analysis is possible but not needed here.

22 A36/B52; compare A48/B65 ('if the object ... were something in itself, apart from any relation to you, the subject') and B67 (outer sense 'can contain in its representation only the relation of an object to the subject, and not the inner [properties] of the object in itself'). Note also A26/B42, A32–33/B49, A42/B59, and B306.

23 B306; see the B67 quote in the last note, A38/B55, A324/B381, and A42/B59 (mentioning the relations, in themselves, that hold among things as they are in themselves).

24 *Prolegomena*, § 13, Note III (Ak. 4, 292–93); see ibid., Note II (Ak. 4, 289 – 'that for us unknown but nonetheless real object'), and § 32 (Ak. 4, 315). Observe also A504 = B532 ('attach a determination of the world, [regarded] as an in-itself real *[wirklichen]* thing') and Bxx ('the thing in itself as indeed real *per se*').

25 At the very least, nothing in the notion of 'object as it exists in itself,' taken by itself, would prevent such a situation from being described as one in which we know the object as it exists in itself. In some circumstances claims about our current failings might still defeat this description (as in the case in which one unwittingly accepts, as accurate, a doctored photograph that by accident presents a situation that actually obtains). But such claims would rest on specific facts about our cognitive situation and not on the mere meaning of 'object as it exists in itself.'

26 A43–44/B50–62. The A44/B62 claim that Kemp Smith translates as 'we do not apprehend them [things in themselves] in any fashion whatsoever' is part of a longer sentence whose relevant part really should be translated 'through our sensibility we do not merely confusedly know [*erkennen*] the nature [*Beschaffenheit*] of things in themselves; through our sensibility we do not know that nature at all.' In this same passage Kant argues that our common-sense intellectual concept of 'right' represents 'a moral property [*Beschaffenheit*] of actions, which belongs to them in themselves' (A44/B61). But if it is true (and hence logically possible) that we grasp such a property of actions, then it is hard to see how Kant could suppose that our having *knowledge* of objects as existing in themselves is impossible given just the very definition of 'object existing in itself.' In *Prolegomena*, § 13, Note I (Ak. 4, 287) Kant even grants that, were we willing to deny his fundamental claim about our *a priori* knowledge of space and time, we could consider the possibility that our senses might represent objects as they are in themselves. Kant rejects this possibility *not* because of

the mere meaning of 'object existing in itself' but because he takes the possibility to run contrary to his claim.

27 In the Inaugural Dissertation, § 4 (Ak. 2, 392–93), Kant argues that sensibility by its very nature can yield knowledge only of objects as they appear and not as they are in themselves; and, as Guyer (1987, 340–41), suggests, related views can be discerned at places in the first *Critique*. But this sort of Inaugural Dissertation argument appeals to a claim about the nature of sensibility and not to any definition of 'object as it exists in itself.' Again, Kant argues that we cannot escape the circle of our representations and know objects as they are in themselves (A104–105; compare A190/B235 and A197/B242 and see Howell, 1981b, 91). But this unconvincing argument (which, taken by itself and without appeal to the Transcendental Aesthetic, does nothing to show that through our representations we cannot know things as they are in themselves) does not appeal to objects as they exist in themselves as being by definition unknowable. Nor does Kant's elaborate indirect argument for transcendental idealism in the Transcendental Dialectic, which essentially proceeds by reasoning, on the basis of various claims about knowledge, that if we take space and time to exist in themselves, contradictions follow.

28 The recent interpretation has been lately defended by Allison (1983, Chapters 1, 2, 5, and 11; and 1990, 3–4). Somewhat similar approaches are suggested in Melnick (1973), Prauss (1974), and Pippin (1982). See also notes 23 and 28 to Chapter Two below, and the remarks in Ameriks (1982b) and Guyer (1987, Chapter 15). Here I consider one main line of reasoning for the Allison interpretation.

29 There is certainly a verbal similarity here to my own view. But, as emerges in Chapter Two, on the fully developed version of my view the situation is actually far more complex, and logically far more delicate, than recent interpreters seem willing to admit. (i) I argue there that besides accepting my above, 'appearing-theory' treatment of the object of knowledge as a thing that exists in itself and also appears through sensible intuition, Kant accepts an 'appearance' theory, according to which there are indeed two entities: the object in itself and the appearance or representation of itself that that object produces in the mind. (ii) I show that, in conjunction with his views on our knowledge, Kant's appearing theory leads to a contradiction that is perhaps most naturally avoided by hypostatizing the spatiotemporal object as it appears and so in the end arriving at two objects. (iii) Familiar branching-merging issues, also noted in Chapter Two, provide a further reason for arriving at two objects.

30 For the material in this and the last paragraph, see Allison (1983, Chapters 1, 2, and 11 – for example, 6–7, 17, 27, 242, 244–45, and 250; and also 1990, 4). Given the above discussion – as well as Allison (1983, 7 and 241) on space and time as the relevant sort of necessary conditions – the recent interpretation also is committed to claiming that things as they exist in themselves are by definition nonspatiotemporal. I reject that claim for the same sorts of reasons that I reject the claim that such things are by definition unknowable by us.

31 Kant of course sometimes talks of our 'taking' objects as appearing and as existing in themselves. But when he does so, he is simply referring to the object in its role of appearing or of existing in itself in independence of its representation to or grasp by a mind. (Compare 'Smith, taken [or considered] as a runner, is good.' This claim merely emphasizes that Smith as a runner – Smith in that role – is good; it makes no

special point about taking or considering Smith.) Kant also holds that in making our claims about objects and knowledge, we theoretical philosophers *think* there to be an object, having existence in itself, that appears to us as the spatiotemporal object that we know (Bxviii–xix note, Bxxvi–xxvii, and the Phenomena and Noumena chapter). Here, however, we simply think there to be a thing in itself in the sense already explained. In other words, we form the thought that there is an object that has an existence ontologically independent of that object's representation to or grasp by any mind; and we take that object to appear to us as spatiotemporal. See, further, Howell (1981b).

32 See Inaugural Dissertation §§ 3, 4, 10, 11, 13, 14–17, and 21.

33 B71; compare A27/B43, B306, and B308–309. In the two latter texts Kant indicates the possibility of many other objects, existing in themselves, that cannot appear to us. See also Kant's letter of August 16, 1783, to Mendelssohn, which includes this sort of possibility in a summary of the main theses of the first *Critique* (Ak. 10, 346).

34 Allison (1990, 250, and 1983, 239).

35 Note, for instance, the first three paragraphs of the Aesthetic and later Aesthetic texts like A22–23/B36, B41, A26–27/B42–43, A35/B51, and B67–69. It is indeed the mind in itself that is here affected (or the sensibility belonging to that mind). It is through the affection of the mind that we first (in logic) *get* outer and inner representations; it is not that we already (in logic) have such representations and that what the object existing in itself affects, in order to appear to us through outer sense, is ourself (or our body), which is already appearing to us *in* outer sense through the outer representations that we already have. Note A26/B42 on 'the subjective condition *under which alone* we can *have* [*bekommen* = get] outer intuition, namely, liability to be affected by objects [*so wie wir nämlich von den Gegenständen affiziert werden mögen*]' (my emphasis); and see B67–69.

36 See Allison (1983, 247–54).

37 Thus consider the affection relation between, say, my mind in itself and an object in itself that now appears to me as a tree. It is not easy to see how the recent interpretation could take that relation to be a phenomenal-world causal relation that holds between my empirical mind and the phenomenal tree, that phenomenal-world causal relation being considered without my referring either to its holding between my empirical mind and the tree or to its meeting the necessary conditions for my knowledge of it. To take that phenomenal-world causal relation in such a fashion would be, for the recent interpretation, to take it to have an existence in itself in which it holds between my mind in itself and the tree in itself. But Kant denies that the causal relation can be taken to apply to things in themselves. Moreover, given the recent interpretation, to take the phenomenal-world causal relation in such a way also would be to suppose that that causal relation, as it holds between my empirical mind and the tree, is *itself* the appearance of an in-itself relation that holds between my mind in itself and the tree in itself. But, for Kant, we cannot regard that causal relation as such an appearance any more than we can regard space as the appearance to us of some in-itself order holding among things in themselves.

38 For texts showing such fallacious reasoning, see, for example, Allison (1983, 7) (the transition from characterizing something *independently* of appeal to sensible

conditions to the conclusion that that thing is by definition *non*sensible); also 241 (when we conceive of things *in abstraction from* space and time, it is 'analytic' that we conceive of them as *non*spatiotemporal); and especially 242 ('what we have is the distinction between a thing considered in a certain relation, in virtue of which it falls under a certain description, and the same thing considered *in abstraction from* this relation, and *therefore not falling* under this description,' my emphasis).

NOTES TO CHAPTER TWO

1 The best reading of Kant's views is indeed that the (pure) categories cannot be shown or known to apply to any but the objects of our possible experience. Observe Avii, Bxxix, B146 (and the heading there), B147–48, B288, A240–44, B300–302 and B302 note, B306–309, A254/B309 ff., A287–88/B343–44; and note the implications of B431–32. Occasionally Kant writes as though he has proved that the categories do *not* apply to any but such objects (thus B149 at the end – but note the last sentence of B149, implying a simple denial of any knowledge of category application outside possible experience, and compare A286/B342 and A696 = B724). However, this latter mode of expression (to the extent that it is not merely a loose way of putting the first reading above, as seems to be the case at B149) makes claims that go beyond what he actually can hope to prove about the categories. (See Chapter Ten on the scope and limits of category application.)

2 See A108; A189–90/B234–35; A197/B242; A250; paragraph three of Kant to Beck, December 4, 1792 (Ak. 11, 395); Kant's marginal note to Beck's letter to Kant of November 11, 1791 (Ak. 11, 311); A116; A117 note; B131 ff.; A370; and A378.

3 See A320/B376–77 as a whole and texts like Reflexionen 2835 and 2836 (Ak. 16, 536–39).

4 For Cartesian ambiguities (overlapping to some extent the list of those noted here), see Kenny (1968, Chapter 5) and Margaret Wilson (1978, 156 ff.).

5 And questions of this sort have exercised earlier commentators. See, for instance, Paton (1936, vol. I, 576, note 1, and vol. II, 289, note 3), Kemp Smith's very different ideas (1962, 270–84), and numerous other writers.

6 See B157–58 note, A278/B334, B146, and Kant to Beck, May 26, 1789, Ak. 11, 51. In this last text Kant says that in order to explain the agreement of our forms of intuition with the categories, 'we should have to have still another manner of intuition than the one we have and another understanding with which to compare our own *and with which everyone could perceive things in themselves*' (Ak. 11, 51; Zweig trans., 153, my italics). The implication of the italicized phrase – compare A278/B334 – is that such intuitions and understanding as we do have exist in themselves in our minds.

7 The fact that the Kantian appearing sentence and the distorting-mirror sentence have a common logical and semantic behavior of course does not imply that the situations expressed by these sentences are completely analogous. (For example, that common behavior is compatible with the fact that appearing in the mirror is not necessary for the objects of our knowledge in the way that, for Kant, appearing in space and time is necessary for those objects.) But the common behavior exists at the points noted.

8 By this phrase and similar phrases in this book I mean those factors that concern what Kant takes to be the faculties, operations, and representations belonging to or

bringing about states of knowledge. I ignore the fact that knowledge states also will involve matters of evidence or justification (or simple reliability). For Kant's views on such matters – whose details he himself commonly ignores – see the discussions of *a priori* and *a posteriori* knowledge and of analytic and synthetic truths in the Introduction, the Doctrine of Method (A709–38 = B736–66, A769–94 = B799–822, and A820–31 = B848–59), and the *Logik* (Ak. 9; § V, 38–39; § VII, 49 ff.; § IX, 65 ff.; and § X, 81 ff.). Phrases in this book like 'states of knowledge' are to be understood, when required, to cover not only states that constitute genuine knowledge in the usual, knowledge-implies-truth sense but also cognitive states that claim, but may not actually be, genuine knowledge. (Such an understanding is required by Kant's use of '*eine Erkenntnis*.' Note the A58/B83 and A709 = B737 talk of 'knowledge that is false.' This talk has been stressed by Meerbote, 1980, 288; 1986, 193–94.)

[9] Kant's use of the term 'experience' ('*Erfahrung*') is ambiguous. By this term he sometimes means (a) empirical knowledge of objects, as here in the texts I have cited within brackets. On other occasions, however, he means simply (b) 'the [unsynthesized] raw material of sensible impressions' (B1; B1 also contains, later in the same sentence in the German, the (a)-meaning). See Beck (1978, Chapter 3, 40–41). I will make explicit in this book which meaning is intended when the context does not make that clear.

[10] For instance, at A93/B126, A109, A116, A121–22, A129–30, B126, B134–35, B137, B137–38, B141–42, B146, and B165 ff.

[11] See Bxxix–xli note, A22–23/B37, B68–69, A107, B139–40, B152–55, B157–58 note, A364, A381–82, B407, B412–13, and B420.

[12] Pistorius (1784, 345), as translated by Kemp Smith (1962, 323). See also 305, 307–308, 323–31, and 467 of Kemp Smith (not all of whose points, however, I accept), and Howell (1979, 341–42).

[13] Observe also A210/B255 ('All increase in empirical knowledge ... is nothing but an extension of the determination of inner sense, that is, an advance in time'), remarked on in Paton (1936, vol. I, 576, note 1, and vol. II, 289–90).

[14] This list of first-*Critique* causal-interaction texts is Adickes' (1929, 5–11), supplemented.

[15] Reflexion 5661 (Ak. 18, 318–20, written sometime during 1788–90; see especially 319) and Reflexion 6319 (Ak. 18, 633–34, written between 1790 and 1795; see especially 633, lines 24–26).

[16] Recall also the support given to this conclusion by our distorting-mirror discussion.

[17] Nothing is gained by objecting to this account that the existence in themselves of our states of knowledge contradicts Kant's view that we cannot know objects as they so exist (given that, for Kant, knowledge implies the possibility of knowing that one knows). The contradiction is clear. But it is simply another difficulty – of a long-familiar sort – for Kant's overall treatment of our knowledge. (The first-*Critique*'s transcendental-knowledge claims about the mental operations that yield us knowledge – operations that we have seen to exist in themselves – already are a notorious source of similar problems.) Nor is anything gained by simply rejecting both the possibility that knowledge exists in itself and the possibility that knowledge has merely the status of something appearing to us, while saying nothing further about how knowledge then is to be treated. (For this rejection and the preceding objection, see Allison, 1987a,

177, commenting on my original presentation of the present issues in Howell, 1979.) Creating mysteries resolves nothing. Moreover, Kant himself suggests no third possibility; and the last two (together with the idea of the joint noumenal and phenomenal occurrence of knowledge, which will not help Kant out of difficulties like those I discuss in this chapter) exhaust the field. (Given my account of existence in itself, either a thing has such existence, independent of the fact that it is represented to a mind, or it does not. Given Allison's recent-interpretation account, any object of knowledge that is considered at all either will be considered without appeal to its relations to the necessary conditions for our knowledge or else will be considered with appeal to those relations.)

18 The importance, for understanding Kant's theoretical philosophy, of distinguishing appear*ing* from appear*ance* theories was first emphasized, to my knowledge, by Prichard (1909, see, for example, 74) in his negative but highly incisive commentary and especially by Barker (1969) in a lucid and illuminating discussion. See also, for standard non-Kantian accounts, Chisholm (1950 and 1957), and for further comments on appearing and appearance theories in Kant, Howell (1979).

19 I also count as appearing theories any theories which, while denying that the object of knowledge has any existence in itself – and while holding, for example, that that object has existence only within the mind – still suppose that that object is logically distinct from any part of the act of mind (or of the representation or other mental entity) via which that object is known. As long as those theories hold that it is *that* object that is grasped and known by the mind, and not simply such an act (or representation), those theories may be said to suppose that that object itself 'appears to' (or 'appears within') the mind via such an act. (Compare the case below in Section 4 of an object, which we know to be fictitious, that is still depicted 'in,' and so 'appears in,' a picture; and note also the Kantian case there discussed.) Here and below I give a working account of appearing and appearance theories, without trying to present a definitive classification of all such theories.

20 Note, for instance, A19/B34 ff., B41, A26/B42 ff., A34/B51 ff., A42–44/B59–61, B67–69, B72, B153, B156, and A190/B235. See also Chapter One, Section 3.

21 Such a way of regarding those intuitions – as being objects (objects also having an existence in themselves) in the forms that *those* objects take as *they* exist in the mind – could itself be treated as a variant of an appear*ing* theory. This variant faces numerous problems. Among them are both the difficulty noted in Section 4 below for Kantian appearing theories and a problem like that noted in Section 4 for Kantian appearance theories. The variant also faces other, subtle difficulties. (For example, and as noted in Section 4, on the Kantian appearing theory the object known is the *same* object as an object having existence in itself, in the spatiotemporal form that that object takes as it appears in the mind. But on the present variant the object known is the same object as an intuition that – given our discussion in Section 2 – exists in itself in the mind. So, on the present variant, the object, having existence in itself, that appears in a spatiotemporal form always *is*, as it exists in itself, simply that very intuition existing in itself *in* the mind. Such a result does not agree with Kant's general view of objects existing in themselves. Nor does it agree with his specific view that the mind exists in itself along with all the intuitions that it has and appears to itself in a temporal form through various of those – inner – intuitions.) This sort of variant theory may well be

active, along with others of Kant's views, in various texts. Its existence might help to explain how, as we see below, Kant can so easily run together (or simply not bother to distinguish) appearing and appearance theories in general. However, I will not consider it further, except for comments in note 36.

[22] For Kant the object known, as appearance, is, strictly, the synthesized intuition. And he has an official distinction, which he commonly disregards, between a [mere] appearance as a '[categorially] undetermined object of an empirical intuition' (A20/B34) and a 'phenomenon.' ('Phenomena' are officially 'appearances [in the A20/B34 sense just given], so far as they are thought as objects according to the unity of the categories,' A248–49.) I disregard such points here.

[23] For the texts cited in the next three paragraphs and the fact that Kant does not sharply distinguish appearing and appearance theories, see Barker (1969) and also note Howell (1979). Defending his recent interpretation, Allison rejects the view that Kant ever identifies the object known with the intuition and thus arrives at two distinct things: the appearance and the thing in itself that produces the appearance in the mind. He admits that Kant talks of objects as identical to representations 'extremely frequently' (1983, 26) but holds that this talk can be explained away or else simply misrepresents Kant's true position. To my mind, this is a bit like interpreting the Bible as an atheist tract into which have slipped extremely frequent misleading expressions of faith.

[24] First quotation in this paragraph: Bxxvii; second: A42/B59; third: *Prolegomena*, § 13, Note I (Ak. 4, 287); fourth: B69 (compare also texts like A22/B37). All italics are mine.

[25] First quotation in this paragraph: A119–20 (observe also *Prolegomena*, § 13, Note III, Ak. 4, 290); second: A45/B62 (see also A43/B60, B164, and A369 ff.); third: A129; fourth: A391. All italics are mine.

[26] A38/B55, my italics; note also the sentences running over from Bxxvi to Bxxvii and from Bxxvii to Bxxviii.

[27] Kant's appearance theory, with its objects existing in themselves and producing appearances of themselves in the mind, is of course also non-Berkeleian. For texts (of both appearance- and appearing-theory sorts) distinguishing Kant's transcendental idealism from Berkeleian idealism, see B69–70; B274; Kant to Beck, December 4, 1792 (Ak. 11, 394–95); and *Prolegomena*, § 13, Notes II and III, and § 32. Note also Bxxvi–xxvii; A252; B306–307; A366–80; Bxxxix–xli note; B274–79; and *Prolegomena*, § 49.

[28] Expounding his recent interpretation and attacking my 1979 statement of the problem below for Kant, Allison (1987a, 168–70 and 177) denies that he or Kant holds any such identity to exist between these objects. This is the same Allison who in the same essay speaks of Kant's distinction 'between a thing as it appears and *the same thing* ... as it is in itself' (ibid., 170, my italics) and who writes that 'on a somewhat more precise specification' that distinction is 'between a *consideration* of a thing as it appears and a *consideration* of *the same thing* as it is in itself' (Allison, 1983, 240, my emphasis on 'the same thing'; see also Allison, 1990, 4). Allison's denial is mystifying. Except in cases not at issue here, like those of qualitative identity or nonstandard quantification, 'same' in 'the same thing' (and '*denselben*' in the original of the Bxvii quotation in the last paragraph) *means* the logical relation of strict identity between genuine objects of reference. (Compare 'the same Allison'

above.) If on his own interpretation Allison does not mean that we take an object of knowledge and then consider that *identical object* without appeal to *its* relations to the conditions for our knowledge of objects, then his interpretation is so far unintelligible. What Allison conceivably has in mind (compare his 1983, 250) is his claim that, for Kant, there is just one group of objects considered in two different ways, not two different kinds or realms of objects. As I have already argued, I take that claim to be mistaken. In any case, it is not a denial of a strict identity but is, rather, an assertion about the kinds of entities to which Kant's theory is committed.

29 By speaking of 'two worlds' I do not mean (whatever others may have meant by such a phrase) two worlds with logically disjoint domains of objects. Rather, and as the main text indicates, I mean (on Kant's appearing theory) two worlds or realms in which the same objects occur in different ways – in one world, in knowable, spatiotemporal forms; in the other world, in unknowable, nonspatiotemporal forms. (For simplicity, I ignore the possibilities, which are easily accommodated within the present discussion, (a) of merely phenomenal objects that have no further, unknowable existence in themselves and (b) of objects that have such an unknowable existence in themselves but that do not occur at all as phenomenal things in knowable, spatiotemporal forms.)

30 As I suggest in Chapter One, note 10, and remark again below in Chapter Five, note 6, one straightforward way of representing the phenomenal world W' is through Hintikka's sort of possible- (or perceptual-) worlds semantics for cognitive and other intensional notions. In such a semantics a claim of the form *H* knows that *p* is regarded as true at the actual world (or, in our present terms, at the Kantian world *W* of objects existing in themselves) just in case, at all worlds W_1, W_2, ... , compatible with what, at *W*, *H* knows, it is the case that *p*. The worlds W_1, W_2, ... , are regarded as, roughly, maximal states of affairs. At each of them the claim is true that the tree is conical (and all the rest of what *H* at *W* knows is true). But at such worlds the truth of claims whose holding or not holding is compatible with, but is not implied by, what *H* at *W* knows will vary. (Thus if, as far as *H*'s knowledge goes, it is not implied that the tree is – or that the tree is not – exactly 75 feet tall, then the claim 'the tree is exactly 75 feet tall' will be true at some of the worlds W_1, W_2, ... , but false at others of those worlds.) Given such a semantics, we can view the phenomenal world W' as, roughly, the result of 'packing together' the worlds W_1, W_2, ... , so as to retain what is true in them all (that is, so as to retain all the truths that *H*, at *W*, knows) while dropping out all the ways in which these worlds differ among themselves (as, for example, while dropping out their differing positions about the exactly 75-foot height of the tree). (Some further refinements would then be required in order to account for the existence of knowers besides *H* and for other points about our knowledge.) See Howell (1973, especially 230; 1979, especially 338 and note 9; and 1981b, especially 111–14 and note 36). See also Hintikka (1962; 1969a, especially 108–109 – a reprint of 1969b; and 1974, Chapters 9 and 10). Although I have found the above semantics very helpful (and intuitively very suggestive), nothing in my treatment of Kant depends on assuming it (or the specific treatment of W' that I just proposed). It should be possible to state my interpretations in terms of any other semantics that the reader prefers, as long as the other semantics is itself adequate and does not conflict with basic Kantian views.

31 See A50–52/B74–76, A67–70/B92–95 ff., and B141–42. As I urge in my 1973

paper, Kant seems to run together that-clause and direct-object knowledge (at least of a perceptual sort). But that point is unimportant here, for on any adequate interpretation of his theory Kant certainly accepts claims like (P) below as expressing knowledge.

[32] See B142, last sentences; A150/B189–90; A821 = B849; *Logik*, Introduction, § VII (Ak. 9, 49–51); and note Reflexionen 2143–45, 2155, 2161, and 2177 in Ak. 16.

[33] In Howell (1979) I describe what I there call 'the' problem for Kant in terms of the appearing-theory horn of the present dilemma, and then in the last section of the article I note the other, appearance-theory horn. The appearance-theory horn has been noted earlier. See, for example, Barker (1969, 288) and G. E. Moore (1953, Chapter 9). As far as I know, however, the appearing-theory horn (and, in effect, the overall dilemma) was first commented on in Howell (1979).

[34] Recall from Section 3 and note 27 Kant's anti-Berkeleian emphasis on objects existing in themselves and appearing to, or producing appearances of themselves in, the mind. Note, for example, Kant's Bxxvi–xxvii comment on not being landed in 'the absurd conclusion that there can be appearance without anything [having also an existence in itself] that appears.' See also A251–52.

[35] A structurally similar hypostatization would take us from (a) the object of knowledge, as that object exists in itself in an unknowable, nonspatiotemporal form in *W* to (b) what we could call the-object-of-knowledge-as-it-exists-in-itself-in-an-unknowable-nonspatiotemporal-form-in-*W*. Compare Prichard (1909, 75, note 1, and 78, note 1) and Howell (1979, note 23). For simplicity, I ignore the possibility of this further hypostatization.

[36] Here note Prichard (1909, 73 ff.). It may be that, through this sort of linguistic confusion, Kant first (in the order of logic) hypostatizes the object, as it appears in a knowable, spatiotemporal form, so as to arrive at the-object-as-it-appears-in-a-knowable-spatiotemporal-form; and then he takes the latter, hypostatized entity to be the appear*ance*, thus arriving at the sort of conflation that we have noticed earlier of appearing and appearance theories. (This way of arriving at that conflation yields a result subtly different from the variant appearing theory in note 21. On that variant, the intuition, the object known, is the *non*hypostatized object – which has also an unknowable, nonspatiotemporal existence in itself in *W* – as *that object* appears to the mind in a knowable spatiotemporal form. It seems conceivable, however, that in places Kant unknowingly accepts both this present way of arriving at such a conflation and the variant theory itself.)

[37] As indicated in note 42, a further reason for adopting some form of Option (III) is provided by the possibilities, discussed below, of 'merging and branching.' However, as observed there, Kant does not focus on these possibilities in any detailed, general way; and they conflict with the appearing-theory identity – which is important for Kant – of the object known with an object that has also existence in itself. So the existence of such possibilities will not support the idea that Kant should regard some form of Option (III) as a part of his own developed view of the object of knowledge.

[38] For simplicity, I continue to ignore the role of rule-governed synthesis in our knowledge and the exact relationship that is supposed to hold between outer sense and inner sense. Those points would considerably complicate, but would not essentially alter, the discussion.

39 The knower can know the time stream of his or her own inner states; and some of those inner states are outer intuitions. See A34/B50–51, A36/B53 ff., and the relevant parts of B66–69.
40 In the situation in which our understanding thinks an object for the intuition, that object is, as it is known through the intuition and that thought, a spatiotemporal thing. (It has all the spatiotemporal features and aspects that are represented by the elements of the manifold of intuition.) So, for reasons already remarked, 'the object is spatiotemporal' will be true at *W*. Hence, given the basic structural parallel between the two situations here noted, the above problem for Kant (or a very similar problem) now arises.
41 In terms of Figure 1, the merging possibility would be realized by distinct objects *o* and *p* in *W* being joined by solid lines to *o′* in *W′* (so that *o* and *p*, although distinct, would thus coincide in *W′*). The branching possibility would be realized by the single object *o* being joined not only by a solid line to *o′* in *W′* but also by another solid line to an *o″* distinct from *o′* in *W′* (so that two distinct objects would thus coincide in *W*). See also note 42 and Howell (1979, note 23).
42 The merging and branching possibilities just noted, like the last question here, evidently reinforce the pressure on Kantian appearing theory to accept some version of the Option (III) treatment of the object of knowledge (or else to accept a related, note-35-style treatment of that object as it exists in itself). In the Paralogisms Kant notes the specific, branching idea that one object existing in itself might appear to the mind as two phenomenally distinct things – namely, as our temporal mind and as our spatial body (A358–59; A379–80; compare B427–28). Less explicitly, an *On a Discovery* footnote (Ak. 8, 209–10; see Ameriks, 1982b, 10, and Allison, 1987a, 169) suggests merging and branching issues. However, Kant does not consider the implications, for his appearing theory, of such issues in any general, detailed way. (Thus for Kant it is always the mind in itself, the thing that *has* the representations of inner sense, that appears *to it*self in inner sense. He never writes as though several distinct minds in themselves might concurrently appear in inner sense – might concurrently appear to *what*? – as one empirical mind or temporal stream of representations. Nor, in his theory of morality, does he talk as though the single mind of which we are aware in inner sense might be the appearance of several distinct minds in themselves that perform mutually conflicting acts of will.) Rather, in different places Kant simply writes in terms of his appearing theory and of his appearance theory. On occasion he seems momentarily to adopt, unawares, an Option (III)-style theory. But he never discriminates sharply among these views.

NOTES TO CHAPTER THREE

1 Thus until the end of Section 5 I allow descriptions of the elements of the manifold as representing spatial parts or general properties of an object that is, as a whole, an extended spatial magnitude that persists through time.
2 Kant actually holds that *each* of our human acts of thought contains or can be accompanied by the *I think*. However, as we see in Chapter Six, it is not obvious that he is justified in holding this point (and his arguments for it fail). So I speak here simply of 'various' of our acts of thought as containing the *I think*.

3 See Chapter Ten, Section 8, for very brief remarks.

4 Textual evidence for the points here and in the next several paragraphs is given below.

5 See *Logik*, § 6. Once our understanding has thus arrived at a concept, it can of course combine that concept with other concepts to create concepts instantiated by no actual objects (note A729 = B757 on the arbitrarily invented concept of a ship's clock). See also *Logik*, § 4 and §§ 100–103. In creating concepts our understanding will in general also make use of *a priori* mathematical concepts and the categories.

6 Thus the *Fortschritte*: 'Now if a concept is taken as a sensible representation, that is, as empirical, it contains as a characteristic, i.e., as a partial representation, something already conceived in sensible intuition, and is distinguished from the intuition of the senses only by its logical form, namely, its general validity' (Ak. 20, 273–74; Humphrey trans., pp. 80–81). Or "Logik Philippi" (Ak. 24.1, 451): '1.... A concept is a general representation; representations that are not general are not concepts. Sensations, let them be taken ever so generally, remain singular [*einzelne*] representations. A singular representation is intuition. 2. For a concept it is required that in the case of the general representation I at the same time comprehend [*einsehe*] [the general representation's] relation in the series of subordination with other representations.' Note especially the second sentence (beginning 'Sensations, let them be taken ever so generally'). See also Ak. 24.1, 452, of "Logik Philippi"; "Logik Blomberg," Ak. 24.1, 251 and 253, and "Logik Busolt," Ak. 24.2, 653. And note Paton (1936, vol. I, 203).

7 Ak. 9, 94; compare also § 7, Note, and § 6, Note 1 (Ak. 9, 94–95). See, further, the relevant parts of Relexionen 2829–84, as listed in note 10, including especially Reflexion 2863 (in which at Ak. 16, 551, Kant says that, as abstracted, distinct concepts 'exist only in thoughts'). Somewhat similar views occur in pre-Kantian Cartesian philosophers like Arnauld and Locke. Arnauld notes that 'All existent things are particulars' but that through abstraction we arrive at general ideas that represent more than one thing (Arnauld and Nicole, 1662, reprinted 1970, Part I, Chapter 6, 86, first sentence of the chapter, Dickoff and James, trans., 1964, 50; see also Chapter 5 and the beginning of Chapter 4). Locke speaks, in a well-known passage, of 'ABSTRACTION, whereby *ideas* taken from particular beings become general representatives of all of the same kind' (*Essay* II.11.9). And he writes that (III.3.11; note also IV.17.8) '*Ideas* are general when they are set up as representatives of many particular things: but universality belongs not to things themselves, which are all of them particular in their existence...' There may well be a circle in Kant's account of giving generality to concepts through the process described above. How can the understanding, through that process, focus attention on a feature common to many objects in order to *give* that feature a generality unless that feature already possesses an inherent generality insofar as it belongs to all those objects? And abstractionist accounts of concept-formation have been subject to much other criticism. (See Geach, 1957.) However, we need not accept Kant's conceptualism in order to appreciate his overall thought or his specific development of the Transcendental Deduction. So I ignore such issues below.

8 Ak. 9, 58.

[9] Ak. 9, 95.

[10] See Reflexionen 2275–88 (Ak. 16, 296–300), paralleling the text of *Logik*, Introduction, § VIII.C, just cited; the relevant parts of Reflexionen 2829–84 (Ak. 16, 553–58), paralleling *Logik*, §§ 1–7 (here note especially Reflexionen 2834, 2839, 2849, 2854, 2855, 2859, 2863, 2865, 2876, 2877, 2881); and Reflexion 2902 (incorporated into *Logik*, § 8, Note). In connection with Kant's idea of a partial concept, note also Reflexion 3921 (Ak. vol. 17, 346): 'Through a predicate I do not represent to myself a part of the thing or have a concept of the part, but I represent to myself the object itself and have a partial concept of it.' The general idea of a partial concept is used by various of Kant's Cartesian predecessors. Thus observe Arnauld on knowledge by abstraction (Arnauld and Nicole, 1662, reprinted 1970, Part I, Chapter 5) and Locke, *Essay* III.6.32. See also G. F. Meier's *Auszug aus der Vernunftlehre*, 1752, §§ 259–62, reprinted in Ak. 16, 549–51, 559–61. (Kant used Meier's book as the official text for his logic lectures, and the Reflexionen on logic are Kant's notes to his copy of the book.)

[11] Compare, both for this point and for our earlier points about concepts, A68–69/B93–94, A78/B104, and A137 = B176.

[12] The language of a concept's 'presenting' a property is of course mine and not Kant's. However, one important strand in Kant's thought is the two-part view that (I) what, to speak strictly, concepts *represent* is the objects falling under them. (Recall, besides Chapter One and note 4 there, the *Logik* and other texts cited above – texts in which concepts, considered as grounds of knowledge, are said to be parts of representations of *things*. And observe A50/B74, which remarks that through concepts we think objects; A69/B94; and Reflexion 3921 in note 10.) And (II) concepts achieve this representation of objects, in part, by displaying to our mind (or being made, by our understanding, to display to our mind) general properties that belong to the objects. In considering this two-part view, I introduce the word 'presents' as a convenient label for the latter, (II)-style 'displaying' operation of concepts. Because Kant does not make the matter clear, I leave it open exactly how concepts carry out this operation.

[13] Ak. 16, 297.

[14] To say that the concept itself *is* the general property raises questions about how such a property can be *in* the mind. Such questions arise also for the Aristotelian or quasi-Aristotelian view (of concepts as 'forms' or general properties that are present in the mind in knowledge) to which the Kantian view is, as I suggest below and in Chapter Ten, related. I ignore such questions here, as they do not affect Kant's argument in the Deduction.

[15] Thus see B12 (in which, at the paragraph-end there, the predicate, weight, is described as a concept that I attach synthetically to the concept of body); *Logik* § 7 and its Note (a concept, 'as a ground of knowledge, that is, as a mark,' Ak. 9, 95); Reflexion 2279 (the partial concept – Kant says – is the mark, Ak. 16, 297–98); and Reflexion 2281 ('concepts ... are marks with a general use,' Ak. 16, 298).

[16] The point here concerns the fact, that, for Kant, it is our understanding's acts of thought that 'refer intuitions to objects.'

[17] Compare a somewhat similar idea in Descartes, *Principles of Philosophy*, Part 1,

§ 59; observe Arnauld: 'the higher degree [in a list of abstractions, say the list: equilateral triangle, triangle, plane figure bounded by straight lines], being less determinate, can represent [*peut représenter*] more things' (Arnauld and Nicole, 1662, reprinted 1970, Part I, Chapter 5, 85, Dickoff and James trans., 1964, 50, with 'can represent' for their 'stands for'); see Locke, *Essay* II.6.32 ('each *genus* is but a partial conception of the *species* comprehended under it'), II.11.9 (as quoted above in note 7, and observe also the II.11.9 claim that 'the same color being observed to-day in chalk or snow, which the mind yesterday received from milk, it considers that appearance [the color] alone, [and] makes it a representative of all of that kind'), and III.3.11 (as quoted in note 7); note also G. F. Meier's *Auszug* (see note 10 above), § 115 (Ak. 16, 296–97), § 117 (through marks we 'represent to ourselves' something as present or absent in a thing and thereby recognize what that thing is or is not, Ak. 16, 305); and compare, finally, the discussion of higher and lower concepts in the passages from Meier's work assembled at Ak. 16, 559–61, in which Meier treats marks or properties like virtue and generosity as general, abstracted concepts.

[18] Recall from Section 1 and note 1 that I here focus on the case of the manifold belonging to intuitions that represent spatiotemporal objects like trees, and until the end of Section 5 I ignore the description that Kant should give when the manifold is considered in the light of the Section 1 minimum Deduction assumption.

[19] Here we touch on many Kantian complications (related, among other things, to the form-matter issues discussed below in Section 5). Kant sharply separates intuition from thought and insists that the matter of knowledge is given through intuition, thought by itself supplying simply ways of synthetically organizing that matter. So he must suppose that intuition-elements, as they are given to our mind in independence of thought, in some way carry or embody that matter. Moreover, he accepts the general representationalist picture according to which mental entities such as intuition-elements carry such a matter by in some way displaying that matter to the mind's inner acts of awareness. So he must suppose (on the appearing theory) that intuition-elements, as they are given to our mind, display to inner acts of awareness the matter in question, so that our thought can then act conceptually on it. It is this showing or displaying function, which Kant does not characterize in any great detail, that I here label 'putting before the mind.' See A19/B33 ff. or A374, cited below: 'perception is that whereby the material required to enable us to think objects of sensible intuition must first be given'; and A246: 'what sort of a thing it is that demands one of these [logical] functions rather than another remains altogether undetermined' by the logical functions and pure categories themselves, that determination requiring a relation of the logical functions and pure categories to sensible intuition. See also *Critique of Judgment*, Introduction, § VII, on 'whatever [in the representation of an object] serves, or can be used, to determine the object (for cognition) is its logical validity' (Ak. 5, 188–89; Pluhar trans., 28) or the § 77 observation (Ak. 5, 406; Pluhar trans., 290) that 'the universal supplied by *our* (human) understanding does not determine the particular; therefore even if different things agree in a common characteristic [*Merkmale*], the variety of ways in which they may come before our perception is contingent ... our understanding is a power of concepts, i.e., a discursive understanding, so that it must indeed be contingent for it as

to what the character and all the variety of the particular may be that can be given to it in nature and that can be brought under its concepts.'

20 The present account of this putting-before-the-mind by the manifold of representations is sharpened below. Relevant examples include A98 ff. on the three-fold synthesis (note also A119 ff.); the discussions of perceiving a house (B162, A190–92/B235–36, A192/B237–38); the cases of perceiving water freezing or a ship first upstream and then downstream (B162–63, A191–93/B236–38); the case of delineating a dog (A141/B180); various examples of recognizing or constructing, in the imagination, geometric figures or numerical representations (A103, A105, A124, B154, A140–41/B179–180, A162/B202 ff., A224/B271); and *Logik*, Introduction, § V, on seeing a house in the distance or seeing the Milky Way (Ak. 9, 33–35; compare Reflexion 1681, Ak. 16, 80–81). Students' lecture notes make clear that Kant was alert to the role, in our knowledge, of matters like points of view and conditions of observation. See "Metaphysik L_1" (Ak. 28.1, 236) and "Logik Blomberg" (Ak. 24.1, 126).

21 This fact is remarked in Parsons (1964), reprinted in Parsons (1983); see Parsons (1983, 99–102, and the addition to note 3 at 100). Parsons' essay raises deep, important questions about Kant's view of synthesis and about the relationship of that view to other aspects of Kant's theoretical philosophy. For earlier Cartesian ideas related to Kant's first claim, note the discussion of knowing the piece of wax in Descartes' *Second Meditation* and Arnauld on our inability to understand composite things unless we consider them part by part [*par parties*] or through their different aspects (Arnauld and Nicole, 1662, reprinted 1970, Chapter 5, 83; Dickoff and James trans., 48); observe Leibniz on distinct ideas ("Meditations on Knowledge, Truth, and Ideas" and *Discourse on Metaphysics*, § 24, Gerhardt ed., vol. 4, 422–26 and 427–63); and compare Locke, *Essay* II.1.3, II.14.3 ff., and II.23.1 ff. Note also Berkeley on an idea as functioning as a sign for another, anticipated idea of the same object (*A New Theory of Vision*, §§ 25, 26, 45; *Principles of Human Knowledge*, §§ 58, 59, 65, 66). And see Wolff (1960) on Hume.

22 I here ignore the distinction, which Chapter Nine will bring out, between the logical functions of thought and the logical forms of judgment.

23 See A70–76/B95–101 and Chapter Nine.

24 See A8; B12; A9–10/B12–14; A69/B93–94; B131, last sentence; the B140 heading of B-Deduction § 19; B143; and Chapter Nine.

25 Kant never makes it clear exactly how to combine (a) his view of a property as first becoming general through the understanding's assigning such a form, or generality, to a matter for a concept with (b) his basic theory of the synthesis of the manifold of intuition, through a judgment, that occurs when we know an object. As a result, he leaves open many questions. (For example, when does the tree's conicality first acquire its generality, and how does that conicality occur before the mind so as to be available for the understanding's assignment of generality?) I disregard these questions as they do not affect the Deduction's main reasoning.

26 Thus the functioning, in our knowledge, of the elements of the manifold of intuition of the tree will correspond to the logical relation, the logical quantity, and the logical quality of the judgment that the tree is conical. That judgment also has an

assertoric logical modality. However, and as A74/B99–100 shows, the logical modality of a judgment concerns not the content of the judgment (the object) but the relation of the judgment to the understanding ('the value of the copula in relation to thought in general,' A74/B100; see also A219/B266). So there is no special functioning of the manifold of intuition, itself, that corresponds to the modal aspect of judgment.

[27] Ak. 4, 289. Note also A68–69/B93–94, A79/B104, and B128–29. The 'given concepts' in various of the preceding quotes must, by our above discussion, be given only as matters for concepts. See also below on Kant's treatment of sensations and elements of the manifold.

[28] Ak. 17, 616–17.

[29] Ak. 16, 298; Ak. 17, 620, 623, 643–47, and 653–57; and Ak. 18, 385–87 (see especially 386, lines 15 to 17).

[30] Ak. 28.1, 404–406. My translation is as literal as possible and does not pretend to an elegance which these lecture notes lack. Guyer (1987, 124 ff.), also considers the present text, for a different reason from mine. See, further, the discussion of related texts – for example, B128–29 – in Chapter Ten. For Kant's view that knowledge of an object is always only through consciousness of predicates of the object, see, also, Lewis White Beck's succinct exposition (1955), reprinted in Beck (1965, 74–91) and in Wolff (1967, 3–22).

[31] See texts like A120, B147, and B160–61.

[32] 'Representation' is here used by Kant in the wide sense remarked in Chapter Two. For perceptions, see also A115–16, A119–20, B147, B160, B207, A225/B272, A367, A374, A378, Reflexionen 2835 and 2836 (Ak. 16, 536–39), and *Logik*, Introduction, § VIII.C (Ak. 9, 64).

[33] Reflexion 3930: Ak. 17, 352; the "Logik Blomberg" quotation: Ak. 24.1, 253. Note also Reflexion 3921 (Ak. 17, 345): 'All our concepts are marks drawn from sensation'; and compare B12, A225/B273, and Reflexionen 2275–2288 (Ak. 16, 296–300).

[34] In the above discussion I have not considered all the subtleties of Kant's views on sensations when those views are taken as an independent topic. (For an interesting discussion, see George, 1981.) Some further issues are remarked below in Sections 4 and 5.

[35] For reasons for holding that intuition-elements, as given, behave in this way, recall note 19, and observe the form-matter discussion below in Section 5. Where Kant uses the one term 'represent' to indicate an intuition's function of displaying an object to the mind, in order to clarify his views, I have now introduced the additional terms (a) of a concept's 'presenting' (on one Kantian treatment of concepts) a property and (b) of an intuition-elements's 'putting before the mind' (on Kant's appearing theory) a matter for a concept (or, as we see in Section 4, a matter for a spatial part). As the discussion has brought out, these three terms designate what, within Kant's picture of knowledge, are logically distinct operations (of whole intuitions, of concepts, and of intuition-elements, respectively).

[36] Note also the talk of 'unanalyzable' [*unauflösliche*] concepts in *Logik*, Introduction, § VIII.C (Ak. 9, 59); and recall that I ignore numerous complications in Kant's views on sensation. While – as I and others have argued – an intuition's representation of an object involves intensionality (and intentionality), the present sort of

'putting before the mind' of a potentially general feature by an intuition-element is not an act of thought and will not involve such a phenomenon. (For nonintentional notions of sensation, see George, 1981. For intensionality and representation, see Chapter Five.)

37 Thus in a famous passage Aristotle writes that 'when one of the undifferentiated things makes a stand, there is a primitive universal in the mind (for though one perceives the particular, perception is of the universal – e.g. of man but not of Callias the man); again a stand is made in these ... it is necessary for us to become familiar with the primitives by induction, for perception too instils the universal in this way' (*Posterior Analytics*, Book II, Chapter 19, 100ᵃ–100ᵇ; Barnes trans., 81). Individual substances, the objects of sense perception, are for him matter plus form, the form being *in* the substance – say the property of being human in Callias. For Locke, '*our senses*, conversant about particular sensible objects, do *convey into the mind* several distinct *perceptions* of things, according to those various ways wherein those objects do affect them. And thus we come by those *ideas* we have of [sensible qualities like] *yellow*, *white*, *heat*, *cold*... .' (*Essay* II.1.3). In perception we notice such ideas in our minds (II.9.1–4), and those ideas function as marks whereby the objects from which they derive may be recognized (II.11.9 and other texts cited in notes 17 and 10). For Meier, see the texts cited in note 17 and his 1752 *Vernunftlehre*: To achieve a correct, clear knowledge of things, we must distinguish the manifold (*das Mannigfaltige*) in the objects we know, by determining what marks are to be encountered in things (*Vernunftlehre*, 482, in Ak. 16, 624). The implication that the manifold (of our representations) of an object is made up of conceptual marks is reinforced and drawn closer to Kant by Meier's claims that 'through immediate experience we acquire only such concepts as are sensations' (*Vernunftlehre*, 417, in ibid., 542) and that what we perceive in a thing are marks of it like bitterness and sweetness (*Vernunftlehre*, 418–20, in ibid., 542–44). Compare Descartes: we know substances only through their attributes (*Principles of Philosophy*, Part I, § 52; and elsewhere); and Arnauld: 'our mind [is] accustomed to know most things as qualified [*modifiée* – as having modes], since it knows them only through their accidents or qualities which strike our senses' (Arnauld and Nicole, 1662, reprinted 1970, Part I, Chapter 2, 74; Dickoff and James trans., modified, 1964, 39).

38 Thus the examples of perceptual synthesis in note 20 are all naturally understood in direct-object terms.

39 For space as such a set of relations, see B67, A265/B321, and A283/B339 ff.; for space as unified thing, see B160 note on space as form of intuition versus space as formal intuition.

40 Because they do not directly affect the Transcendental Deduction, I ignore complications about outer intuitions on Kant's appearance theory. (For example, and despite Kant's own practice, on that theory it would seem most accurate to identify a particular outer-intuition element not with a potentially definite spatial part itself but with, roughly, a potentially definite spatial part as observed under given conditions from a specified point of view.)

41 See, further, A42/B59–60, B66–67, A76–77/B102–103, A99 ff., A101–102, B160 and B160 note, B202–203, A162/B203 ff., B207–208, A263–64/B319–20, B457 note, and A494/B522. Observe also George (1981, 238–41).

42 This third aspect emerges because Kant takes each sensation to have a degree or magnitude and links that fact, through his views on the logical quality that a judgment has, to the possession, by the object judged about, of intensive qualities. See A143/B182–83, B207–208, A166/B208 ff., and Chapter Ten. As Kant's views here are obscure and rather peripheral to the argument of the Deduction, I largely ignore them below.

43 Observe the characterization of a mark (as a 'ground of knowledge of this thing itself') given in the *Logik*, Introduction, § VIII.C (Ak. 9, 58) text quoted in Section 2 and also the B12 passage quoted in Section 2 shortly after that text; and note Reflexion 3921 (Ak. 17, 346), quoted in note 10.

44 Thus given B201–202 note, matters for concepts should yield heterogeneous elements of the manifold, whereas matters for spatial parts should yield homogeneous elements of the manifold. As the discussion below shows, (ii) raises complex issues. In stating (ii), I here try simply to present one plausible interpretation of Kant's views.

45 See A523–24 = B551–52 and, for the present point and the point in the preceding sentence, Parsons (1964), as reprinted in Parsons (1983, especially 98–104). The present point about objects as given as wholes in intuition seems to connect also (in ways whose details are not immediately clear) with A25/B39 and A31–32/B47 on space and time as being given as single, unique things and as not being built up out of previously given, constituent spaces and times. See also A438 = B466, A512–13 = B540–41 and A169–70/B211–12.

46 See A512–13 = B540–41, A523–24 = B551–52, and Parsons (1964); and compare A163/B204 and A169–70/B211.

47 See note 19.

48 This sort of fact *might* be held to pose no such problem. However, I doubt that the line of thought below really resolves (a) satisfactorily. As we note in Chapter Four, Kant seems committed to the position that any relations that we can know to hold are 'works of the mind' that do not actually obtain among entities without activities of thought. But the relations here noted are ones that we can know. So Kant really should not grant that such matters stand in such relations in the way suggested below. (Nor can he argue that all that holds between such matters, as they are given, is the potentiality for standing in actual relations. If *a* and *b*, as given, potentially stand in *R*, then *a* and *b*, as given, actually stand in the relation *R'* of being such that it is potential with respect to them that they stand in *R*.)

49 Parsons (1964) makes a subtle and interesting proposal (in terms of a distinction among levels of complexity in the appearances of objects) that would let one eliminate the idea in question. And some main parts of my interpretation of the Deduction can be made independent of that idea. (These parts include, for example, Chapter Five on intensionality and Kantian thought-consciousness, Chapters Six and Seven on unity of apperception and its consequences, various parts of Chapter Eight on the concept of an object in general, and various parts of Chapters Nine and Ten on the logical functions of thought, judgment, and the categories.) However, and as noted below in Section 5 in connection with an adaptation of Parsons' proposal to regress issues, at least in its adapted form that proposal seems to require giving up some of Kant's own claims. (Parsons himself observes that the distinction underlying his proposal is not fully present in Kant. See Parsons, 1964, reprinted in his 1983, 102 ff.;

and note also his 1983 addition, at 100, to his footnote 3.) Moreover, and as remarked immediately below, important Transcendental Deduction claims seem to lead, given Kant's picture of knowledge, to the idea that through the elements of the manifold we take in individual parts and properties separately. So I will not develop Parsons' proposal here.

50 For the Kantian grounds underlying the points made here, see also Chapter Four.

51 See the Transcendental Deduction example of drawing a line in thought (B154, B137–38, A162–63/B203; compare B162, B292, and A102) and A162–63/B203–204 on extensive magnitudes ('the representation of the parts makes possible, and therefore necessily precedes, the representation of the whole'; 'all appearances [and so all spatial objects] are ... intuited as ... complexes of previously given parts').

52 We have been reading the present conclusion in the style of (ii) above, as holding that individual elements of the manifold put spatial parts of objects, in potential forms, before the mind. A different interpretation is suggested by Kant's comments on spatial objects as given as wholes in intuition and on our arriving at the parts of space through the introduction of 'limitations' (A25/B39; compare B419, A412–13/B439–40, and A438 = B466). For this different interpretation, through outer sense the potentially unitary space occurs in the mind. Through sequential thought-acts of attention limitations are introduced into this space so as to yield sequentially presented intuition-elements. Each of these elements provides to the grasp of one such thought-act an individual, presumably potentially definite, part of space (or part of a spatial thing). (And then synthesis makes all these elements function as a single intuition representing parts of the one, overall definite space – or parts of one definite spatial thing.) This different interpretation yields problems like that in the first sentence of (b) above. It also seems to conflict with the many texts like A120, in which Kant says that the elements (perceptions) in the manifold, as they are given, occur in the mind separately and singly; for this interpretation makes the actual existence of a multiplicity of such elements dependent on acts of attention in thought. (There are, however, many subtle issues lurking here; see A524 = B552 and Parsons, 1964, at 103 of Parsons, 1983.) Moreover, this interpretation leads to essentially the same sort of treatment of the Deduction (in terms of the subjection of a manifold to unity of apperception) as does our (ii)-style account. So I ignore it hereafter, despite its intrinsic interest.

53 See also A124, B137–38, B162, A162/B202 ff., and A224/B271.

54 Ak. 16, 298.

55 See note 20.

56 See A22–23/B37, A33–35/B49–51, A98–99, and especially B66–69 and § 24 and § 25 of the B-edition Transcendental Deduction. The interpretation of the fine details of Kant's account of inner sense is quite difficult and for the most part irrelevant to the parts of the Transcendental Deduction on which we focus in this book. In regard to such details I follow the lead of Wolff (1963, 191–202, especially 199–201), except as observed in note 57. On Wolff's type of interpretation, initially outer sense presents a spatial manifold. In order to become conscious of this manifold, our understanding seeks out (apprehends) its elements in acts of attention. (Apparently, also, our understanding does the same sort of thing in regard to our noncognitive, and non-outer-sense, representations of pleasure, pain, and the will – compare B66 – although

Kant is not clear on this point and it is not discussed by Wolff.) Through these acts of attention, our mind affects inner sense in such a way that these elements take on a potentially definite temporal order. When these elements are then (in the order of logic) acted on by imagination (and by other synthetic operations), they take on an actual, definite, time order of the sort mentioned above in the main text. (Here I accept a suggestion that Wolff makes but rejects at 201–202. I also add '(and by other synthetic operations)' to indicate that, insofar as we recognize and can know this latter sort of (subjective) time order, it must itself involve concept application by the understanding as well as activities of mere imagination. Here note B151, in B-Deduction § 24, on the relevant transcendental synthesis of imagination as being figurative synthesis that is 'directed to' unity of apperception and the categories.) See also Kemp Smith (1962) and Weldon (1958), as discussed by Wolff; and note Ameriks's criticisms (1982a, 243 ff.). (At least some of those criticisms do not seem to apply to my present version of Wolff's interpretation. The others can, I think, be met.)

[57] Aside from the reasons for my view already presented, see B154: 'Inner sense ... contains the mere form of intuition, but without combination of the manifold in it, and therefore so far contains no *determinate* intuition, which is possible only through the consciousness of the determination of the manifold by the transcendental act of imagination (synthetic influence of the understanding upon inner sense).' Wolff has been an eloquent defender of the interpretation of inner sense that I here reject. He argues that, despite demands within Kant's overall theory for my type of view, a 'coherent, non-metaphorical account' of synthesis requires seeing synthesis as a temporal activity that applies to entities given to us, through inner sense operating by itself, as occurring in a definite time order (Wolff, 1963, 202; see 200–202 and 126–32). Wolff is right to worry about the difficulty of making clear, detailed sense of atemporal acts of synthesis as helping to create actual, definite time orders. But I do not think this difficulty is clearly insuperable. (For a defense of the general intelligibility of talk at least of atemporal experience, see Walker, 1978, 34–41.) In any case, in order to come to grips with Kant's own account of knowledge, one must appeal to such ideas. Doing so certainly need not rule out offering temporal analogies for those ideas or developing, for contemporary philosophical purposes, temporal reconstructions of them. Moreover, and as urged in Chapter One, in any event Kant has *not* demonstrated the atemporality of entities existing in themselves.

[58] Moreover, and as I argue below in Chapter Five, in the cases of interest in the Transcendental Deduction we *should* in fact (while accepting my present view of inner sense) take the elements in question to occur mentally, through imagination and other synthetic operations, in an actual, definite sequential order before thought-consciousness.

[59] Lewis White Beck in fact implies that this distinction has a claim to stand as part of a postcard-length summary of Kant's system (Beck, "Kant's Strategy," reprinted in Beck, 1978, 17–18; see also Beck, 1969, 458).

[60] I thank H. S. Thayer for comments on the form-matter topic. He is not responsible for the uses to which I have put them.

[61] The use, here, of this explicit-implicit terminology is mine, not Parsons', as are various of the points below. See Parsons (1964), reprinted in Parsons (1983,

especially 100–102). Parsons identifies three levels of complexity in the appearances of objects in perception and (in terms of the first two levels) distinguishes between what is explicitly and what is only implicitly given. He does not talk of intuitions themselves as being explicit or implicit, and he raises a number of issues that, for simplicity, I ignore.

62 Or so I here proceed, supposing that if, at a later stage in the perception of an object, parts of the object come to our attention that are not displayed by the implicit intuition, then that intuition is replaced by an explicit intuition. One might also suppose that an implicit intuition, as displaying a part of the object that is implicitly complex, itself somehow involves an implicit complexity or manifold that at a later stage may become explicit. (Compare Parsons, 1964, 102.) But how can such an implicit intuition, as one actual part of an overall, synthesized intuition (as, in general, it will be) involve such a manifold without itself involving an actual – or merely an implicit(?) – synthesis of that manifold? Yet to accept an actual such synthesis seems to re-create something like the preceding regress (or else it returns us to non-manifold-containing implicit intuitions). And I am not clear how to develop the idea of an implicit synthesis in a satisfactory way. One might suggest – compare A524 = B552 – that an implicit intuition contains a manifold in the sense that if attention in thought were paid to that intuition, then it would divide into a manifold of subelements each representing an object-part. (The implicit synthesis would then amount, roughly, to this counterfactual fact's holding with respect to the implicit intuition.) However, either (a) the elements of the implicit intuition actually exist in the intuition prior to the counterfactual division's actually being carried out or (b) they do not so exist. (a) fits texts like A513 = B541. But if the elements are actual then they stand in actual combination relations (for example, of all occurring in this one implicit intuition). Yet, for Kant, and as noted in Chapter Four, the holding of actual combination-relations seems to depend on actual acts of thought. (b) avoids this problem but seems not to fit texts like A513 = B541. (b) also conflicts with what looks like the clear fact that if a spatial object is given then all of its parts are actually given too.

63 That is, and as can be seen from Chapters Six to Ten, main Deduction points about unity of apperception, about the concept of an object in general and its role in synthesis, and about the logical functions of thought, judgment, and the categories, are independent of views like the claim that every intuition representing a spatial object contains a manifold of elements that represent spatial parts of that object.

64 Questions like these last last two are raised, along with other problems, in Schrader (1958), reprinted in Wolff (1967, 134–55; see 136–37).

65 This statement of the strong and the weak readings is sharpened in Chapter Eight. As I state the weak reading here and below, the notion of intuition that that reading involves evidently should be taken in a sense corresponding to the wide sense of 'representation' remarked in Chapter Two.

66 For a serious problem about the use of the Deduction to show this sort of point, see, however, Chapter Eight, note 40. (To the extent that – as in the first half of the B-Deduction – the argument concerns only an 'intuition in general,' this use of the Deduction also would have to appeal to facts about the *a priori* forms of our human intuition in order to show the point about spatial parts.)

NOTES TO CHAPTER FOUR

[1] Objects simply of inner sense (like our own sequences of representations) are of course also supposed to be subject to various categories; we return to that topic later.
[2] Thus note the implications of A97 ff., A118, and A120.
[3] I ignore various fine points about the role, in relation to Kant's notion of the given, of operations of synthesis. In the light of our discussion here and earlier, it should be clear how to accommodate, in the above account of the given, Kant's talk of objects, as well as of intuition-elements, as being given (for example, at A19/B33, A50/B74, A43/B60, and A62/B87).
[4] The relating together of course turns out to involve synthesis, for Kant. See B130–31: 'the concept of combination includes, besides the concept of the manifold and of its synthesis, also the concept of the unity [*Einheit* – oneness] of the manifold. Combination is representation of the *synthetic* unity of the manifold.' Compare A99, A117–18, A120, and B138. We may ignore the distinctions that can be drawn among act, concept, and representation of combination. In principle, and without attempting an analysis, I think that the idea of entities as making up one thing should be taken as liberally as possible, since we are concerned with general conditions on knowledge of objects. (Thus any set will be 'made up' of its members.) Eventually Kant narrows the idea, as it applies within experience, to relations definable in terms of the categories.
[5] There is a complication here that ultimately is not immensely important but should be noted to avoid confusion. Kant talks, in the ways indicated, of given intuition elements as occurring in the mind without standing in relations that make them form combinations (A120: 'different perceptions ... occur in the mind separately and singly'; B134: 'Combination does not ... lie in the objects, and cannot be borrowed from them'). But this talk cannot be literally correct, for any given intuition elements will stand in *some* relations and form *some* unities, whether or not those relations or unities are grasped by the mind. (Thus, and whether or not I recognize the fact, my inner-sense elements *m* and *n* are related together as both belonging to me, and *m* and *n* jointly make up *one* two-element collection of my inner-sense representations.) The best way to deal with this problem is this. By claiming that combination cannot be given, Kant should mean, strictly, that no combination and no relations can belong to the given elements of the manifold in such a way that that combination and those relations are, simply through the givenness of those elements, made available to the mind (in outer or inner sense or in thought-consciousness) so that the mind is able without further processing to become aware of that combination or those relations as belonging to those given elements. To simplify the discussion below, I will proceed in Kant's own terms, but I will assume that the arguments and discussion have been implicitly adjusted to agree with this reformulation of the combination-cannot-be-given claim.
[6] There is a further, fourth, point, connected to Kant's view that a *necessary* combination of the manifold cannot be given. But this further point is best discussed later, after our account of the necessity of unity of apperception.
[7] See A98–99; compare B207–208 and A167–68/B209–10.
[8] Here especially the adjustments indicated at the end of note 5 need to be made to Kant's views, for he clearly allows that relations (unknowable by us) may hold among objects existing in themselves. (Thus note Chapter One on the affection relation.) In

particular – compare note 5 – there will be relations holding among given intuition-elements. What Kant must therefore be taken to mean is that no relations (or properties) that we can know to hold belong to given intuition-elements and then are made available to our mind for it to recognize as holding. Rather, any relation we can know to hold depends for its existence or holding on activities of our thought and understanding.

[9] See, for example, Kant's comments on such Humean experience at A90/B123, A102, A111, A112, A121–22, and B134, and in his May 26, 1789, letter to Herz (Ak. 11, 48–55; see especially 50 and 51–52 on the difference between being conscious, separately, of each individual representation and being conscious of all these representations as together representing an object).

[10] See also the final summaries of the Deduction at A130 and B168–69 and the emphatic statement at *Metaphysical Foundations of Natural Science*, Ak. 4, 476 note.

[11] In the Deduction, Kant uses 'objective validity' (*objektive Gültigkeit*) and 'objective reality' (*objektive Realität*) as stylistic variants. Thus note the shift from 'objective reality' (A84/B117) to 'objective validity' (A87/B119–20, A88/B121, A89/B122, A90/B122, A93/B126) in the introductory section; the same shift from A95 to A97 and again from A109 to A111; the B137, B140, and B142 talk, in the first half of the B-Deduction, of objective validity; the B145 (§ 21) comment that § 26 of the second half of the B-Deduction will show the 'validity of the categories with respect of all the objects of our senses'; the shift at B148 (§ 23) from talk of validity to talk of objective reality (and B150, in § 24, on objective reality); the § 26, B161, assertion that the categories are 'valid *a priori* for all objects of experience,' and the B168, § 26, talk of objective validity. Outside the Deduction Kant also frequently shifts from one term to the other – for example, at A28/B44 (quoted in the next paragraph below) and A155–56/B194–95.

[12] *On a Discovery* (Ak. 8, 191) and *Fortschritte der Metaphysik* (*What Real Progress*) (Ak. 23, 279); compare A223/B270 and A220–21/B268. (However, in these last two texts Kant is considering, specifically, concepts and possibility.)

[13] See also later in A94/B126, A125–26 (quoted above), A92 at the end, and B147.

[14] A93/B126; compare A736–37 = B764–65.

[15] (A) is the minimum that Kant means and needs, in the Deduction; however, as we see later, his views ultimately lead to stronger, *de re* necessity claims that go beyond (A).

[16] See A115–17, A119, A121, B135, B136, B138–39, B145, and B168–69.

[17] A95 claims the categories are contained in the concept of a possible experience and are conditions of a possible experience. By itself, that claim is ambiguous between (A) and (C). But the context shows that (A) is meant. (Note the immediately following A96 paragraph: 'a concept which ... expresses such a formal and objective condition of experience [not: such a condition of the possibility of there being experience].' And note A96, again, on the categories as the 'concepts which thus contain *a priori* the pure thought involved in every experience.')

[18] See especially Wolff (1963, 52–55), and his comments on the synthetic, progressive character that Kant intends the reasoning in the first *Critique* to have. For present purposes, I take p to be a necessary condition of q just in case q implies p, expressing that implication truth-functionally where relevant (as in (A)).

[19] Modern work on presupposition goes back to van Fraassen, Strawson, and ultimately Frege. For an application to Kant, which has inspired various of my suggestions below, and further references, see Brittan (1978, 35–42; compare 112 ff.).
[20] Besides the texts that we interpreted in a necessary-condition way in discussing (A), note, for example, the earlier A93/B125–26 quotation (in the paragraph tagged by note 13) and its clause 'antecedent conditions *under which alone* anything can be ... thought as object in general' (my italics). The conditions mentioned in this clause are clearly amenable to – and, to my mind, cry out for – treatment as straightforward necessary conditions.
[21] See A809–18 = B837–46 and A825–30 = B853–58.
[22] First quotation in this paragraph: *Prolegomena*, § 4, Ak. 4, 274; second: § 5, Ak. 4, 279; third: § 4, Ak. 4, 274; fourth, § 4, Ak. 4, 275.
[23] See Wolff (1963, 44–56); Brittan (1978, 30 ff., 112 ff.); Hintikka (1973, Chapter 9).
[24] More specifically, the first *Critique* deductively infers the objective validity of the twelve categories and the truth of each of the principles of the understanding. (Kant supposes that these results establish also that there will be some specific synthetic *a priori* principles of mathematics and dynamics, but by themselves they do not demonstrate the truth of any particular such principles. See A162/B201–202 and the Preface to the *Metaphysical Foundations of Natural Science*, Ak. 4, 467–79; compare ibid., 482.)
[25] In the Preface; Ak. 4, 474–76.
[26] Ak. 20; see especially 271–72.
[27] For the development of the Deduction, see Vleeschauwer (1934–37) and his historical summary (1962); Beck (1969, 457–81); Guyer (1987, parts I and II); Carl (1989); Beck (1989).
[28] The quotation above from the 1772 letter is at Ak. 10, 130; Zweig trans., 72. Later in the same paragraph, Kant asks 'how my understanding may formulate *real* principles concerning the possibility of [*a priori* concepts concerning qualities], with which principles *experience* may be in exact agreement and which nevertheless are independent of *experience*' (Ak. 10, 131; Zweig trans., 72, last emphases mine); this sentence, which Carl (1989) cites in his note 2, leads me to my '[which may be a sensible object]' reading of the original quotation. It may well be, however, that Kant's central concern in the letter is how *a priori* concepts can relate (as the Inaugural Dissertation supposes) to noumena, as Beck (1989) forcefully argues. (Beck also suggests an alternative interpretation of the Ak. 10, 131, sentence above.) If so, then the main problem of the 1772 letter is better described as an ancestor (or structural analogue) of the first-*Critique* § 13 problem than simply as 'close to' it.
[29] Ak. 4, 476.
[30] Walker (1978, 77) says that the 'late form' of the Deduction (which for him includes the *Fortschritte*, Ak. 20, 271, statement noted above) 'turns entirely on the notion of a judgment.' But at Ak. 20, 271, Kant connects unity of apperception and unity of consciousness with judgment in the same sort of way that he does in B-Deduction § 19. See also, for example, Kant to Tieftrunk, December 11, 1797, Ak. 12, 222–23.
[31] That Kant is in effect generalizing from the case of an arbitrary given sensible

intuition can be seen from his language. Note, for example, B129 ('the manifold ... given in *an* intuition which is purely sensible'), B135 ('the manifold ... given to me in *an* intuition'), B137 ('*a* given intuition'), B138 ('all my representations in *any* given intuition'), B159 (the categories 'as *a priori* modes of knowledge of objects of *an* intuition in general'), A96 ('pure *a priori* conditions of *a* possible experience'), A116 ('the unity of knowledge necessary for *a* possible experience'), A125 ('the form of *an* experience in general'), A127 ('unity of apperception in respect to *a* manifold of representations'), A129 ('the mode in which *the* manifold ... belongs to one consciousness'), and many other places (all emphases are mine). The generalization is presented more clearly in B- than in A-.

32 The main condition that *is* set in the A-Deduction is of course that all the relevant intuitions are in time, the pure form of inner sense (A99, A115, A123–24). But in introducing the categories in the Metaphysical Deduction, Kant also notes space as providing, along with time, 'material' for the categories (A76–77/B102); and he mentions space along with time at A99–100 and A101–102 in the threefold synthesis. A few cryptic A-Deduction remarks even suggest something like the B-Deduction reasoning noted below from (a) category application to the object of an intuition in general to (b) category application specifically to space and time and thence to all objects of human sensible intuition. See A77/B103 (synthesis is pure if the manifold 'is given *a priori, as is* [for example] the manifold in space and time,' my italics); A95 ('intuitions *in general* ... constitute the field ... of possible experience,' my italics'); A118 (synthesis of the manifold in imagination is transcendental 'if *without distinction of intuitions* it is directed exclusively to the *a priori combination [Verbindung*] of the manifold,' my italics). Except for the points noted below about judgment, the main structure of the B-Deduction thus may conceivably have been available to Kant, although not yet clearly expressed, in the A-Deduction. (Observe also the various emphases on '*ein*' or on the idea of a single representation or object at A92/B125, A99, A110 (twice), and A129; and compare the comments below on '*ein*' and the B-Deduction.)

33 Kant's overall use of '*überhaupt*' supports the present reading of 'intuition in general' (*Anschauung überhaupt*). (Thus note the analogous talk, at A93/B126 and many other places, of a 'concept of an object in general' – a concept of an object in the most general sense of 'object'; and see Howell, 1981b, 90.) For Kant's B-Deduction use of the notion of an intuition in general (with his idea of abstracting, in that notion, from all questions about the specific sensory 'mode' in which actual human intuitions are given), see A79/B105, B144–45, B148, B150, B151, B154, B159, B161, B162, B163; and compare A247/B304 and A254/B309.

34 See A11–12/B25, A13–15/B27–29, A55–57/B80–82, A65/B90 ff., and A76–77/B102.

35 Kant's procedure here is thus consistent with the texts cited in the last note and so with his attempt to determine the exact origin, scope, and objective validity of the categories (as applying to the object of any sensible intuition in general but – as seen below – as yielding us knowledge only of the objects of human sensible, empirical intuitions). For further comments on the structure of the B-Deduction, see the remarks below on Henrich's interpretation.

36 Note that materials for this logical-function-structuring line of argument are

already present in A-texts. (Compare A245 with B128–29; and observe A67–70/B92–94 and A78–79/B104–105.)

[37] Bxlii already licenses consultation of the A-text.

[38] Besides the important essays by Henrich (1968/69 and 1989) considered below (and the discussions of his work there noted), the following contain representative or suggestive views on the B-Deduction's structure and remarks on its second half: Kemp Smith (1962, 284–91); Vleeschauwer (1934–37, vol. 3, 13–41, 85–296); Vleeschauwer (1962, 89–114); Paton (1936, vol. I, 499–546); Ewing (1950, 114–31); Grayeff (1970, 131–204); Ameriks (1978); Nowotny (1981); Pippin (1982, Chapter 6); Allison (1983, Chapter 7); Baum (1986, 1987); Allison (1987a); and Aquila (1989, Chapter 5).

[39] Allison (1983, Chapter 7, especially 134 ff. and 146 ff.) argues that the first half of the B-Deduction establishes merely the 'objective validity' of the categories (their application to objects in a very broad, 'logical' or 'judgmental' sense of 'object'), whereas the second half establishes the 'objective reality' of the categories (their application to actual objects of human experience or possible experience). Allison suggests that this strategy is marked by Kant's use of '*Objekt*' in the first half and of '*Gegenstand*' in the second half of the Deduction. I do not see this interpretation, however, (i) As observed in note 11 above, in the Deduction Kant himself uses 'objective validity' and 'objective reality' as stylistic variants. (Note Allison's own qualifications, 1983, 134.) (ii) Kant makes it clear that in the B-Deduction first half he establishes category application specifically to objects of sensible *intuition* in general and so not to objects in some very broad, 'logical' sense. (Note Kant's usage, in describing what the first half has done, at B137, B138, B143 in several places, B144–46, B148, and B159; and see also this same sort of criticism in Förster, 1985, 736–37.) (iii) As Förster (ibid.) points out, in the B-Deduction Kant uses '*Gegenstand*' where, if Allison were right, he should use '*Objekt*.' (See B137, B138, B144 note, and B146 – 'the thought of a *Gegenstand* in general, by means of a pure concept of the understanding.' In B155 note, 'motion of an *Objekt* in space,' Kant does the reverse.) One's suspicion that the *Objekt-Gegenstand* difference in the B-Deduction is merely stylistic is reinforced by the fact that in other new B-texts (B69, B306) Kant shifts from one term to the other for what seem purely stylistic reasons.

[40] Category application to space and time does not directly imply category application to individual objects in space and time. But Kant's B-Deduction argument for category application to space and time is supposed also to establish category application to specific, determinate spaces and times and so to the objects that occupy those spaces and times. (Note B138, B154, B156, B160, and B162–63.)

[41] This point (and my whole *ein* line of thought) fits the B-Deduction as Kant presents it. In his presentation, however, Kant does not initially acknowledge the two different readings of the minimum Deduction assumption noted in Chapter Three (the strong reading, according to which a single object is known through, and is distinct from, the elements of *i*; and the weak reading, according to which the object known through those elements may amount, as far as this assumption by itself goes, to no more than those elements themselves, as they are presented to the mind). As his argument progresses, however, it becomes clear – as we see in Chapter Eight – that he means to emphasize and start with the strong reading. He then moves on, in a way that he does

not make explicit, to the weak reading. The idea that *i* functions for the mind as *one* unified intuition representing *one* object, and so my present interpretation of Kant's emphasis on *ein*, fits the strong but not the weak reading. But because he focuses at the start on the strong reading and does not make the weak reading very explicit (and because of the way that he ultimately uses the weak reading in his argument), that fact is not really evidence against my interpretation. The importance of Kant's B-Deduction emphasis on '*ein*' was first stressed by Henrich (1968/69).

[42] Note also B144, and compare B140 and A129.

[43] Compare A92/B125, A99, A110, and A129.

[44] Henrich (1968/69). For the points below, see 645–46, 650–53, 653–55.

[45] Henrich (1989, 35; see 30–39 generally).

[46] Ibid., 39, 252.

[47] For other reactions to Henrich (1968/69), see Brouillet (1975); Wagner (1980); Robinson (1981); Nowotny (1981); Allison (1983, 351–52); and Allison (1987a). My views were reached independently of these discussions. Each of these authors notes the relevance, to the B-Deduction, of the idea of an 'intuition in general.' In this connection, see also Vleeschauwer (1934–37, vol. 3, 154, 158, and 232), and Paton (1936, vol. I, 528–29 and 541–42).

[48] Besides the reasons for this conclusion given above, observe B136 note. Kant there argues that 'space and time ... are *intuitions*, *therefore* [*mithin* – my emphasis] singular representations,' just as my account would require. At B136 note Kant also refers us to the second half of the B-Deduction (to § 25, presumably a mistake for § 26), where at B160 he says again that 'space and time are represented... as *intuitions* ... and *thus* [*also* – my emphasis] with the determination of the *unity* of [their] manifold.' Note also the reference in B160 note to space 'represented as *object*'; recall that intuitions by definition represent single objects; and compare A129 (through the unity of consciousness 'the manifold is thought as belonging to *one* [*Einem*] object') with B144 note and with B161 before the asterisks.

[49] See notes 34 and 35 and the text they tag. A further reason for Kant to proceed in this systematic way, from category application to intuition in general to category application to space and time and thence to category application to the objects in space and time, is that by doing so he makes very clear *how* it is possible for the understanding and the categories to relate to sensibility. (See B159 and Henrich, 1968/69, 650–53.) Note also that because Kant wants to provide a systematic, detailed account (for the reasons just given) of the application of the categories to the objects of our senses, the second half of the B-Deduction has philosophical work to do. *Deductively*, Kant is in a position to infer that application directly from the § 20 conclusion that the categories apply to the object of any sensible intuition in general; he can argue simply that all the § 15 to § 20 points about apperception, judgment, and the unification of the manifold carry over to human sensible intuitions. But such a simple deductive argument would leave unanswered the question of how exactly such application is possible given our particular human mental capacities. It also would not give a thorough transcendental account of the scope and limits of our *a priori* knowledge through the categories. (Note § 22 and § 23.)

[50] The worry would be, for example, that Kant's appeal to 'origins' in justifying category application could be explained simply in terms of his acceptance of the

Cartesian tradition (as exemplified by, for example, Descartes, Leibniz, Locke, Hume, Tetens, and others) of investigating the scope and limits of our knowledge by studying the operations of our minds, that acceptance of course being acted on by Kant's sensibility-understanding distinction and his views on transcendental knowledge (see note 34 and the text it tags). The legal model would then enter merely as a convenient expository device. Of course it also could be that Kant's views, and Kant's acceptance of the tradition, are acted on – in their contents as well as in their form – by the legal model (or that all these things interact). The later parts of Henrich (1989) may suggest the latter possibility.

[51] For a brief, persuasive account of the explanatory side of the B-Deduction, and the observation that the *whole* B-Deduction may be read as explaining the possibility of relating the categories to intuition, see Henrich (1968/69, 650–53). Henrich makes a convincing case for distinguishing the B-Deduction task of explaining that possibility (B159) from the A-edition subjective deduction. (He also is absolutely correct to reject the Adickes-Paton view that B-Deduction §§ 15 to 20 constitute an objective and §§ 21 to 26 a subjective deduction.) For the B-Deduction, the proof/explanation distinction thus cannot be identified with the specific Axvii objective/subjective-deduction distinction. For the Axvii distinction, see Wolff (1963), Paton (1936), Vleeschauwer (1934–37), and Kemp Smith (1962).

NOTES TO CHAPTER FIVE

[1] For additional evidence that the B-Deduction assumption concerns a being like us, see the § 15 description of the knower's faculties and B138–39, B145, and B148–49.

[2] Compare such § 19 points with the § 15, B131, observations on seeking the unity of the manifold in 'that which contains the ground of the unity of diverse concepts in judgment.'

[3] I first applied this idea, and intensional logic, to Kant in Howell (1973), which provides a general framework that was later refined in Howell (1979, 1981a, and 1981b). I was stimulated by Hintikka's work (1969) on intensional logic and by the comments in Wolff (1963, 109–10) and Parsons (1964, in Parsons, 1983, 96) linking intuitions with the philosophical notion of intentionality. Subsequently I discovered that in 1949 Beck suggested a connection between Kant's ideas and that notion (Beck, 1965, 105–107; Beck also cites a 1925 work by Günter Jacoby). Other work connecting Kant to issues bearing on intensionality or intentionality (or both) includes Hintikka (1974, Chapter 10); Kirk Wilson (1978); Posy (1981, 1983, and 1987); Aquila (1983 and 1989); Massey (1986); and Meerbote (1987).

[4] Certain failures of existential generalization also mark the presence of the intensionality phenomenon, but that fact may be ignored here.

[5] In this book '*H*' and '*i*' are, within my interpretation of the official argument of the Deduction, variables subject to appropriate generalization. However, to simplify the exposition I sometimes use '*H*' and similar terms, as here in (T), as though they were constants naming specific individuals ('*j*' also is such a constant). As below, I also sometimes talk, for example, of sentences like (T) where, strictly, I should introduce metalinguistic variables and talk of sentences of the relevant form. Other points of usage should explain themselves.

[6] As observed in Chapter Two, note 30, in a possible- (or perceptual-) world

semantics for claims like (T), *W′* really should be replaced by a set of worlds W_1, W_2, ... compatible with what, at *W*, *H* thinks. But for simplicity, and to retain the expository connection with Kant's idea of a single 'field of appearances' (A40/B57) or phenomenal world, I proceed in terms of *W′*.

[7] As noted in Chapter Two, it also would be possible to allow *H*'s states of knowledge – and, as one can see, *H*'s thoughts – to belong to *H*'s mind as it appears, so that claims like (T) then would hold true at *W′* as well as at *W*. Arguments similar to those below then would show the intensionality of (T) considered as true at *W′*; and that intensionality then could be appealed to (much as I use points about intensionality below) in interpreting the Deduction and Kant's overall theoretical philosophy. But for the reasons given in Chapter Two, I work in terms of the truth of claims like (T) at *W*.

[8] (T) holds at *W*, and the state of *H*'s knowledge – of which the thought expressed in (T) is a part – exists in *W*. So, for reasons given in Chapter Two, *at W* the claim is true that 'the object before *H* has the property of being conical.' Hence the designation at *W* of 'the property of being conical' is the property of being conical. Hence also the above identity claim between that property and the property which is such that *J* thinks that that property is interesting holds at *W*, and the two singular terms in that identity claim are coreferential at *W*. Similarly for the identity claim in the next paragraph. (Throughout this discussion, I ignore the Chapter Two problem for Kant and the issues that it raises.)

[9] We are in fact here discussing what Kant in the above *Logik* passage calls distinctness (and indistinctness) in concepts.

[10] The present argument can of course also be given when the concept is taken to present, but not to be identical with, the general property, for when the property is presented to *H* by the concept, *H* may deny the truth of the identity given above, and the truth of (T) at *W* need not then be preserved by the substitution indicated.

[11] The reasoning here (and the reasoning in the preceding paragraph) is not successfully attacked by urging that, for Kant, the cited property identity does not hold, the ground for this attack being that in making the concept or property of being conical distinct, *H* in *H*'s thought simply *replaces* that concept with a different concept. Kant makes it clear in many texts that the original concept or property that is clear but not distinct has exactly the *same* content, in terms of the general properties that are thought through it, as does the distinct (in his sense) concept that is arrived at by analysis. He indeed speaks over and over as though analysis of the original concept simply reveals what we already thought in it and thus as though it is one and the same concept that occurs with less and then with more distinctness before consciousness. See *Logik*, Introduction, § VIII, especially Ak. 9, 64 (when I make a concept distinct, the content of my knowledge remains the same, and I simply 'learn to distinguish ... with greater clarity ... what was *already* lying in the given concept,' my emphasis); and compare *Logik*, Introduction, § V, Ak. 9, 33, 35; A5–6/B9–10, A7/B11, A7–8, B12, A43/B60–61, A718 = B746, A721 = B749, and A730–31 = B758–59. Note also the implications of Kant's remark, in his November 25, 1788, letter to Schultz, that if '3 + 4 = 7' were analytic (as for Kant it is not), then 'I would have to think *exactly the same thing* by "3 + 4" [which expresses a concept] as by "7" [which also expresses a concept]' (Ak. 10, 556; Zweig trans., 139, my emphasis).

[12] Indeed, in an early expression of the idea that objects always 'transcend' our

individual experiences of them, Kant holds in "Logik Blomberg" that all objects known by us through experience have only some of their parts clear, the rest remaining obscure (Ak. 24.1, 125–26).

[13] See the second half of the B-Deduction at § 24, B150 ff., and § 26, B161 ff.

[14] Note the comments in the first half of the B-Deduction, at § 18 and § 19, on inner sense and association, and observe also § 17, B136 note, and B137–38.

[15] These elements of course actually occur before *H*'s mind interspersed with feelings, thoughts, other intuition-elements, and so on. The details can be ignored here.

[16] Thus consider Kant's puzzling A90/B123 claim that 'appearances can certainly be given in intuition independently of functions of thought,' in his introductory Deduction statement of the problem of how the subjective conditions of thought can have objective validity. I read this claim as expressing, at the start of the Deduction, what is possible – that is, what is not ruled out as impossible – given merely the minimum Deduction assumption and Kant's distinction of sensibility from understanding. (Compare the notorious problem-sentence running over from A90 to A91, at B123 – 'since intuition stands in no need whatsoever of the functions of thought,' etc.) Subsequently, in the Deduction, Kant will introduce unity of apperception and will argue on its basis that while intuitions certainly are *given* independently of thought, they cannot function for the mind as single, unified representations of genuine objects of knowledge (synthesized appearances) save through the application to them of the functions of thought. Nor can we (for whom unity of apperception holds) be aware of intuitions except in the context of their so functioning.

NOTES TO CHAPTER SIX

[1] I use '*I think*' to refer to the mental act and '*I*' to refer to its first-person component (both in italics). In the relevant places, 'I,' without italics, refers to the thinker. Any deviations from this policy should be clear from the context. Kant does not always bother to distinguish the *I think* from its component *I*. I sometimes follow him in this practice where it causes no confusion.

[2] See B132. Kant there describes the *I think* as 'an act of *spontaneity*.' It is spontaneously produced by the understanding; like all such acts, it is an inner determination of the mind (and so a representation in the wide sense of Chapter Two); and, because of its first-person function, it is a representation (in the narrow sense of Chapter Two).

[3] Compare *Anthropology*, § 1 (Ak. 7, 127).

[4] I omit this qualification below, except where it is immediately relevant.

[5] See B134 and especially B135 of § 16, and also B136–37 and especially B138 of § 17.

[6] Confusion exists about what, exactly, the B135 'principle of the necessary unity of apperception' *is*, the principle that B135 and B138 say is analytic. Allison (1983, 137–39) suggests a number of claims, including both the opening sentence of § 16 and what I have just noted as being, by B135 and B138, the correct answer. Guyer (1987, 115–16) identifies the principle with claims like 'only insofar as I can grasp the manifold of the representations in one consciousness do I call them one and all *mine*'

(B134). Now Kant does not explicitly identify the principle at B135, where he first mentions it and calls it analytic. But at B138 he observes that 'this proposition ... is, as already stated, itself analytic. For it says no more than that all *my* representations in any given intuition must be subject to that condition under which alone I can ascribe them to the identical self as *my* representations and so can comprehend them as synthetically combined in one apperception through the general expression *I think*.' As B135 is the only earlier B-Deduction passage that says any proposition is analytic, this B138 text thus gives the precise content of the principle of the necessary unity of apperception. (Compare the similar, B136, statement of the 'supreme principle of the possibility of all intuition' in its relation to understanding; and the sentence running from B132 to B133.) This B138 claim is logically distinct from the 'synthetic proposition' – 'that all the variety of *empirical consciousness* must be combined in one single [or unified] self-consciousness' – that at A117 note Kant says is the 'first and synthetic principle of our thought in general.') At B407–409 Kant states various other claims involving 'I' that he holds are analytic. These other claims are themselves logically different from the B138 principle. See also note 32.

[7] The qualification here, that the principle is obviously and trivially true *given* that unity of apperception holds with respect to all my representations, is important. If unity of apperception has not been shown to hold with respect to all my representations, then it is not obvious or trivial that all my representations *can* be accompanied by my *I think* and so are subject to the condition in question. (Moreover, taken by itself without such a qualification, the *sentence* 'all my representations satisfy the condition under which I can accompany them all by the *I think* and can take them all to be my representations' is not any sort of analytic truth.)

[8] See also B138–39 and B145–46.

[9] Recall Chapter One. An intellectual understanding would be able, out of its own resources, and so without the aid of any sensory information, to generate thoughts that would grasp objects as single, individuated things. In so doing, it would, Kant holds, bring those objects into existence. Besides B135 and B138–39, see B72, B68, B307, B308, A544 = B572, A770 = B798; *Prolegomena*, § 34, footnote; and *Critique of Judgment*, § 76 and § 77.

[10] *H* and *i* may also be supposed, in (K), to satisfy any other conditions the first *Critique* has imposed on intuitions and beings like us prior to the Transcendental Deduction.

[11] The logical behavior of first-person thinking and linguistic statements – which express what Lewis (1979) calls attitudes *de se* – was first explored in detail by Castañeda (1966, 1967, and 1968), who notes connections with Kant. Subsequent work by Kaplan (1989), Perry (1977, 1979), Lewis (1979), Chisholm (1981) (who also notes connections with Kant), and others suggests further ways of sharpening our grasp of this area. For other details on Kant, see also Howell (1981a), in which many of the points made in the present chapter and in Chapter Seven were first presented.

[12] For the *I think*-accompaniment form of unity of apperception, see B131, B137, B138, B140, and compare A341/B399 ff., A347–48/B405–406, A398 ff., B406, B418 ff., B422–23 note, B428 ff.; for the *I* accompaniment form, see B68, A117 note, A123, B135, B278, A345/B403 ff., A349 ff., A363 ff., A381–82, A398 ff., B407 ff., B412 ff., B418 ff., B422–23 note, B430, A443 = B471, A788 = B816; for the *have* or

possession form, note the implications of A106–108, A110 ff., A116 ff., A120 ff., A127 ff., and see A122, A129–30, B132–33, B134, B135, B138, A352. Occasionally Kant also uses a representation-*I am* accompaniment form. See B138, B157, B277, A405, and Bxl note.

[13] The formalization below of (S) could be sharpened. Thus we could replace the talk of '*y*, *z*, and ... elements of *i*' by talk of the *I think* accompaniment of all members, taken together, of a sequence of elements. We also could eliminate the implication that *H* as simple knower is explicitly aware of the act or representation *I think* as accompanying representations. (Such theoretical awareness really is proper only to *H* as Kantian philosopher reflecting on what is involved in *H*'s awareness as knower.) To do so, read (S) as claiming roughly '$(\exists T)(\exists R)[T =$ the *I think* & $R =$ the accompaniment relation & (y) (z) ... [y, z, and ... are the elements of $i \supset H$ is or can become conscious in thought that $(TRy$ & TRz & ...$)]]$.' In addition, we could consider the question (which we need not answer here) of the exact relation Kant should take to hold between the *I think* itself and the awareness that *H* has of the *I think*.

[14] At A117 note Kant says that it does not matter whether the representation *I* is clear or obscure or even actually occurs at all; all that matters is the relation of knowledge to apperception as a faculty. However, this claim obviously does not answer the present difficulty but, rather, in effect simply asserts that the difficulty does not exist. Since I pointed out this '*can* become conscious' difficulty in Howell (1981a), it has been noted in Castañeda (1990), who also observes points similar to some made later in my (1981a).

[15] In Chapter Eight, we discuss the unity-of-apperception-required necessity of *H*'s thinking the union of the manifold of *i* in the concept of an object, and we consider what Kant takes to be the nongivenness of this necessity. As that discussion shows, Kant would take the present supposition – about *i*'s elements, as given, as standing in the required combination – to clash with the necessity that he attributes to unity of apperception and its required combination. But as the Chapter Eight discussion indicates, Kant's views on these matters need further argument.

[16] Such views are found in the classic work of Strawson (1966). Bennett (1966, § 29 and § 30; 1970, § 18) also attacks Kant's view of combination as created by acts of synthesis. See, further, Henrich (1976, 21–23, 92, 103–105) and Harrison (1982). The nonsynthesizing view of the Deduction has recently been defended by Guyer (1987) with force and clarity. C. I. Lewis's Kant-influenced 'pragmatic conception of the *a priori*' (1929) also should be mentioned here.

[17] See the opening paragraph of § 16, for example.

[18] Note, however, that (a) evidently assumes that the elements of *i* stand together before a single act of thought via which *H* knows. As we see in Section 4, that assumption is compatible with only some readings of the minimum assumption in (K).

[19] Lewis (1979) suggests roughly that when *a* believes that *p*, *a* self-ascribes, in a first-person or in an equivalent way ('*de se*'), the property of inhabiting a world where it is the case that *p*. One might apply such a theory to support (S) (an application not at issue in Lewis's paper). But the original suggestion needs defense. Otherwise one simply makes the assumption that there are – and presumably can be – no cases of belief or knowledge by a being like us that are not first-person (or the equivalent) self-

ascriptions (for example, cases of a simple impersonal belief to the effect that it is, in the actual world, the case that *p*). Lewis gives systematic grounds for his suggestion but also grants that other theories might be proposed. Since the assumption just noted does not seem obvious, further argument is needed before one accepts Lewis's suggestion. (There also are complications in applying that suggestion to reach (S), for (S), unlike the suggestion, concerns the knower's or believer's first-person awareness of having mental elements or experiences – the elements of i – that are supposed to be part of the having of the relevant belief.) Chisholm (1981) develops a theory somewhat like Lewis's, which he relates to Kant. He does not use that theory to defend anything equivalent to (S), but a defense might be mounted thus. Assume that H's having i_1 and H's having i_2 are 'self-presenting' (Chisholm, 1981, 79–80) and that H can consider the question whether he or she has i_1 and also has i_2. Then by principles Chisholm accepts (1981, 80, 88), one can infer that it can be certain for H that he or she has i_1 and has i_2; and one can argue thence to (S). However, it is not obvious without further argument that we should here assume that H's having i_1 and H's having i_2 are indeed self-presenting, for that assumption involves something tantamount to the supposition that H can, in the first person (or in an equivalent way), consider *himself* (or *herself*) as having i_1 and as having i_2. (Note Chisholm, 1981, 80 and 29.) Moreover, it also is not obvious that in every case in which a being like us *can* consider whether he (or she) has x and *can* consider whether he (or she) has y, that being also *can* consider whether he (or she) has x and y. (Considering x or y separately might, for example, already exhaust the limits of one's comprehension.) 'Blindsight' is suggested by Pollock (1988) and Castañeda (1989, 1990) as an actual case of awareness without first-person (or the equivalent) self-awareness.

20 In stating (NCA) and (NUA) below, I ignore fine logical details not now relevant.

21 In Howell (1981a, note 10), I suggested a different logical route from (K) to a claim like (NUA). But the generalization I have argued in Chapter Four that Kant makes on the case of a sensible intuition in general strongly supports the present argument. (Note, incidentally, that while I have couched this argument in modern logical terms, the underlying reasoning that it formalizes turns on straightforward generalizations and appeals to necessity that any intelligent eighteenth-century reader could accept.)

22 Since this fact is introduced into the third argument through (K), (K) is still itself appealed to in that argument.

23 Most likely Kant would agree immediately that (W) and (S) are different, if he were asked; and (S) rather than (W) is what the Deduction requires. But he introduces (W) – or does not bother to distinguish (W) and (S) – simply because he supposes one can easily pass from (W) to (S). For texts showing his recognition of (S), see B132 ('stand together in one universal self-consciousness'), B133, B136–37, B138, A109–10, A112, A116, A117 note ('collective unity'), A129 (a complete unity of appearances 'in one and the same apperception'), and A354 ('I think (the manifold in a representation)'). For texts ambiguous between (W) and (S), observe the start of the opening § 16 sentence, B135 ('I call them one and all *my* representations'), A122 ('I ascribe all perceptions to one consciousness'), A123 (the I 'forms the correlate of all our representations'), B408 ('in all the manifold of which I am conscious I am

identical with myself'), and A784 = B812. (W) claims (sometimes connected with (S) claims) occur in the end of the § 16 opening sentence, B133 ('my accompanying each representation with consciousness'), B137 (the given representations have in common the *I think*), A120 (relation of perception to a consciousness), A354 (the *I think* belongs to each [*jeder*] experience), A382 (the *I* can accompany the two kinds of representation), B407 (the I in each [*jedem*] thought). I note below places where it seems Kant means to move from (W) to (S).

[24] I also set aside other minor, variant arguments for (S) (or (W)) that one may perhaps see Kant as presenting.

[25] See Kant to Herz, May 26, 1789 (Ak. 11, 48 ff., especially 51–52). Compare the 1762 "Mistaken Subtlety," § 6 (Ak. 2, especially 59–60) and *Anthropology*, § 1.

[26] A similar clause is present toward the start of A117 note; see also A116 and A120. Given especially these last two texts, Kant's 'impossible' point might perhaps be that (i) we would have no evidence for the existence of a representation if we could not be first-person conscious of it. Or else – as better fits the tenor of A116 and A120 – he means that (ii) a representation would not be a representation (in the narrow, object-representing sense) unless we were first-person conscious of it. I doubt Kant is holding (i), given the possibility of inferring representations of which we are not immediately conscious and given also his own explicit acceptance of animal representations and of such inferred, non-immediately-conscious representations in our human case (*Anthropology*, § 5). (ii) is not helpful to the present argument for (S), for (ii) allows that a representation (in the wide sense, in which it is simply some inner determination of the mind) could exist unaccompanied by the *I think* and yield us a Humean knowledge of itself.

[27] See A116, A120, and A346/B404. A116 and A120 support the idea that intuitions are nothing to us until they are taken up into consciousness.

[28] The 200-element case turns on the fact that any finite being like us will have some upper limit to the number of elements it can hold before consciousness. There also might be beings like us that on certain occasions, owing to some mental peculiarity, simply could not achieve an 'I think (i_1 and i_2)' thought even though they could achieve separate 'I think i_1' and 'I think i_2' thoughts. (Suppose the i_1-i_2 combination were too disgusting or horrifying to be thought.) Since the Deduction is meant to apply to *all* beings like us, such examples cannot be eliminated by arguing that *H* is some sort of ideally rational or competent being.

[29] Note also, as justification for this interpretation, the first sentence in the second quotation and the beginning of the quotation from the A-Deduction.

[30] In thus explicating the B132–33 passage (and in my immediately following comments on arguing in a similar way to (S) by way of (W)) I ignore (r) above. In (r), Kant says that since all my representations are mine, they must conform to the condition under which alone (s) can hold – that is, to the condition under which alone I can accompany them all by the *I think*. That condition, not given in this passage, is the fact of my synthesizing these representations, a fact introduced in the next paragraph, at B133. If we ignore (r) and this condition, then the argument clearly does move, as I have claimed, from (t) to (s). This movement also can be easily seen in the first sentence of the second quotation above, from B134.

[31] While the above B132–33 argument most naturally fits the first, direct-to-(S)

interpretation that I gave, it thus allows this (W)-to-(S) interpretation if we suppose that Kant first derives (W), which he immediately takes to imply – or else simply identifies with – (S).

32 Note Kant's B408 claim that 'the proposition, that in all the manifold of which I am conscious, I am identical with myself, is ... implied in the concepts themselves, and is therefore an analytic proposition.' If this proposition, which on one interpretation simply comes to (d), is analytic for Kant, then it is hard to see why he would not also take (d) as analytic (and so necessary). (See also Howell, 1981a, note 23.) I say 'trivial or analytic' above because of qualms that some may have about the existential presuppositions (or pragmatic conditions on 'I') in (c) or (d) and qualms that others may have about Kant's account of analyticity. Nothing in the third argument for (S) really turns on the specific idea that (c) or (d) is analytic, as should be clear. (Note also that, as our earlier discussion shows, the 'principle of the necessary unity of apperception' that B135 and B138 say is analytic is not (c) or (d) but, rather, something on the order of the entire B132–33 claim quoted first above.) If we do apply an analytic-synthetic distinction to first-person claims like (d) (and to do so, some fine tuning of that distinction is needed), then later third-argument claims like (e) and (S) itself should emerge as synthetic. (Note my comment below on the problem with (e).) Again – and to ignore additional qualms and details – Kant will take (c) and (d), as necessary, to be known *a priori*. Similarly, and roughly, (e), (f), and (h) will self-ascribe different forms of that *a priori* knowledge or awareness. (S) itself (and (g) or (i)) Kant will take to involve necessity (see Chapter Seven) and to be known *a priori*. Because, however, he infers (S) (and (g) or (i)) fallaciously (see below) and these claims really are not necessary, the status of such claims as *a priori* knowledge cannot be justified, I think. Accepting that status, however, Kant will take such claims to express transcendental knowledge that, like all such knowledge, is certain (Axv, A822–23 = B850–51). The *I* or *I think* that such claims involve (although not the claims themselves or Kant's general view of what the *I think* can achieve) of course derives from Descartes. For related issues about apperception, analyticity, and the *a priori*, see, on analyticity, Guyer (1987, 134 and 438, note 2); and, on the *a priori*, Henrich (1976, 58–59, 64–65, 69–70, 86–87), Guyer (1979 and 1987, 86–87 and 131–54), and Allison (1987a).

33 Actually to get from this last claim (after the colon) to a possession form of (S), some further inferences are required (involving the claim that my consciousness of possession is *de re* with respect to i_1 and i_2 and that i_1 and i_2 are the sole elements of i). The details are unimportant here.

34 We have to be able to deduce (e) from (K) (and not make *H*'s ability to think as in (e) a separate assumption of the Deduction) in order (i) to deduce (S) from (K) in the way required in the third argument for (S) and then (ii) to continue the overall Deduction reasoning according to the pattern explained earlier (infer the conditional whose antecedent is (K) and whose consequent is (S), generalize on that conditional, and so on). The mere fact *H* is assumed to have and use a first-person thought-consciousness does not by itself guarantee that *H* is able to accept the specific truth in (d). Nor does the fact that (d) itself is 'trivial or analytic' guarantee that fact. The triviality or analyticity of (d) means merely – and roughly – that the *sentence* (d), evaluated with respect to *H* or any other being like us as the referent for 'my' in (d),

comes out true by virtue of its structure. It does not mean that *H* himself or herself needs to be able to grasp the truth expressed in that sentence. Compare the case of a ten-page-long truth-functional tautology involving the sole constant 'George.' Replacing 'George' in that tautology by appropriate grammatical forms of 'I' will yield another 'trivial or analytic' claim true with respect to any being like us; but few beings like us will be able to grasp the truth expressed in that claim.

[35] And there is no reason to suppose I have to know that identifying fact, for it concerns all my representations whatsoever, and there is no reason to think every being like us can list the totality of his or her representations and say that they, and no others, make up that totality. Moreover, to infer (g) from (f) by way of knowledge of the identifying fact, my knowledge must take the first-person form 'I am or can become conscious in thought that the sum total of my representations = *r*, *s*, *t*, and so on.' But to assume that I have such knowledge is so close to assuming (S) as to be question-begging.

[36] Various slightly different ways exist of formulating each of the *de re* and *de dicto* claims given in this and in later chapters. The formulations that I provide are those most directly relevant to our interpretation of the Deduction.

[37] (g) in the first version, from which that version infers (S), is clearly to be read as a *de re* claim about the individual representations *r*, *s*, *t*, and so on; whereas (f), from which (g) is (fallaciously) inferred, must be *de dicto* if (f) is plausibly accepted as true. I noted the operator-shift fallacy in the second version of the third argument already in Howell (1981a); but I did not consider, separately, the first version of the third argument or note its problems. Guyer (1987, Chapter 5) considers some related issues but focuses on necessity-operator shifts and ignores what I think is clear evidence that Kant's third argument for (S) (which is one of the fundamental arguments of the Deduction) centrally involves intensional fallacies, including an operator-shift fallacy, that turn on 'I am or can become conscious that' or the equivalent. (Kant's overall reasoning also does involve necessity-operator shifts, as we see in Chapter Seven and later.)

[38] See Henrich (1968/69, 653–55). Note that Henrich himself sometimes speaks in an (S)-style of *all* representations as bound in a unity of consciousness (653, 654) and sometimes in a (W)-style simply of *a* representation as being taken up into my consciousness (654). Although if I am right the fallacy here turns on matters of intensionality rather than on an ambiguity in 'mine,' Henrich deserves full credit for uncovering this fallacy.

[39] See also B133–34 note on the analytical unity of consciousness that belongs to general concepts. These texts and A-texts like A103–104 and A116–17 note, as well as Kant's theory of general concepts as being arrived at by a process of comparison, reflection, and abstraction, support the interpretation of analytic and synthetic unity of apperception offered here and below.

[40] The argument would appeal to *H*'s need to use concepts in order to recognize that the actual (i_1, i_2) sequence occurs before the act of thought through which *H* knows.

[41] For more details of this type of argument, see Howell (1981a, 401 ff.). I here simply sketch the main points. A prime inspiration for such reasoning is Kant's view of concepts as rules for synthesis, which Wolff (1963) especially has stressed. See the familiar passage at A106: 'All knowledge demands a concept ... a concept is always,

as regards its form, something universal which serves as a rule. The concept of body ... as the unity of the manifold which is thought through it, serves as a rule in our knowledge of outer appearances. But it can be a rule for intuitions only insofar as it represents in any given appearances the necessary reproduction of their manifold, and thereby the synthetic unity in our consciousness of them. The concept of body, in the perception of something outside us, necessitates the representation of extension, and therewith representations of impenetrability, shape, etc.'

42 This sort of point about reproduction is made in the A-Deduction. But it is ignored in the B-Deduction and plays no role in Kant's basic argument from the holding of unity of apperception with respect to *i* to category application to the object of *i*. (As far as that argument goes, the claim with regard to reproduction simply is that that holding requires the imaginative reproduction and further synthesis of *i*'s elements to proceed in such a way that those elements come to represent a category-subsumed object.) So for the most part I ignore reproduction in imagination hereafter.

43 By A352, thoughts are, essentially, 'internal accidents belonging to a thinking being'; the subject of thoughts, this thinking being, is 'known only through the thoughts that are its predicates' (A346/B404). So each thought necessarily has a single thinker. See also A349; A355; B407, paragraph (1); B411–12 note; and *Prolegomena*, § 46. Of course the above A352 quotation comes from a paralogistic argument, which Kant criticizes, for the simplicity of the soul. But the view which that quote expresses, when that view is stripped of its paralogistic implication that the thinking being in question is properly taken by us to be a simple, enduring substance, is one that he endorses. (The thinking being is in fact the formal subject of thoughts, he holds.)

44 The reasoning to that form of (S) would be similar to that suggested earlier for (S), with the sorts of inferences remarked in note 33 also being needed. Some may think that Kant's views on self-awareness require only on existential quantifier, rather than a uniqueness quantifier, in (j). However, as that point does not particularly affect the discussion below, I ignore the issues it raises.

45 Other reasons could be given (see, for example, Howell, 1981a, note 53).

46 As in effect emphasized by Castañeda and more recently by writers like Perry and Lewis, the converse of this claim does not hold, for *H* may have non-first-person modes of *de re* awareness of himself (or herself).

47 One might object that such a knower *can* regard the above position as true at least in the sense that were that knower's intelligence and knowledge suitably improved, then that knower would grasp that position and regard it as true. Discussion of this objection is complicated by the fact that Kant is not absolutely clear about the sense in which the *I think can* accompany all *H*'s representations. Moreover, the objection raises general questions about abilities and 'can' that go well beyond the scope of this book. However, it still seems clear that, contrary to the objection, the fact that a being like us *would* acquire the relevant abilities through study or some science-fiction-like manipulation of the being's intelligence does not imply that the being *currently has* the ability. (It also might be that, in the case of some beings like us, manipulation of intelligence would lead to their rejecting rather than accepting the position!) Furthermore, and as our earlier discussion of the 'can become conscious' problem shows, to appeal to the possibility of such manipulation is to make actual category application in some cases dependent on the very remote condition that such manipula-

tion actually occurs. And that result is a far cry from the Deduction's goal of showing actual category application, under all conditions, to all the objects of knowledge of any being like us. (See also Howell, 1981a, 420–22.)

[48] In Howell (1981a, 411–17), I reject the idea of defending the equivalent of the fourth argument for (S) by reading the first-person 'he himself' (or 'she herself') in (k) (and 'the *I think* accompanies' in (S)) as abbreviating the uniqueness quantifier in (j). Such a defense is akin to that suggested here via the purely existential, (j)-like form of (S). For reasons given below, that defense is inadequate to the Deduction's goal of showing category application to all objects known by all beings like us. However, some of my 1981 objections to the uniqueness-quantifier form of unity of apperception are unsatisfactory. For example, the claim '*H* does or can think: I have *m* & *H* does or can think: I have *n*' does not deductively imply the claim '*H* does or can think: I have *m* and *n*.' So, contrary to Howell (1981a, 411–12), a first-person form of the holding of unity of apperception is not better than a uniqueness-quantifier form for showing (as Kant desires) that all the objects of *H*'s knowledge are relatable within a single experience of *H*'s.

[49] Howell (1981a) amounts to an extended defense of this point. Guyer (1987, and in earlier work) also has criticized Kant's views on unity of apperception.

NOTES TO CHAPTER SEVEN

[1] This formalization could be sharpened, but that is not necessary here. The present conclusion can be reached from (K) given the assumptions just noted and the readings of (K) that we saw the fourth argument to accept. The reasoning would be similar to that already indicated for (S) (and would include inferences like those mentioned in Chapter Six, note 33).

[2] We could also consider a purely existential accompaniment form of (S) to the effect that *H* is or can become conscious in thought that the representation *that there is a single thing such that* ... accompanies all of *i*'s elements taken together. But for simplicity I focus just on the purely existential form of (S).

[3] As observed in Chapter Six, note 48, and in Howell (1981a, 411–13), developing the Deduction along these lines deprives Kant of one way of arguing that all the objects of *H*'s knowledge are relatable together within a single experience of *H*'s. But, as that note shows, first-person forms of unity of apperception like (S) really are no better for that purpose than is the present way of developing the Deduction.

[4] Recall Chapter Two. Note also B157–58: 'In the original synthetic unity of apperception, I am conscious of myself, not as I appear to myself, nor as I am in myself, but only that I am. This *representation* is a *thought*, not an intuition; ... although my existence [of which I am conscious via the *I think*] is not indeed appearance (still less mere illusion), the determination of my existence can take place only in conformity with the form of inner sense... . Accordingly I have no *knowledge* of myself as I am... . The consciousness of self [via the *I think*] is thus very far from being a knowledge of the self; ... for knowledge of myself I require, besides the consciousness, that is, besides the thought of myself [via the *I think*], an intuition of the manifold in me, by which I determine this thought.' Observe, further, B423 note: 'The *I think* precedes the experience which is required to determine the object of

perception through the category in respect of time; and the existence here [the existence of the self referred to or expressed via the *I think*] is not a category… . [What is expressed via the *I think* is] something real that is given, given indeed to thought in general, and so not as appearance, nor as thing in itself (*noumenon*), but as something which actually [*in der Tat*] exists, and which in the proposition *I think* is denoted [*bezeichnet*] as such. … the *I* in [the] proposition [*I think*] … is purely intellectual, because belonging to thought in general.' And also see B429: 'If [via the *I think*] I would be conscious of myself simply as thinking, then since I am not considering how my own self may be given in intuition, the self may be mere appearance to me, the I that thinks, but is no mere appearance insofar as I think; in the consciousness of myself in mere thought [via the *I think*] I am the *being itself* [*das W e s e n s e l b s t*], although nothing in myself is thereby given for thought.' For A-texts, note the first-person wording at, for example, A122.

5 Chapter Two, Section 2; and note 12 there.

6 B67–68, B155, B157–58, B277, B422–23 note, and B428–29. Regarding the *I think* as giving us such a *de re*-like awareness *of* our self in itself does not infringe the Paralogisms position that we cannot take the self of which we are aware via the *I think* to be an ontologically simple substance that is a persisting person. In regarding the *I think* in the above way we imply nothing about the nature, in itself, of our self. We imply only (and Kant would agree) that, as far as our *I think* awareness *of* our self in itself goes, our self is one thing (it is in that sense logically self-identical), of whose further properties we are not aware (it is logically simple), that has all of the relevant representations *r*, *s*, *t* (it is the logical subject of thoughts). In implying these things we allow that, as far as our awareness *of* it goes, our self in itself could be, for example, a plurality of impersonal, dependent particulars momentarily acting in concert. (See either version of the Paralogisms. For similar comments about the *de dicto*-like view of the *I think* – noted below – and Kant's position in the Paralogisms, see Howell, 1981a, 406–409.)

7 A346/B404: 'Through this I or he or it (the thing) which thinks, nothing further is represented than a transcendental subject of thoughts = X. … Consciousness in itself is not a representation distinguishing a particular object, but a form of representation in general, that is, of representation insofar as it is to be entitled knowledge; for it is only of knowledge that I can say that I am thereby thinking something.' A350: 'The *I* is indeed in all thoughts, but there is not in this representation the least trace of intuition, distinguishing the *I* from other objects of intuition.' B407: 'That the *I* of apperception, and therefore the *I* in every act of thought, is *one*, … and consequently signifies a logically simple subject, is something already contained in the very concept of thought.' B426: The implication is that all that the *I* gives me, taken by itself, is 'the completely undetermined concept of a thinking being in general.' Note, further, A355: 'In attaching the *I* to our thoughts we designate [*bezeichnet*] the subject of inherence only transcendentally, without noting in it any quality whatsoever – in fact, without knowing anything of it either by direct acquaintance or otherwise. It means a something in general (transcendental subject), … that which is represented through the concept of a mere something.' And see B429: When I take the *I think* merely as an act of pure thought, 'I think myself only as I do any object in general from whose mode of intuition I abstract.' These A355 and B429 passages sound ambiguous. They

may express simply a Kantian reconciling way (see below) of getting a *de re*-like awareness of self out of a *de dicto*-like *I think* by applying the *de dicto*-like *I think* (roughly, the concept of 'something in general' or of a 'thinking being in general') to sensible, inner intuition. Or, as the A355 talk of 'designates' suggests (and the similar talk at B430, following on B429, where Kemp Smith's 'distinguish' translates '*bezeichnet*'), A329 and B429 may well express Kant's first-person, *de re*-like *I think*, in which the *I think* 'designates' but does not 'determine' the self in itself. (See the discussion below of the B-text.) That A355 and B429 still sound *de dicto*-like ('it means a something in general') would then reflect Kant's apparent lack of clarity about the differences between the *de dicto*-like view of the *I think* (and his reconciling way of deriving a *de re*-like awareness of self from that view) and the first-person, *de re*-like *I think*.

[8] Nor does the Kantian suggestion, indicated below, of a way of reconciling the *de dicto*- and *de re*-like views of the *I think* answer Pistorius. That suggestion does not show why such nonphenomenal existence must belong to the self *of* which we are made aware via the *de dicto*-like *I think* when – according to that suggestion – that *I think* becomes *de re*-like through its application to sensible intuition. (Even if, according to that suggestion, that self is held to appear to us via an inner-sense intuition, why must that self have a nonphenomenal existence going beyond its mere existence as it appears?) Simply to assume that it must, or simply to invoke some 'no-appearance-without-a-nonphenomenal-thing-appearing' principle, is to beg the question against Pistorius.

[9] See Howell (1981a) for further discussion of such pressures and related issues. (This 1981 discussion needs to be read in the light of the attempted Kantian reconciliation, noted below, of Kant's two views of the *I think*.) By holding that the first-person, *de re*-like *I think* yields an awareness *of* the nonphenomenally existing self in itself to which *all* appearances appear, Kant gives himself an excellent reason (within his own framework) for counting that *I think*, taken by itself, as pure and so as nonempirical. Observe the texts cited in note 4, as well as B132 and B422; and compare A116 note, A123, and A107–108.

[10] 'Problematic' in the sense of A74/B100 and A218/B265 ff., not in the sense of A254–55/B310–11. I have avoided Kant's problematic-assertoric terminology because I do not see how the problematic *I think* (as accompanying all representations in knowledge) can avoid 'containing existence' if Kant is to have any answer to Pistorius. And, as noted below, in texts like B429 that concern the *I think* taken by itself (and so the problematic *I think*, it seems), Kant seems to agree.

[11] Or the specific assertion (like the preceding assertion, using the assertoric *I think*) that 'I exist thinking.' See B418–20, B428, B429–30, and compare B426.

[12] Thus B426–27: 'I think myself on behalf of a possible experience, at the same time abstracting from all actual experience'; but (Kant says) it would be a mistake for me to 'conclude therefrom that I can be conscious of my existence even apart from experience and its empirical conditions.' See also further at B427.

[13] See, for example, A247/B304, B306–308, and A252–53.

[14] Conceivably one might attempt to accommodate this Kantian claim within the reconciling way as follows. Suppose that, via abstraction, we form the idea of the existence in itself of the particular self *of* which we are made aware via the *de dicto*-

like *I think* as that *I think* is applied to sensible, inner intuition. Then, once we have formed that idea, we can carry it along even in our use of the *I think* merely to accompany all representations in knowledge. This attempted accommodation fails, however, for it does not explain why, in *all* uses of the *I think* to accompany representations in knowledge, we *must* (rather than we simply *can*) carry along the idea of the existence in itself of the particular self in question. But to accommodate the present Kantian claim and to answer Pistorius, that point needs explanation. Moreover, as I observe below, in several B-texts Kant indicates that (i) the *I think* (or *I*) is a pure act of thought that can precede *any* experience that determines an object (and so can precede *any* experience of our empirical self from which we could then abstract in the above reconciling way). And Kant indicates that (ii) the *I think*, as such a pure act of thought in (i), 'designates' – and so yields a *de re*-like grasp of (even if no knowledge of) – the knowing self in itself. The present attempted accommodation cannot accept (ii), however, given (i), for (ii) implies that the *I think*, as a (i)-style pure act of thought, *already* grasps that self in a *de re*-like way.

[15] See the B429 quote in note 7 for this statement.

[16] See the B429 text cited at the end of note 4.

[17] See B157, B422–23 note, B429; compare A355. Note that '*bezeichnet*' is translated by Kemp Smith as 'designate' at A355, as 'denote[d]' at B423 note and A382, and as 'distinguish' at B430. B422–23 note especially shows Kant desperately trying to respond to the intellectual pressures he is under and to find some way to combine his immiscible views of the *I think*. For further comments on B422–23 note and other texts showing Kant's response, see Howell (1981a, 425–29 and note 68).

[18] For our doubts about Kant's claim, recall especially Chapter One. The recent ideas about first-person self-awareness (and first-person reference via 'I') suggest that such awareness requires sense experience of (or some sort of non-*a-priori* acquaintance with) oneself without requiring one to grasp any Fregean sense of – or any special conceptual criteria of application for – first-person terms like 'I.' See Perry (1977, 1979) and Lewis (1979). The general view that first-person self-awareness occurs without one's having to grasp defining or identifying criteria of oneself is (or at least sounds) akin to Kant's view of the 'designating' but not 'determining' role of the *I think*. Strawson (1959), Castañeda (1967), and Chisholm (1981), among others, comment on the relation between Kant's view and modern ideas. See also Howell (1981a).

[19] Instead of stipulating (S), we could of course proceed in terms of assumptions needed to make one of the Chapter Six attempted proofs of (S) succeed. Such stipulations have the same general effects on the Deduction as does the stipulation of (S).

[20] Note B-Deduction § 18 and A122–23.

[21] See the Second Analogy and A760–68 = B788–98.

[22] See, for example, the Transcendental Deduction reasoning about the affinity of the manifold and the necessary unity of nature at A108, A111–14, A121–22, A125–28, B138–40, B161–65, and B168.

[23] In addition, I concentrate on *I think*-accompaniment forms of such Kantian necessity claims, and I do not consider the possession-forms that could be introduced. (Recall Chapter Six.)

24 The formal version of (N_1) would be the same as the Chapter Six formal version of (NUA), with the 'it is necessary that' operator moved directly before '*w* is or can become conscious in thought that' within the formal version of (NUA). To give examples of (N_1) or logically similar claims (with my emphases): At B136–37, Kant in effect asserts that *given* – and so *if* – something is known via certain representations, *then* those representations '*must* allow of being combined in one [*I think*-governed] consciousness'; at A104, he says that 'such [unitary] *consciousness* [of unity of synthesis] ... *must* always be present,' otherwise knowledge of objects is impossible; at A108, he comments on the '*necessary* consciousness of the identity of the self [and so of the *I think*-accompaniment]' with respect to the manifold through which we know; at B138, he asserts that synthetic unity of consciousness (which by B137 involves representations having 'in common the act of apperception *I think*') is something that *must* occur with respect to every intuition that is to 'become an object [of knowledge – note the B137–38 emphasis on knowledge] for me.' See also claims at A116–17 and A117 note. Various other examples could be given, but as many of Kant's texts are ambiguous (as between, say, (N_1) and (N_3)) or impossible to analyze briefly, the above will suffice. (A similar comment applies to the (N_2) and (N_3) examples given below.)

25 Thus (N_1), unlike (NUA), imputes a conditional, relational, essential property to particular beings like us and the individual, particular sensible intuitions in general through which such beings know. Note that (N_1) (like (N_2) and (N_3) below) also should be *de re* with respect to the *I think* itself. But as this point raises complicated questions, not considered by Kant, which do not affect the basic Deduction argument, I will not pursue it here.

26 The formal version of (N_2) is exactly the same as that of (N_1) (as described in note 24) except for the required changes in the antecedent and consequent. B-texts showing (N_2) or logically similar claims include (with my emphases) the opening sentence of § 16 (all my representations are such that, necessarily, I am or can become conscious that the *I think* accompanies them); a later B132 claim (since, as Kant says there, all my representations '*must* conform to the condition under which alone they can occur in one universal self-consciousness,' all my representations are such that, necessarily, I am or can become conscious of their *I think* accompaniment); and the B138 statement of the principle of necessary unity of apperception (which is like the last B132 claim). For A-texts see, besides the A121–22 passage just noted, A117 note (all representations have a necessary relation to a possible empirical consciousness, and all empirical consciousness has a necessary relation to the consciousness of myself).

27 That is, are of the form, roughly, of the formal version of (NUA) in Chapter Six, with the relevant changes to antecedent and consequent and with the 'it is necessary that' operator occurring between '*w* is or can become conscious in thought that' and 'the *I think* accompanies (*y* and *z* and ...).'

28 Note also B135: 'I am conscious of the self as identical in respect of the manifold of representations because I call them one and all my representations... . *This amounts to saying* that [in (N_3) style] I am *conscious* to myself *a priori* of a *necessary* synthesis of representations ... under which all representations that are given to me must stand' (my emphases). For other texts exemplifying (N_3) or a logically similar form, see (with my emphases) A111–12 ('in original apperception, everything must *necessarily*

conform to the conditions of the ... unity of self-consciousness'; and that necessary conformity must, in logic, derive from our (N_3)-style consciousness in thought that, necessarily, the *I think* accompanies all the elements of the manifold); A112 (we meet with – and so we are *conscious in thought* of – a *necessary* unity of consciousness, and so a necessary *I think*-accompaniment, in the manifold); A116 ('we are *conscious a priori* [and hence, given Kant's account of the *a priori*, we are conscious of the *necessity*] of the complete identity of the self in respect of all representations which can ever belong to our knowledge'); A118 (transcendental unity of synthesis is *represented* as – and so we are *conscious* of it as – *a priori necessary* in relation to original unity of apperception – and so unity of apperception itself is a *consciousness* of a *necessary* relation of representations to the *I think*); B133 (I represent to myself, and so am *conscious in thought of*, the identity of consciousness in the representations given in intuition, an identity of consciousness that Kant evidently takes to be *necessary*); and B144 ('a manifold, contained in an intuition which I call mine, is *represented* [to my *consciousness*] ... as belonging to the *necessary* unity of self-consciousness'). Note that in (N_3)-style claims we have *de re* claims involving the necessity-modality embedded within the consciousness-in-thought modality.

29 Such is the tenor of well-known 'distinct ideas are separable' passages like *Treatise* I.3.3 (Selby-Bigge ed., 79–80); I.3.6 (ibid., 92); compare also I.4.6 (on personal identity, ibid., 253, 259–60) and Appendix (ibid., 634, 635); *Enquiry concerning Human Understanding*, § 4, Part 1 (Selby-Bigge ed., 29, 30); § 4, Part 2 (ibid., 35, 26–37, 38); § 5, Part 1 (ibid., 42). Such passages (at least if we exclude the *Treatise* texts on personal identity and its Appendix) are centrally concerned to show that no relation *R* that does or may hold between two distinct ideas *a* and *b* holds necessarily between *a* and *b*. But (as parts of these passages and especially the *Treatise*'s personal-identity discussion and its Appendix indicate) Hume's use of the 'distinct ideas are separable' principle implies the possibility of the extreme situation noted above. Of course, for Hume, a being like us will in general take at least some relations of succession or resemblance to hold between distinct ideas *a* and *b*. (For another, detailed, discussion of Kantian claims that run contrary to Humean views of mental states, see Patricia Kitcher, 1982.)

30 Although I have expressed doubts, in note 3 above and in Chapter Six, note 48, that Kant can successfully reason in this way, there is no doubt that he believes he can show, and that he means to show, such results. See, for example, A110, A113–14, B164–65, A230/B282 ff., A582 = B610, and Howell (1981a, 411–17). (These comments apply also to the relevant points in the next paragraph below.)

31 Even if one wanted to defend (S) and (NUA), it is hardly clear that one should accept this present claim of *de re* necessity. This claim implies that *if* in actual fact I happen to know via *i*, then it is *necessary* that I self-ascribe i_1 and i_2; and so if in actual fact I happen to know via *i*, then I self-ascribe i_1 and i_2 in every possible world in which I have i_1 and i_2, including possible worlds in which I do *not* know via *i*.

32 Of course it remains possible that some further, recondite lines of reasoning could establish such non-(NUA) claims. But I have not been able to construct any convincing such reasoning; and, for the reasons just given, I think it unlikely such reasoning can be provided. See also note 31, and recall Chapter Six on proving (S).

33 That is, Kant argues (in any of the ways suggested in Chapter Six) to (NUA). He

then moves 'it is necessary that' across both the implication signs in (NUA) (in the Chapter Six formal version of (NUA)) so as to reach an (N_1)- or (N_2)-style claim that is like (NUA) but has 'it is necessary that' directly prefixing the '*w* is or can become conscious in thought that' consequent of (NUA). (The effect of this move is to create a double *de dicto*-to-*de re* operator-shift fallacy, binding '*v*,' '*w*,' '*y*,' '*z*,' and so on, within the scope of 'it is necessary that,' to quantifiers outside that scope.) That Kant moves from the necessity of the whole conditional to the necessity of its consequent can be seen at B136–37. As observed above in note 24, he there in effect asserts that *if* something is known via certain representations, *then* those representations '*must* allow of being combined in one [*I think*-governed] consciousness' (my emphasis). His underlying reasoning, which places this '*must*' where it is, evidently is that because (i) it is *necessary* that (*if* something is known via certain representations, *then* those representations are combined in one consciousness), therefore (ii) *if* something is known via certain representations, *then* it is *necessary* that (those representations are combined in one consciousness). Observe also the other texts cited in note 24, where he reasons similarly. A106 already graphically illustrates his general tendency to move from the necessity of a whole conditional to the necessity of its consequent. He there writes that 'the concept of body, in the perception of something outside us, necessitates the representation of extension,' thus in effect moving from the true *de dicto* necessity claim that 'it is necessary that (if we apply to a thing the concept of body, then we represent that thing as extended)' to the highly dubious *de re* necessity claim that 'if we apply to a thing the concept of body, then it is necessary that (we represent that thing as extended).'

34 To see the details here, return to Chapter Six, Section 4.C, prefix (c) or (d) of the third argument for (S) with 'it is necessary that,' and then carry through either version of the argument in the way sketched in this paragraph. Because Kant does not explicitly offer every step of the third argument, no texts show all the detailed reasoning here indicated. But a number of texts support the idea that he reasons in this general way. Thus take the B135 quote in note 28. If we understand the opening sentence of that quote in the way proposed in Chapter Six, Section 4.C (in the context of the B132–33, B134, and A129 quotes there considered), then we should take that sentence to involve the claim that 'I am conscious that all my representations are mine.' But to infer from that claim the necessity-involving, (N_3)-style conclusion that is drawn in the rest of this quote (see note 28), Kant must be treating my consciousness that all my representations are mine as being or implying my consciousness that *it is necessary that* all my representations are mine (compare a similar expression of necessity at A129, quoted in Chapter Six, Section 4.C). And he presumably must then be taking my consciousness of that necessity to derive from my recognition of the necessity of the simple claim 'all my representations are mine.' (Here note again the A129 claim just mentioned.) A similar interpretation could be defended for A129 and especially for B408, where Kant asserts that 'the proposition, that in all the manifold of which I am conscious, I am identical with myself, is ... an analytic [and so a *necessary*] proposition. But this identity of subject [and so the *necessity* of this proposition] ... I can be *conscious [of]* in all my representations' (my emphasis).

35 Thus A109 refers to 'the unity that *must* be *met with* in any manifold of knowledge which stands in relation to an object' (my emphases) and so seems, on (i)-style

grounds, to be inferring that knowledge through a manifold implies the necessity of my being conscious in thought that an object-signifying unity relation holds among the elements of that manifold. But then at A109 Kant continues by remarking that 'this unity *must* be regarded as [itself] *necessary a priori* – otherwise knowledge would be without an object' (my emphasis), thus seemingly transforming, in a (iii)-style, the previous necessity of my consciousness of the object-signifying unity relation into a – still (on (i)-style grounds) *necessary* – *consciousness* of a *necessary* object-signifying unity relation. He then emphasizes this result at A110 by writing that 'all appearances, insofar as through them objects are to be given to us [and so known], *must* [on (i)-style grounds] stand under those *a priori* [and, by the (iii)-style transformation just noted, *necessary*] rules of synthetical unity' (my emphases). But such rules are, for him, consequences of the holding of unity of apperception. So he asserts, finally, that 'appearances in experience *must* [on (i)-style grounds] stand under the conditions of the *necessary* [on (iii)-style grounds] unity of apperception.' (The necessity of the rules derives from the necessity of unity of apperception; note A106–107 and Chapter Eight below.) Other texts where Kant introduces multiple necessity-operators include A125 ('this unity of nature *has* [*soll*] to be a *necessary* one,' my emphases) and B127 ('the understanding must think concepts ... as being necessarily connected in the object'). See also the B135 quote in note 28.

36 Kant's A593/B622 ff. distinction between the absolute necessity of things (which he rejects) and the absolute necessity of judgments (which he accepts) can be plausibly interpreted as introducing the equivalent of a *de re-de dicto* distinction. (I ignore interpretive complications. For example, Kant is of course not here giving any fully general such distinction. Moreover, if the absolute necessity of a thing is indeed *de re*, then that necessity must involve the claim that there is some thing that necessarily has a certain property and not just the claim that necessarily there is some thing that has a certain property.)

37 Recall the example of Berkeley from Chapter Six, Section 4.C. Modal fallacies of such sorts have been committed down into the twentieth century.

38 See note 28.

39 My views about Kant's treatment of the necessity of unity of apperception go back to the first draft of this book. Except for comments on the (*de dicto*) necessity of claims similar to (NUA) and a brief (and not very helpful) mention of necessity-operator shifts in note 26 of Howell (1981a), I did not publish those views, however – although in Howell (1981a) I emphasize structurally similar shifts on the consciousness-in-thought operator. Harrison (1982) suggests a Kantian confusion between the necessity of the consequence ('$N(p \supset q)$') and the necessity of the consequent ('$p \supset Nq$'); and Guyer (1987) focuses both on the move from the former to the latter and on Kant's related move from (again with my abbreviations) *de dicto* claims like '$N(x)(Fx \supset Gx)$' to *de re* claims like '$(x)(Fx \supset NGx)$' (see, for example, 121–24). I must protest, however, Guyer's implication (437) that I (Howell, 1981a) miss Kant's 'insensitiv[ity] to the difference between the necessity of a conditional and the necessity of its antecedent,' which (Guyer holds) enables Kant to infer from the necessity of the conditional the necessity of its consequent. I did not miss that insensitivity for the simple reason that, as far as I can see, it does not exist. If it did, Kant would move easily from the necessity of 'everything which thinks, exists'

(Guyer's B422 note example) to the necessity of 'something thinks' or from the necessity of 'if we apply to a thing the concept of body, then we represent that thing as extended' (my A106 example of note 33 above) to the necessity of 'we apply to a thing the concept of body.' But neither at B422–23 note or A106 nor elsewhere does Kant make such a move. Rather, and roughly, he simply confuses a necessary conditional with a conditional with a necessary consequent. (Given this confusion and the truth of the antecedent, he can then infer the necessary consequent.) As he does so, he also often illicitly shifts the consciousness-in-thought operator. (Guyer himself at 123 introduces a perplexing 'awareness of necessary regularity' instead of the 'necessary awareness of regularity' that one expects given his own discussion of Kant.)

NOTES TO CHAPTER EIGHT

[1] Another harmless simplification here is that it is unlikely *H* would ever come to know the spruce in exactly the above sequential fashion.

[2] See also A77/B102 ff., A97, A119 ff., and B151–52 and B162–63. For reasons given in Chapter Six, note 42, I will not try here to sort out Kant's views on reproductive versus productive imagination in any detail. *Roughly*, imagination used productively does for pure space and time, in a logically prior (and category-governed) way, what reproductive imagination does in the normal perception of empirical objects. (Note A79/B102 ff. on the pure spatiotemporal manifold to be synthesized; A98–104, A118, A121–22, A123, B150–52, B154–55, B160–63, and A140/B179 ff.) Reproductive imagination leads the mind 'to reinstate a preceding perception alongside the subsequent perception to which it has passed, and so to form a whole series of perceptions' (A121; see A99 ff. and A119 ff.). In reproductive imagination (see A100 ff. and A121) perceptions become so linked that, given the occurrence of one perception, the mind is led to reproduce relevant earlier perceptions (and the mind is led to anticipate, by imagining them, future perceptions). Given A102 (on obtaining a 'complete representation') and A121, the single, compound representation that this process yields occurs as one representation – though not necessarily in one piece – before thought-consciousness. And so Kant should hold, since (as we see below) unity of intuition derives from the fact that one act of thought (of thinking there to be one object to which belong the features presented by the intuition's elements) is involved in the intuition's synthesis. That position raises complex questions, since ordinary experience requires that we regard the one act of thought as itself occurring sequentially in the mind. None of these questions directly affects the Deduction argument from unity of apperception to category application, so I ignore them here.

[3] A50/B74, A80/B106, A92–93/B125. Reasons for this assertion are noted in Section 3. Briefly, combination (and so the idea of the various properties and spatial parts as belonging to *one* object) is not given via sensible intuition. Hence *H* must use *H*'s only other cognitive faculty, that of discursive thought, to *think* there to be a single object that has the properties and spatial parts presented by *j*'s elements.

[4] Chapter Three, Section 2. Observe also B5–6, A6–7/B10–11, A8, B11–12, and A9–10/B13–14. As one can see, all the points below can be duplicated, with minor changes, if we adopt Kant's alternative view that concepts present properties from which they are distinct.

5 The above account of synthesis is that of A98–104 (and A119 ff.), taken in conjunction with concept-of-object-in-general and transcendental-object texts like A104–10 and A50/B74 ff., A92–93/B125, B137, B146–47, B158, A190–91/B235–36, and A197/B242–43. See also A113 on the notion of a rule, A245, A247/B304, B139, A129 (in the summary), B145–46, A414/B441; recall Reflexion 4643 and "Metaphysik Volckmann," as quoted in Chapter Three, and the other Reflexionen there cited. For the specific points about P_1, P_2, and the concept of a thickly needled red spruce, see especially A101 and A106; and observe A7–9/B11, A20–21/B35, A43/B61, and B131 note. In the case of the points about s_1, s_2, and the concept of being conical, complications emerge since Kant is not wholly clear about how the conceptual synthesis of the spatial-parts manifold occurs. I avoid these complications by talking simply of s_1 and s_2 as jointly 'specifying' the presence of the relevant concept. See A77/B102–103, A102, A105, A124, B137–38, B162, A162/B202 ff., and A224/B271. Throughout this chapter I abstract from the role of the logical functions and specific categories in synthesis.

6 Kant's overall use of '*überhaupt*' supports the present reading of 'concept of an object in general' (*Begriff von einem Gegenstande überhaupt*). Kant speaks at A93/B126 of concepts of an object in general because, I take it, he is there thinking of the various categories as being realizations of the concept of an object in general – that is, as being themselves concepts of an object in general. Besides A93/B126 on the concept of an object in general, see B137, B146, B158, A245, and A247/B304. For my general view of the much-disputed issue of Kant's concept of the transcendental object and its relation to the concept of an object in general, see Howell (1981b), from which many of the points in the present chapter derive.

7 There are various slightly different ways of formulating each of the claims in this chapter. The ones that I give are most directly relevant to our present goals. Note that (i) one could of course add a uniqueness condition to the quantifiers in (T*j*) and (T*i*) below. (ii) For reasons emphasized later in connection with (T*i*), the terms 'P_1,' 'P_2,' 's_2,' and 's_2' in (T*j*) really should occur in *de re* ways within the scope of '*H* is conscious in thought that.' (iii) (T*j*) really should concern the overall, imagination-reproduced presentation of what j_1 to j_4 individually put before the mind and not the individual presentations that are effected by j_1 to j_4 themselves. No harm results, however, if we ignore these points, and their analogues for (T*i*), hereafter.

8 Note the A105 quotation above ('the concept of this unity is the representation of the object = X'); B137 ('an object is *that* in the concept of which the manifold of intuition is united,' my emphasis); B146 ('knowledge involves two factors; first, the concept, through which an object in general is thought (the category); and, secondly, the intuition, through which it is given'); B158 ('for knowledge of an object distinct from me I require, besides the thought of an object in general (in the category), an intuition by which I *determine* that general concept,' my emphasis); A245 (apart from their relation to sensibility, the categories 'are *not* concepts through which an object is known and *distinguished from others*, but only so many modes of *thinking an object* for possible intuitions,' my emphases); A247/B304 ('Thought is the act which relates given intuition to an object. If the mode of this intuition is not in any way given, the object is merely transcendental, and the concept of understanding [of the object] ... expresses *only* the *thought* of *an object in general*, according to the different modes [logical functions],' my emphases). Note also A104. Together with the reasoning just

given, these texts show clearly that *H*'s *thought*, via the concept of an object in general, of the object that has the intuition-presented properties and spatial parts is the *de dicto*-like thought of *an* object but of no object in particular. See also Howell (1981b, 90–95 and 100–104).

9 We could get this understanding of the clause in question by rewriting the relevant part of (T*j*) as '$(\exists z)[z = j_1$ & *z* puts before *H*'s mind (or is) P_1 & *H* is conscious in thought that (*z* is an experience & *z* occurs & *z* grasps (or is of) P_1 ...)].'

10 In considering (T*i*) I continue the sorts of simplifications made in connection with (T*j*). I speak in (T*i*) simply of 'features or aspects' since *i* is a sensible intuition in general from whose sensible mode – and so from whose presentation of anything like spatial parts – we are abstracting.

11 Note B137–38, B145, B146–47, B150, B158, B160–61; and see the relevant parts of the A- and B-Phenomena and Noumena discussions and Paralogisms.

12 Indeed, as suggested in Howell (1981b, 104–14), where the present problem is discussed for the case of human intuition, one can argue that, for Kant, the first fact is to be analyzed in terms of the second fact. (For difficulties with such an analysis, see ibid., 109–10. The analysis of course needs to be restricted to cases in which one's thought *of* a single, individuated object is not mediated by an earlier *de re*-like grasp *of* that object or by inference.)

13 Such a *de re*-like thought can be expressed by moving the existential quantifier in (T*i*) in front of '*H* is conscious in thought that.'

14 I here follow Kant's reasoning, with some changes and additions, in texts like A104–105 (where he says that 'the unity [in the manifold] which [relation of the manifold to] the object makes necessary' is a certain unity among the elements of the intuition that puts us 'in a position to say that we know the object') and B137, first full paragraph (quoted below in Section 4). For further discussion of this complex topic, see Howell (1981b, 90–95).

15 To be made exact, this claim needs qualification. The specific features of the precise object to which *i*'s elements are related (and, given the need for the above individuating supposition, the definite, individuated character of the object itself) derive from the thought-independent character of those elements rather than from *H*'s *thinking* there to be a single object to which the features belong. Moreover, and as was noted in an earlier chapter, Kant's Deduction discussion of knowledge abstracts from matters of evidence or reliability.

16 Formally, this strong reading is of course just like (T*i*), except that in it the '$(\exists x)$' quantifier occurs at the beginning, and '*H* is conscious in thought that' is replaced by '*H* knows that.'

17 In the present section, I also bypass the need for the individuating supposition about *i*, returning in Section 5 to the issues it raises.

18 Thus in the B137 quotation below from S17, Kant's claims about the relation of intuition to an object are presented as concerning the relation of intuition to the object known. Note also the emphasis on knowledge in relation-to-object A-texts like A104–105 and A108–10.

19 In the heading and opening paragraph of § 17, Kant asserts that synthetic unity of apperception is the supreme principle of the understanding. However, our Chapter Six, Section 2, discussion shows that the § 17 principles about synthetic unity of

apperception are really no more than the § 16 principle already discussed in Chapter Six and there seen not to be required by the strict argument of the Deduction. And in fact the distinctive contribution of § 17 is not made in its repetition of the § 16 principle but in the present B137 remarks.

20 Indeed, given especially (s_1), Kant holds that that knowledge through *i* 'consists in' the occurrence of the (T*i*) thought or in the relation that that thought establishes to the object. But in this section we are ignoring such points about knowledge.

21 For this reading of 'unity of consciousness' in (s_3) and (s_4), see B137–39, especially the full paragraph at B138 (beginning 'Although this proposition').

22 Notice also the B134 implication that because unity of apperception holds and hence synthetic unity exists in the manifold of representations, I do not have 'as many-colored and diverse a self as I have representations of which I am conscious to myself.' Such an implication seems clearly to imply that, even on the weak reading of *H*'s knowledge (when, as far as (K) goes by itself, all that *H* knows might be *H*'s diverse representations, taken separately), the holding of unity of apperception requires synthesis of the manifold in a (T*i*)-style thought. A similar implication is present in the many A-Deduction texts arguing that the holding of unity of apperception rules out the possibility that knowledge might be about no more than disconnected representations 'crowd[ing] in upon the soul' (A111) in the form of 'merely a blind play ... less even than a dream' (A112).

23 In the A-texts Kant begins by working with the strong-reading idea that knowledge is a connected whole of representations which is governed by unity of apperception and whose existence requires a (T*i*)-like synthesis of those representations. He then in effect moves to the claim that on the central, weak reading, the holding of unity of apperception requires the same sort of synthesis. For the strong-reading start of this reasoning, see the relevant parts of A92–93/B125–26 (as discussed already in Chapter Four), A97, A99, A102, A103, A104–108, A111–12, and A119–23; and, for versions of the strong-to weak-reading movement, and the requirement, by the holding of unity of apperception, of a (T*i*)-like synthesis, see relevant parts of A105–108, A111–13, A119–23, A125–26, and A129. Note also the interpretation below relating A106 ff. to B137. There is of course a very great deal in the A-Deduction that I ignore here.

24 This inference – apparently a devastating logical error at the heart of the B-Deduction – has puzzled many writers. To take two recent examples, Allison (1983, 146) tries to escape the problem by arguing in effect that Kant means merely that unity of apperception is a sufficient condition for *H*'s thinking an object in the 'broad conception of an object' that Allison supposes is at work in the first half of the B-Deduction. But, as I have argued in Chapter Four, note 39, Allison's view of that half is mistaken. Guyer (1987, 117–18) holds that while § 17 starts out to reason from unity of apperception to conditions for knowledge of objects, § 17 in fact simply identifies, without argument, conditions that Kant independently claims to hold for knowledge of objects with conditions for unity of apperception. As I urge below, however, an interpretation is available that allows Kant's fundamental § 17 reasoning to be seen not as that sort of elementary blunder but as part of a unitary, if not ultimately a convincing, argument that is present also in the A-Deduction.

25 As we see below, such reasoning, when developed, appeals to the 'originality' of unity of apperception and its necessity. While § 17 and Kant's above B137 argument

do not mention the necessity of unity of apperception, the 'necessary synthesis of representations – to be entitled the original synthetic unity of apperception' – *is* mentioned directly before § 17 begins, at the end of § 16. B136, B136 note, and B137 of § 17 then themselves refer to that original synthetic unity (which § 16 of course has discussed at B132–33). Moreover, as we have seen, at B138–39 of § 17 Kant clarifies the general 'principle of synthetic unity' (B136) that § 16 introduces. Thus in terms just of the § 17 text, it is highly plausible to appeal to the originality of apperception in interpreting Kant's B137 argument. That appeal then is simply reinforced by the A106 ff. reasoning, remarked below, and its parallels with § 17. To note such parallels, A106–107 appeals to 'transcendental apperception' as the 'original and transcendental condition' that grounds the necessary synthesis of representations in the concept of an object. The A106 comment that 'this object is no more than that something, the concept of which expresses such a necessity of synthesis' resembles the B137 definition of an object (quoted above; compare also A108). And the last paragraph at A107 begins by talking of apperception in solely necessary-condition terms in a way that parallels B137 above in (s_1) and (s_3), moving then at the end to the claim that the unity of apperception grounds, and so is sufficient for, all concepts. This movement is similar to the B137 (s_4) inference that unity of consciousness implies the (T*i*)-style relation of representations to an object. (It then simply is that, unlike B137, A106–107 has earlier made clear the reasoning that underlies this move.) A107 also mentions space and time, just as does § 17; and note the parallel between the A107 claim that 'by relation to [unity of consciousness] representation of objects is alone possible' and (s_4) in B137.

[26] In the A-Deduction (A108, A111–12, A116–19) Kant argues that the holding of unity of apperception is necessary and so requires a necessary synthesis of the manifold of intuition. Because, by the Metaphysical Deduction, the categories provide the rules for necessary synthesis, Kant concludes that unity of apperception requires a category-governed synthesis of the manifold that (as Chapter Ten shows) requires *H* to think a (T*i*)-like thought. This sort of reasoning is not, however, at work at B137, for the Metaphysical Deduction really shows, at best, only that the categories are required in judgments; and the B-Deduction introduces judgment and the categories only in § 19 and § 20.

[27] Compare also the paragraph running over from A125 to A126.

[28] To add details: at A106 Kant argues that all knowledge demands a concept. The application of that concept yields a necessary reproduction of the manifold and hence the (necessary) synthetic unity in our consciousness of appearances. This necessity has a transcendental ground – a ground of 'the unity of consciousness in the synthesis of the manifold of ... intuitions, and consequently also of the concepts of objects in general, ... a ground without which it would be impossible to think any object for our intuitions.' Hence that ground – identified at A107 with transcendental apperception – is a ground for the necessity of our thinking a (T*i*)-like thought. Kant then concludes A106 by giving the (s_2)-reminiscent definition of an object remarked in note 25; and in A107 (as also observed in note 25) he goes on to assert that only through relation to unity of consciousness is representation of objects possible.

[29] Note, for example, Reflexion 5655 (Ak. 18, 314), dated as from 1788–89: 'Trichotomy. All relation of representations through concepts has a threefold

dimension: 1. the relation of a representation to consciousness; 2. of a different representation to consciousness; 3. unification of both together in one consciousness. Thereby the unification of representations with each other first becomes possible *(connexa uni tertio sunt connexa inter se)*.' Observe especially the last, parenthetical claim. On the present line of thought, i_1 and i_2's relation to the *I think* thus constitutes their original or logically first combination before *H*'s consciousness. The *I think* (or relation to it) 'is what, by adding itself to the representation of the manifold [the given intuition-elements i_1 and i_2], first makes possible the concept of the combination' (B131, in the final § 15 paragraph leading up to § 16). See also the B132 quotation in note 31.

30 Here recall the main Chapter Six arguments for unity of apperception and its necessity, especially the first (by appeal to the opening sentence of § 16) and the third (the § 16 appeal to the possibility of my calling all my representations mine). The (fallacious) third argument is a main ground for the necessity of unity of apperception. As that necessity is, on my present interpretation, crucial to the reasoning that underlies the central B137 argument for (T*i*) that we are here considering, I therefore think, contrary to Henrich (1969, 655), that Kant does not confuse himself about the B-Deduction's structure by an uneasy reliance on the third argument. Rather, that argument (although unfortunately fallacious), or some other argument yielding the necessity of unity of apperception, is *required* if the B137 reasoning is to go through. (See also Chapter Four above on Henrich, 1969.)

31 Here and in the last several sentences I interpret Kant's B132 remark that 'I call it [the representation *I think*] ... *original apperception* because it is that self-consciousness which, while generating the representation *I think* (a representation which must be capable of accompanying all other representations, and which in all consciousness is one and the same), cannot itself be accompanied by any further representation.' According to a proposed emendation (Schmidt, ed., 141b) 'accompanied [*begleitet*] by' should here be 'derived [*abgeleitet*] from.' For 'original,' note A544 = B572: 'an *original act*, such as can by itself bring about what did not exist before' (at B132, what did not exist before is a mental unity of i_1 and i_2, which is brought about by their relation to the *I think*). Compare also B72. Because the *I think* accompanies – and so surely is itself accompanied by – other representations, Kant's above B132 remark is misleading and should, I believe, be read along my above lines in the main text. The 'derived from' emendation is not required by my interpretation but can easily be incorporated into it. (A further reading, in terms of which my basic sort of B137 interpretation also could be developed, is to take 'the *I think* cannot be accompanied by any further representation' to mean 'the *I think* cannot be the mental object of any further representation.' Here observe views in B157–58 note.) I ignore the minor issue of whether at B132 it is the *I think* or the self-consciousness that generates it that is ultimately original.

32 Kant does not present the argument that follows. But such an argument is certainly allowed by his view of the *I think* as necessarily capable of accompanying all other representations and as, through its unity, being a source of *a priori* knowledge (B132) and of various necessities (A106–107, B135).

33 The necessity of synthesis that, by A106, is expressed in the concept of an object in general (and that belongs to *H*'s thinking (T*i*)) is a necessity that follows from the fact

that 'all knowledge demands a concept' (A106). That is, it is necessary that *if H* knows through *i* (and *if* other Kantian points hold), *then H* applies the concept of an object and thinks (T*i*). Compare A105 on 'the unity which the object makes necessary' and the claim that 'it is only when we have ... produced synthetic unity in the manifold of intuition [through the use of the concept of an object in a (T*i*)-style thought] that we are in a position to say that we know the object.'

34 Deductive relations because the knowledge-assumption (K), given the strong reading, and taken in conjunction with other Kantian points, logically implies (T*i*), as we have seen above. Given that logical implication, the present necessity-of-the-consequent follows immediately and requires no further explanation in terms of special acts of *H*'s mind.

35 That is, a necessity expressed by prefixing (T*i*) – or '*H* is conscious in thought that' in (T*i*) – by 'it is necessary that.'

36 I am of course thinking here of Kripke's work. My talk below of necessities known *a posteriori* refers to such suggested examples as the claim (supposed to be known *a priori*) that if person *P* comes from gametes *g* and *h*, then necessarily *P* comes from *g* and *h*; hence given that I know – *a posteriori* – that *P* comes from *g* and *h*, I know – *a posteriori* – that necessarily *P* comes from *g* and *h*.

37 See Chapter Seven, Section 2.

38 Note, for example, 'one single experience in which all perceptions are represented as in thoroughgoing and orderly connection' (A110).

39 This formulation of the weakened result is, for example, suggested by B278–79: 'inner experience in general is possible only through outer experience in general.' Other formulations could be proposed.

40 To the extent that, for the reasons given here and above, we doubt that Kant can give such a proof, we also must doubt that in the Deduction Kant can achieve a hope indicated in Chapter Three. That was the hope of showing that, on the weak reading of the knowledge that they yield, the elements of *i* – and so of any sensible intuition through which a being like us does or can know – put before the mind general properties of a single, category-subsumed object.

41 Thus the B274 ff. Refutation of Idealism, traditional interpretations of Wittgenstein's private-language argument, and views like those in Strawson (1966, 101–102), all can be made to imply something like this claim about the object.

42 Just such a replacement in effect occurs in the one first *Critique* text in which Kant explicitly considers something like the weakened result, the B274 ff. Refutation of Idealism. That text comes well after the Deduction, and its argument (in terms of knowledge of the existence of outer, spatial objects as necessary for consciousness of representations as occurring in a definite time order) is quite different from the Deduction's appeal to unity of apperception as by itself leading to category application. Similarly for such modern defenses of the Deduction as Strawson's influential 1966 work (97–112), which, while very interesting, replaces the Deduction argument with reasoning of Strawson's own. (Strawson of course discusses Kant's view of the Deduction.)

43 Kant could of course reply that i_1 and i_2 become known as objects (and are category-subsumed) insofar as the subject seeks to acquire knowledge of itself as empirical mind. But that reply still leaves a gap, for i_1 and i_2 are, just through their *I*

think-accompaniment, known entities. And the *Deduction* reasoning *itself* does not then show that i_1 and i_2 are, in that accompaniment, empirically known and so category-subsumed; it shows only that the single object known *through* them in the *(Ti)*-thought is category-subsumed.

NOTES TO CHAPTER NINE

1 I also ignore other § 18 points not now relevant – for example, B140 on 'the pure form of intuition in time, merely as intuition in general.' (I take Kant here at B140 to continue to speak from our human point of view, regarding time as an example of an intuition in general subject to unity of apperception. As we saw in Chapter Four, and as B148 and the first sentence of § 24, at B150, show clearly, Kant's § 20 results about category application extend beyond our human modes of intuition to the objects of any sensible intuitions in general.)

2 Because they are not relevant to the main argument of the Deduction, I set aside the presumed differences among the B139 subjective unity of consciousness, the B139–40 empirical unity of consciousness, and the B140 empirical unity of apperception. (Roughly, I take the subjective unity – as in the main text above – to be the unity that intuition-elements have in virtue of their associative organization. The empirical unity of consciousness or of apperception arises when intuition-elements, as they occur associatively organized in the mind, are grasped by thought-consciousness.)

3 A real proof would involve, among other things, complex issues about appearing and appearance theories that do not directly affect the Deduction. Note also that the object should be distinct from any merely associative organization of i_1 and i_2 in *H*'s mind.

4 Observe that – to mention a point discussed briefly in Chapter Eight and again below in Section 5 – this last reasoning does not show that the object in question is distinct from all intuition-elements whatsoever of *H*'s, taken separately or in sequence.

5 I see this implication in the closing remarks on objective validity in § 18 and § 19. For the *Prolegomena*, see especially § 18 and § 19, where Kant asserts that 'objective validity [of the unity of apperception, the sort of validity provided by the first sort of objective unity above] and necessary universal validity (for everyone) are ... identical concepts' (Ak. 4, 198; Lucas trans., 57). The *Prolegomena* § 18 and § 19 remarks (and those in B-Deduction § 19) are couched in terms of judgment, but that fact is easily accommodated to our present account. (I ignore here the notorious *Prolegomena* § 18 distinction between judgments of perception and judgments of experience, a distinction that prima facie conflicts with the B-Deduction § 19 insistence that all judgments are objective. I will note, however, that Kant's idea of a judgment of perception – as not needing a category – seems confused and is not required by anything in the basic Transcendental Deduction argument itself.)

6 This last point is trivially true just because (and in the sense that) it amounts to saying that because Kant's above position implies that *H* thinks the (T*i*)-thought, therefore Kant's above position implies that i_1 and i_2's forming an objective unity of the second sort itself implies that *H* thinks the (T*i*)-thought.

7 Will Kant understand the *p*, *q* (T*i*)-like thought to be about numerically the same

object as the *m*, *n* (T*i*)-like thought? In *Prolegomena* § 18, last paragraph, Kant argues explicitly for the present implication of the second (universal-validity) objective unity by the first (relation-to-a-distinct-object) objective unity, and his argument strongly suggests that the numerically same object *is* at issue ('if a judgment agrees with an object, all judgments about the same object must agree with one another'; Ak. 4, 298, Lucas trans., 56). This question raises complex points but can be ignored here, however, since the reasoning that I have just given works whether the *p*, *q* and *m*, *n* (*Ti*)-like thoughts concern numerically the same object or only objects of the same kind.

[8] Of course it does not follow either that what is claimed in that judgment goes beyond what is claimed in *H*'s (T*i*)-thought or that that judgment, if it does go beyond what is claimed in that thought, yields *H* knowledge. Indeed, in Chapter Eight we considered only the basic Kantian view that, given the holding of unity of apperception with respect to *i*, it follows that *H* thinks the specific (T*i*)-thought and that that thought yields *H* knowledge. I return to such points below.

[9] In his A132/B171 claim, Kant focuses just on categorical, affirmative, subject-predicate judgments. But the extension to other sorts of Kantian judgments, which is not difficult, is not needed for the point here.

[10] The terms 'simple,' 'basic,' and 'compound' are mine, not Kant's.

[11] In the case of the logical function of relation (A70/B95 ff.), *all* basic judgments have the one categorical, subject-predicate logical function; the other two relational logical functions (hypothetical and disjunctive) belong to compound judgments. Compound judgments have, directly, none of the other, nonrelational, logical functions but are made up of basic judgments that themselves exemplify those other logical functions in various ways.

[12] Of course, as examples below illustrate, this judgment may also be about other objects too. Furthermore, as we see in Chapter Ten, this present claim does not solve all the problems about the application of all the categories to *i*'s object.

[13] For Kant's own acceptance of the idea that (T*i*)-like thoughts can be involved in further judgments, note, for example, the A68–69/B93 example of how, in the judgment that all bodies are divisible, 'the concept of the divisible applies to various other concepts, but is here applied in particular to the concept of body, and that concept again *to certain appearances* that present themselves to us.' (My emphasis. Kant changed 'appearances' in his copy of the first *Critique* to 'intuitions'; see Schmidt, ed., 109.) Of course in this example Kant's acceptance is made in the context of Aristotelian logic and its assumption that the subject-terms of categorical judgments are nonempty.

[14] The trancendental unity of apperception has been argued earlier to hold with respect to i_1 and i_2. But it has not been explicitly argued to hold either with respect to what i_1 and i_2 put before the mind or with respect to the concepts (or further judgments) that occur in the judgment associated with the (T_1)-thought. (Nor has it been argued to hold with respect to any other concepts or further judgments of *H*'s.) This is one of the complications we bypass until Chapter Ten.

[15] The concepts or further judgments are what Kant calls the logical matter of the judgment, and their relation together through the logical functions is the judgment's logical form. See A266/B322, and compare Reflexionen 3046, 3039, 3042, 3044, 3045, 3050, and 3060 in Ak. 16; and *Logik*, § 18 ff.

[16] There are other difficulties, which I ignore here.
[17] Thus suppose that *H* judges, as in the (T*i*)-thought, that there is a single object that has *F* and *G*. Then *H* can also judge that all *FG* are *F*, that all *FG* are *G*, that all *FG* and *F* are *F*, that all *FG* and *G* are *F*, and so on; that all *FG* and *F* and *G* are *F*, and so on; that if all *FG* are *F*, then all *FG* are *G*; that if (if all *FG* are *F*, then all *FG* are *G*), then (if all *FG* are *G*, then all *FG* are *F*); and so on (including disjunctive claims). Because such judgments involve different combinations of the relevant logical functions (and so of the concepts and further judgments involved), they are distinct judgments. There are in principle an infinite number of such judgments, most of which *H* cannot make, as they outrun the storage and comprehension capacities of *H*'s (finite) mind. But there are obviously many such judgments – like the ones just listed – that *H can* but need not make while making the specific (T*i*) judgment. And that fact is all that is needed to create the problem that I note below for (P).
[18] The judger makes *z*, with its specific logical form; and by (P) the logical form of a judgment consists in or derives from the holding of objective unity of apperception with respect to the concepts occurring in that judgment.
[19] The judger at least contemplates the contents of these judgments, whether or not the judger actually asserts them. Given Kant's own account of the assertoric and apodeictic logical functions, if either of those logical functions is involved the judger may well actually make the judgment.
[20] One might read § 19, B141 in this way: 'a judgment is nothing but the manner in which different modes of knowledge are brought to objective unity of apperception.' Note also *Prolegomena* § 22 (Ak. 4, 305; Lucas trans., 64): 'the logical moments [functions] of all judgment are so many possible ways of unifying representations in a consciousness,' and § 39 (Ak. 4, 323; Lucas trans., 86): judging is 'the act of the understanding which contains all the rest and is only differentiated by different modifications or moments.' (By A70/B95, the functions of thought fall 'under four heads, each of which contains three moments.')
[21] To this difficulty it might be objected that even if, say, the subject-predicate logical function is itself *a priori*, it is nevertheless wholly contingent that that function holds with respect to particular concepts or features presented by the manifold of intuition. So there is no necessity here to be explained through the holding of unity of apperception; and thus the present difficulty collapses. Although it may seem plausible in its own terms, this objection is not, however, one that Kant himself can easily offer. (a) In his (P) view, Kant regards the holding of unity of apperception as responsible for the holding of the individual logical functions with respect to the concepts or features presented by the manifold of intuition. And his reason for so proceeding is surely that he takes the latter holding itself to be in a certain way necessary and *a priori*. (b) In Chapter Ten we see that, for Kant, when individual logical functions are applied to intuition-presented concepts or features in such a way that the associated categories are then applied to the objects judged about (as happens in every judgment yielding knowledge), an element of necessity *is* involved in this application. (That fact is true even when the judgment is contingent.) And Kant will suppose that this element of necessity requires a source in the holding of unity of apperception.
[22] Note also the Chapter Eight discussion of the § 17 'sources' argument for (T*i*).
[23] Besides the B140 heading to § 19, which shows the depth of Kant's adherence to the fundamental Deduction idea here noted, see B131 of the introductory § 15, where

Kant clearly implies that the holding of unity of apperception is the 'ground of the unity of diverse concepts in judgment' and hence (it would seem) is the ground or source of all the specific, determinate sorts of logical combination and so of logical form that occur in judgments. Note "Metaphysik L_2" (Ak. 28. 2, 1 at 548): a 'logical ground is the relation of [pieces of] knowledge, how one is inferred from the other'; a 'ground is that through which something else is posited'; a 'ground is that upon which something follows in an entirely necessary way'; Kant to Reinhold, May 12, 1789: 'a ground is (in general) that whereby something else (distinct from it) is made determinate'; 'if the ground is posited, the consequent is determined' (Ak. 11, 35; Zweig trans., 138).

[24] Kant also has to show that the object is distinct from the relevant intuition-elements. But that point is independent of his views about the logical forms of judgment and unity of apperception.

[25] Of course to the extent that Kant in § 19 means to claim anything about a knowledge-yielding judgment involving but going beyond the (T*i*) judgment, Kant extends his explicit § 17 view that unity of apperception implies *H* thinks the knowledge-yielding (T*i*) thought.

[26] See B128–29, *Metaphysical Foundations of Natural Science* footnote to the Preface (Ak. 4, 474 ff.), and Reflexion XLII (Ak. 23, 25).

[27] One can also generate related problems. For example, how exactly is the necessary holding of unity of apperception – which is expressed in what is presumably the judgment that, necessarily, the *I think* accompanies all my representations – itself the source (on Kant's 'sources' view) of *every* judgment's logical form or range of possible logical forms? I will not try here to discuss such worries (which raise further questions about the 'sources' view), beyond drawing attention to the general comments below.

[28] The three quotations here (all with my emphases) are from the beginning of Reflexion XLII (Ak. 23, 25); *Prolegomena* § 22 (Ak. 4, 305); and *Metaphysical Foundations of Natural Science*, footnote to the Preface (Ak. 4, 475). Note that it is not implied in this last quotation that in order for concepts or properties like *stone* and *hard* to have these uses the judger must consciously think them to have such uses.

[29] I talk here of 'concept (or property)' in order to bypass complications postponed until Chapter Ten.

[30] Even though it is necessary that if (a) holds, then (b) holds, the present example shows that it does not follow that if I think that (a), then I think that (b). This fact illustrates again our Chapter Five points about intensionality and thought-consciousness.

[31] Here note the texts cited in the third paragraph of this section; observe that even in the texts cited later, in the paragraph tagged by note 28, Kant does not suggest the specific idea just mentioned; and notice that when he argues, in *Anthropology*, § 5, for the existence of unconscious mental activities and representations, the activities that he holds can be unconscious are simply those of association in imagination. (Compare also the well-known A78/B103 comment on imagination.) The fact that it is just associative (or productive) activities of imagination that Kant argues may be unconscious shows that while he is an important ancestor of recent cognitive theories of the mind, he assumes nothing like the range of unconscious higher-order mental

processes that many such theories routinely postulate. (Commentators like Kemp Smith who attribute to Kant an acceptance of wide-ranging unconscious mental processes to my mind simply impose on the actual theory of the first *Critique* their own views.)

32 For such views see, for example, Arnauld and Nicole (1662), Part 2, Chapter 3, first three paragraphs. Note also Locke, *Essay*, III. 7. 1, and Kant's own early view of judgment as a comparison of a thing, the subject of the judgment, with a mark, the judgment's predicate, in "False Subtlety," § 1 (Ak. 2, 47).

33 For other first-*Critique* remarks on the copula, see A74/B99–100 and A598–99 = B626–27 (where Kant in effect distinguishes the copulative from the existential use of 'is'). Note also *Logik*, § 24.

34 And in Section 2 we adopted a similar understanding of the relation-to-a-distinct-object type of objective validity.

35 Or taken as having any other merely associative organization in the mind. (This qualification applies throughout, below.)

36 Nor is it necessary here to discuss the *Prolegomena* § 18 distinction between judgments of perception and judgments of experience. (For a brief comment, see note 5.)

37 Here recall Chapter Eight, especially Section 5.

38 This comment is of course made from the standpoint of what Kant himself should try to argue, given his belief (criticized in Chapter Eight) that in § 17 he proves that unity of apperception with respect to *i* implies that *H* thinks and knows an object through the (T*i*)-thought.

39 Wolff (1963, 163–64 and ff.) emphasizes strongly a difficulty resembling the present difficulty for Kant; his discussion has influenced mine here and earlier.

40 Such a supposition resembles (and perhaps lies behind) the *Prolegomena* § 18 idea of judgments of perception. (Compare note 5.) To avoid what I think are insuperable difficulties with that idea, the present supposition would need to take the direct-inspection judgments themselves (unlike judgments of perception) to introduce the categories. Even if the categories could be introduced, for the reason given below this supposition will not work.

41 As we noted in Chapter Seven, Kant's B142 points do not rule out his expressing the holding of necessary unity of apperception by an (N_1)- or (N_2)-style claim to the effect that if a being like us knows through (or has) a given sensible intuition, then it is necessary that that being is or can become conscious in thought that the *I think* accompanies the elements of the intuition taken together. Such a claim is compatible with the contingency of the judgment that is made about the object of that intuition. (For this reason, among others, I reject the claim in Guyer, 1987, 120, that the B-Deduction argument up to § 20 'collapses into the assertion that empirical judgments of objects are actually *claims of necessity*.' Kant's reasoning here is much subtler than that.)

42 B142; compare the last sentences of § 18.

43 See the last sentence of *Prolegomena* § 18 (Ak. 4, 298). (Kant's language there admits of the ambiguity about 'same object' that was remarked in note 7; the criticisms below can be applied to either way of taking that ambiguity.)

44 That is, we cannot here plausibly infer such a fact about *m* and *n* without introduc-

ing Kant's position that the holding of unity of apperception implies that the knower thinks, with respect to the relevant intuition-elements, a (T*i*)-like thought. (Of course if we introduce that position, then we can infer that *m* and *n* are involved in a judgment about a distinct object. Recall Section 2.)

NOTES TO CHAPTER TEN

[1] Although I suppress the initial quantifiers and conjuncts in this simplified claim, they are still officially present; and so '*F*' and '*G*' are still to be understood in a *de re* fashion.

[2] As Chapter Nine in effect indicates, '*R*' occurs *de re* in (C*i*) below. Thus (C*i*) really is to be imagined as arrived at by adding '(∃*R*)[... *R* is a logical function ... ' to the other quantifiers and conjuncts before '*H* is conscious in thought that' in the Chapter Eight (T*i*).

[3] By our Chapter Eight discussion, *i*'s object is a single, individuated object that is brought under the category. This fact, which I will not mention again, holds throughout this chapter. (As noted in Chapter Eight, the effect of this fact is that '(∃*x*)' in (T*i*) and similar claims may be moved in front of '*H* is conscious in thought that.')

[4] I ignore many fine points. In connection with (a) and (b), recall from Chapter Three that both the appearing and the appearance theory are logically compatible with the idea that concepts are properties taken as general.

[5] 'Syntactic' and 'semantic' are somewhat anachronistic here but are useful in understanding Kant's position. I consider below how far any sharp, modern sort of syntax-semantics distinction will apply to Kant's account of judgment and the logical functions.

[6] Categories, by B128, 'are concepts of an object in general, by means of which the intuition of an object is regarded as determined in respect of one of the logical functions of judgment.'

[7] In these comments I ignore fine distinctions, which can be drawn in several different ways for Kant's work, among the act of judgment, the judgment judged (and its content), and the means or constituents of each.

[8] See Chapter Three, Section 2.

[9] I use 'modern' throughout this chapter simply to recall positions commonly or recently held that yield an instructive comparison with Kant's own ideas. As it happens, I accept many (although not all) of these modern views; but it is not necessary to endorse any of them to appreciate the comparisons.

[10] That is, *F* as subject is identified with – or is taken to have an unexplained relation to – the object *x* in the form that *x* takes as *x* is thought through this judgment (and represented by *i*). (The object *x*, in this form, then is to be treated in either appearing- or appearance-theory terms.) This point applies throughout this chapter.

[11] Ak. 2, 47. Of course the texts that I cite here and below come from different periods, and Reflexion 3921, quoted below, is an unpublished note. But I see no evidence that Kant changed his basic views on the matters now under discussion; and the texts, taken together, support the overall position I have advanced.

[12] Ak. 17, 346, my emphasis; observe also Reflexion 4643, as quoted in Chapter Three, Section 3.

[13] Ak. 4, 475; Ellington trans., 13, with slight changes to follow exactly the textual emphasis of the German.
[14] Ibid. Substance (and accident) is of course the category associated with the subject (and predicate) logical function.
[15] Note from Kant's language that it clearly is the feature of [being a] *stone* – which Kant here describes as a concept – that is thought as the subject, and not simply the object having that feature. Kant does not here explain the relation of the feature, as thought as the subject, and the object. However, in writing that 'I represent to myself *in the object* as *determined* that the *stone* ... must be thought only as subject,' he talks in a way that makes it natural for me to speak, later below, of the 'subject-element in the object.'
[16] The pure concept of what can be thought only as subject (and never as predicate of something else) is the pure category of substance (A147/B186), and that pure concept is schematized through the notion of the 'permanence of the real in time' (A143/B183). Given this schematization, something in experience is thought only as a subject if and only if, roughly, that thing is (thought as) permanent or unchanging (and it is never thought as predicate of something else).
[17] See note 16, and compare A285/B341 and Arnauld and Nicole (1662), Part I, Chapter 2: Because our mind knows most things only 'as modified,' our mind 'often divides the substance itself in its essence into two ideas, of which it regards one as the subject, the other as the mode' (1970 edition, 74, my translation).
[18] Ak. 2, 48.
[19] Both the *Logik* and "False Subtlety" quotes are presented in connection with the logical principle of the *dictum de omni et nullo*, 'the supreme principle of affirmative syllogisms' ("False Subtlety," Ak. 2, 49). According to "False Subtlety," § 2, this principle runs 'Whatever is universally affirmed of a concept is also affirmed of each concept that is contained under it' (Ak. 2, 49). According to *Logik*, § 63 Note, it claims that 'what belongs to or contradicts the genus or species belongs to or contradicts all the objects that are contained under that genus or species' (Ak. 9, 123). The fact that Kant puts this principle indifferently in terms of concepts or of genus and species itself suggests his identification of concept with general feature.
[20] As noted in Schmidt, ed., 109, in Kant's copy of the first *Critique* 'appearances' here is changed to 'intuitions.'
[21] For example, in *Essay* IV. 8. 4 Locke holds that, in the case of definitions, any part of the definition is to be regarded as predicated of the word defined, so that in the definitional proposition *lead is a metal* what is signified by 'metal' is predicated of the species (and so of the general feature), *lead*, that comprehends the individual pieces of lead. See also Meier's *Auszug*, § 363 (Ak. 16, 715) and Aristotle, *Categories*, 1^b10 ff.
[22] In this paragraph I simply sketch a standard modern idea in order to throw light on Kant; I am not trying to give any comprehensive account of modern views of syntax and semantics.
[23] The modern ideas above of course do not logically *require* the present account of judgment, but they can make that account seem very natural, given especially the view that concepts present features from which they are distinct.
[24] Note also the "Metaphysik Volckmann" text in Chapter Three, Section 3.
[25] Here and below I continue to focus on the idea of concepts as presenting properties from which they are distinct. That idea, along with texts like the A55/B79 comments

on 'logical form' quoted in the last paragraph, makes it natural for the present sort of modern reading of Kant to proceed in terms of a syntactic notion of logical form. There are of course also modern model-theoretic, semantical conceptions of logical form. I do not here endorse any specific analysis of that notion (if any one analysis is possible) but simply indicate how one modern approach to Kantian judgments can develop.

26 In noting that the holding of objective unity of apperception is the only way this *specific* logical form (or merely *some* logical form) can exist in the concepts, we raise issues considered in Chapter Nine, Section 3. Those issues also obviously affect our present question about the relation between the establishment of logical form in a judgment and the referring of the judgment's concepts to the object. In a full-scale study of Kant on judgment, these matters should be investigated thoroughly. But I must ignore them here.

27 The general idea that a judgment will have a purely syntactic organization (as well as semantic relations to the world) is something that most modern readers would expect, whether or not they would also suppose (as has the sort of modern reader whom we have been discussing above) that 'logical form' itself is a matter purely of syntax.

28 Here see especially the second quotation from B128–29, and also our discussion below.

29 This classification is in various ways rough (and some texts below may be argued to fall under several headings). But it is useful here.

30 Because of the 'insofar as' qualification here, this B143 text may well be better classified as a (c)-style view than as a plain (b)-style claim that the categories are the functions of thought. (This (c)-style classification would then fit the (c)-style classification of the central B128 definition of a category that Kant added to preface the B-Deduction.) Given the B128 context (a preface to the B-Deduction) and this 'insofar as' qualification, Guyer's claim (1987, 120) that B143 'absolutely violates Kant's original constraint that the categories cannot merely be identified with the logical functions' is wrong. Although the B143 *wording* can be argued to fall under (b), B128 and Kant's other characterizations of the categories show that, as my interpretation below brings out, Kant at B143 is *arguing* on the basis of a (c)-style treatment of the categories that does not identify them simply with the logical functions.

31 Ak. 4, 324; Lucas trans., 87.

32 Ak. 17, 620.

33 Ak. 4, 302; Lucas trans., 61; the 'moments' here are the moments of A70/B95, which give the twelve specific logical functions.

34 Ak. 4, 474–75; Ellington trans., 12–13.

35 Ak. 18, 392.

36 In stating these two ways, I idealize Kant's descriptions somewhat, for clarity. The central B128 and *Metaphysical Foundations* texts fit my account here quite accurately; other (c) texts resemble it in relevant ways.

37 By making this point about R, I here assume the identification noted in Section 3 of (i) the logical-function organization of concepts c_1 and c_2 in the *judgment* that the object has F and G with (ii) the organization that occurs within the *content* of that judgment, the fact that the object has F and G.

[38] Recall from note 10 that this identification or unexplained relation is with respect to the object in the form that that object takes as it judged about and represented through *i*.
[39] Thus A183/B227: 'In all appearances the permanent is *the object itself*, that is, substance as phenomenon' (my emphasis); A181: 'All appearances contain the permanent (substance) as the object itself, and the transitory as its mere determination.' Observe also the "False Subtlety" and Reflexion 3921 texts cited in the Section 3 discussion of features as standing in the subject-predicate relation. And note A399 in that same discussion ('if I am to declare a *thing* to *be* substance in the appearance,' my emphasis).
[40] This description is a bit rough, for the (SP*i*)-thought may be only a part of the overall judgment that *H* makes, and thus still other concepts and logical functions may be involved.
[41] The points made in note 26 are relevant to any exact understanding of the way in which the two parts are inseparable.
[42] See also the Section 7 discussion below of necessity and B128 on category-determination of the object. To note one point that is not clear, suppose that we adopt the present, abstractionist view of general logic and the logical employment of the understanding. Then even if we need to use the concept of an object in general to determine *F* to function as subject, why cannot we abstract from that use, once it is established, and focus in our abstraction on the determined *F* as having a definite, well-defined, 'purely syntactic' function of subject? (This sort of question also casts doubt on Kant's view, indicated below, that in general logic we consider simply *possible* ways that concepts can be determinately organized into judgments.) I ignore all such worries below.
[43] See Section 3; recall also the *Metaphysical Foundations* footnote quoted there and the "Metaphysik Volckmann" text given in Chapter Three, Section 3.
[44] Ak. 18, 390; my emphasis.
[45] Ak. 9, 101; my emphasis.
[46] Mellin (1797–1804), vol. 4, 107–108, from the article 'Materie'; all but the last emphasis mine.
[47] The first and third of the above texts focus on the subject-predicate case. One can work out plausible extensions to other cases.
[48] Once we bring in the other relevant logical functions, this concept of a single, actual *x* is of course not simply the category of substance anymore; it is the combination of that category with the other categories involved.
[49] And for many interpretive purposes we will be justified in bringing to bear this model of *H*'s thought-consciousness, as we did in Chapter Nine. We are thereby enabled to apply a sharp, clear logical apparatus that lets us understand many of Kant's views in a comprehensive way without distorting their major aspects. However, at the present point Kant's views diverge from our model. And though his present views can be detached from the overall argument of the Deduction, it is important to note them here in considering his own conception of category application.
[50] Besides the preceding grounds for thinking that Kant does not maintain this sharp distinction (or the various distinctions that a modern reader might draw here), note *Prolegomena*, § 21 (Ak. 4, 302; Lucas trans., 61). Kant there seems to assimilate

intuitions (and so, I take it, objects) to judgments: 'concepts of intuitions in general which are determined in themselves as judgments.' Note also ibid., § 21[a] (Ak. 4, 304; Lucas trans. 63): 'those judgments which the understanding makes solely out of sensible intuitions.'

51 An identification that seems to occur in the A79/B104 text quoted above, which I would therefore count as falling into group (b) of my Section 4 classification of the categories. If we do not identify concept with property, then the subject-predicate organization of features in the object is only analogous to the subject-term/predicate-term organization of concepts in the judgment. But Kant accepts that identification at A79/B104; he there clearly means to take numerically one and the same logical function to give unity both to the object (intuition) and to the judgment. (Note the A79/B104 language and the texts cited earlier.)

52 For the sake of absolute clarity: The representations unified in the judgment are the concepts c_1 and c_2; the representations unified in the intuition i, through the use of the same logical function in the same act of mind, are the intuition-elements i_2 and i_2 that put before the mind – or are – the features F and G with which c_1 and c_2 are identified.

53 I thank Julius Moravcsik for suggesting to me in 1973 that the above sort of puzzling features of Kant's treatment of judgment and the categories might be explained by appeal to Aristotelian theory. He is not responsible for the explanations below. My goal is of course not to argue that Kant adopted the features noted above directly from Aristotle, without being influenced by subsequent authors. Rather, I simply indicate a constellation of points in Aristotle's work that are paralleled, directly or indirectly, in those features. Given the undoubted influence of Aristotelian logic on Kant, the parallels then offer a general framework for explaining the existence of those features in Kant's texts. The detailed working out of this influence would require further study.

54 As observed below, Aristotle on occasion speaks of forms in the mind as 'likenesses' or 'images' (*De Interpretatione*, 16ᵃ3 ff.; *De Anima*, 431ᵃ8 ff., 431ᵇ2 ff., 432ᵃ3 ff.) that are mentally contemplated (*De Anima*, 432ᵃ3 ff.; compare 431ᵇ2 ff.). But the roots of early modern representationalism appear to lie in Hellenistic and scholastic philosophy.

55 I here follow *Categories* 2ᵇ29 ff. ('so the species and genera of the primary substance stand to all the rest: all the rest are predicated of these'; Ackrill trans., 8). There are questions of interpretation here, and Aristotle's actual position is considerably subtler than my comments (and the quote just given) suggest. (Thus note Ackrill on 2ᵇ29 ff., at 84 of his translation.) I am, however, now trying simply to indicate Aristotelian parallels with Kant and not to develop Aristotle's views in full detail.

56 For the material in the present paragraph, see, for example, *Categories*, Chapter 5.

57 For the views reported here and in the next sentence, see *De Anima*, Book II, Chapter 12, and Book III, Chapters 4 and 8.

58 See note 54. Although Aristotle speaks of the mind as contemplating images, he apparently does not suppose (as would a genuine representationalist at this point) that forms in the mind are the mental objects of a special inner consciousness to which those forms represent the objects that possess them.

59 Thus *De Anima*, Book III, Chapter 6 (Hamlyn trans., 60–61): 'Where there is both falsity and truth, there is always a combination of thoughts as forming a unity. ... For falsity always depends upon a combination; for even if someone says that white is

nonwhite he combines white and nonwhite. ... And that which produces a unity is in each case the intellect.'

60 This point is not clear; but see the texts cited in note 54.

61 Thus *Metaphysics*, Book VI, Chapter 4: 'the true judgment affirms where the subject and predicate really are combined, and denies where they are separated, while the false judgment has the opposite of this allocation' (Ross translation, in McKeon, ed., 782); Book IX, Chapter 10: 'This depends, on the side of the objects, on their being combined or separated, so that he who thinks the separated to be separated and the combined to be combined has the truth' (ibid., 833); 'As regards the '*being*' that answers to truth and the 'non-being' that answers to falsity, in one case there is truth if the subject and the attribute are really combined, and falsity if they are not combined' (ibid., 834).

62 The texts cited in note 61 certainly seem to allow for the same sort of subject-predicate combination to occur both in reality and in the mind. But they could, it seems, also perhaps be read as requiring no more than a structural parallel between the relevant types of combination. One need not try to decide this matter here.

63 Commenting on *Metaphysics* Book VI, Chapter 4, 1024^b17 ff. in his translation, 199, Kirwan suggests that Aristotle was, perhaps, not free of a confusion between a thought of pale Callias and the thought that Callias is pale. See, further, his 199 comments on 1027^b25 and taking 'Callias' being pale' as equivalent to 'that Callias is pale.'

64 See, for example, Hamlyn in his translation of *De Anima*, 104 and 113–114.

65 Ideas like Descartes' notion of a 'principal attribute' – which I would suppose has ultimately an Aristotelian origin – would also help to make it natural to think of features of objects in this way (*Principles of Philosophy*, Part I, § 53). Observe also the Arnauld-Nicole view quoted in note 17.

66 Kant of course does not follow all the details of Aristotle's views. Thus he does not describe the subject-feature as inhering in the matter of the stone in the way that Aristotle would. Rather, Kant regards the substance of the body as itself being 'matter,' in the First Analogy sense of that notion.

67 Note also Moravcsik, "Aristotle's Theory of Categories," in Moravcsik (1967), at 133.

68 In the comments below, I ignore points important to the details of Aristotle's theory (for example, the distinction between things present in and things predicable of a substance) but not directly relevant to my comparison with Kant.

69 Kant also introduces other modifications of traditional Aristotelian theory, for example in his much-attacked notion of infinite judgments.

70 In thus describing Aristotelian categories, I follow the discussions by Cook Wilson (in Moravcsik, ed., 1967, especially at 86) and Moravcsik ("Aristotle's Theory of Categories," in ibid., especially at 135–36, 141, 143–44).

71 It is not, of course, that Kant is logically incompetent or has nothing of philosophical importance to say about logic. To mention only three points, Kant's views on the nature of general logic, its distinction from mathematics, and the role of logical form in specifying the categories of thought are philosophically fascinating and have been deeply influential. (For a forceful corrective to the view of a logically incapable Kant gripped by ossified doctrine, see Brittan (1978), especially viii–ix.) However, the version of Aristotelian logic Kant accepts is restricted to Aristotelian syllogisms and immediate inferences, with some ability to deal with hypothetical and disjunctive

claims but without any developed, general propositional logic or any consideration of the possibility of a general quantification theory. (See Parsons (1969), part II, for example. If, in response to such points, one simply identifies Kant's logic with, say, monadic predicate logic – perhaps in a free-logic form – then the problems noted below about moving from logical structure to category application open up.)

72 See for example, Strawson's standard objections (1966, 79–82).

73 Nevertheless, neither I nor anyone else that I know of has *disproved* the existence of such inferences, and the possibility remains open (although it seems not great) that someone could infer the above sort of conclusion and therefore vindicate a central Kantian tenet.

74 For simplicity, I here continue the assumption, which we have seen to be mistaken, that Kant *can* show (T*i*) (and then can apply his theory of judgment to the judgment in which *H*'s (T*i*) thought is involved). If that assumption is dropped, the comments below should be presented in terms of ideas like the Chapter Eight modified-(T*i*) claim or the weakened (T*i*)-style result.

75 As observed in Chapter Three, note 26, the logical function of modality does not concern the content of the judgment (the object) but the relation of the judgment to the understanding. (That relation in turn concerns whether the affirmation or negation in the subject-predicate judgment is taken as possible, actual (true), or necessary. See A74/B99 ff.) In consequence, the modal logical functions have no direct effect on the synthesis of the manifold into objects but instead concern, roughly, how those objects are taken to relate to experience. The effect of the logical functions of quality is noted in Section 8.

76 Arguing against the present sort of objection about hypothetical judgments, Allison (1983, 121–22), says that while such judgments always involve the Kantian pure ground-consequence relation, they need not involve the cause-effect relation between events in human experience. Allison underestimates the difficulty of defending Kant's views here, however. Allison is right that invocation of the pure ground-consequent category does not of itself introduce the phenomenal-world cause-effect relation. But (a) in texts like B-Deduction § 20, Kant wants to establish category application on the basis of the determination of the manifold of intuition by the logical functions of judgment. In turn, application of the pure ground-consequent category associated with the hypothetical logical function is supposed to yield, through schematization, application of the phenomenal-world cause-effect relation. But in that latter relation, a cause is a roughly a *ratio essendi*, a reason for the being and not merely for our inferring of the effect. (See A144/B183, A189, B233–34, A193–94/B238–39, and A198/B243.). Hence (since it is obviously not introduced by the process of schematization itself) the reason-for-being feature of a cause must derive from the fact that in the pure ground-consequent relation, the ground is a reason for being of the effect. (Allison, op. cit., seems to understand 'ground' simply in an evidential, reason-for-inferring sense; Kant's own characterizations of 'ground,' as cited in Chapter Nine, note 23, seem to shift between reasons for being and reasons for inferring.) Therefore if texts like B-Deduction § 20 are to prove category application just by appeal to determination of the manifold by the logical functions, hypothetical judgments must *all* involve a ground-consequent logical function in which the situation in the antecedent is meant as a *ratio essendi* for the situation in the consequent. But that is not so for the smoke-fire or Smith-grandmother cases. (b) Given Kant's views in the Metaphysical Deduction, in texts like B-Deduction § 20, and in the Schematism, it is

not at all clear why, for Kant, a commitment to a phenomenal-world cause-effect relation between antecedent and consequent situations will not automatically be made in hypothetical judgments about phenomenal objects and the events involving them. But no such commitment occurs in the Mary-Bill case.

77 In response to an objection related to the second one above, Allison (1983, 119–21) argues that Kant does not suppose that in every categorical, subject-predicate judgment we apply the actual, traditional category of substance to the judgment's subject. Rather, within each such judgment, we think the subject of the judgment to be – as far as *that* judgment goes – a subject that is not itself a predicate. And (Allison could hold) within some other judgment we can think that subject as a predicate. Again, however, Allison underestimates the difficulty in defending Kant. (a) (and here compare (a) in note 76) the argument of B-Deduction § 20 is meant to prove the application of the pure category of substance (the notion of 'a something which can be thought only as subject, never as a predicate of something else,' A147/B186). Hence since that argument proceeds simply by appealing to the determination of the manifold of intuition by the logical functions, Kant must hold that in *every* knowledge-yielding categorical judgment the pure category of substance *is* applied. (Of course that application ultimately involves the Schematism, but that point is not at issue here.) Given Kant's language at B128–29, and in the footnote to the *Metaphysical Foundations* Preface, when we apply that category in making some specific categorical judgment, we commit ourselves to holding that the subject of that judgment is *always*, and with regard to *every* context of knowledge (including other judgments), a subject and not a mere predicate. However, and as seen above, in other knowledge-yielding judgments the subject of our original categorical judgment will be a predicate and *not* a subject. This last fact contradicts the commitment that we made in the original categorical judgment and so creates a real problem for Kant. Or else (and as suggested by the discussion below), this last fact forces us to the conception of an object-feature that we *have* to treat *both* as subject *and* as predicate, which returns us to the second objection. (b) As just seen, if the B-Deduction § 20 argument is to succeed, Kant must hold that in every knowledge-yielding categorical judgment the pure category of substance is applied. But, as is shown by Bennett's 'his amiability cloys' example (1966, 183) that Allison cites, this view is quite implausible.

78 The latter sort of claim is necessary in the (N_1)- or (N_2)-style noted in Chapter Seven. As remarked in Chapter Seven and in Chapter Nine (note 41), this necessity is compatible with the contingency of the judgment *H* makes.

79 See A245: categories are 'representations of things in general, so far as their manifold *must* be thought through one or another of these logical functions' (my italics); *Prolegomena*, §§ 20–21; Reflexionen 5854 (Ak. 18, 369–70), 5931 (Ak. 18, 390–91), and XLII (Ak. 23, 25).

80 By *Logik*, §§ 52–53 (Ak. 9, 118), the judgment that something divisible is a body is the 'converse by limitation,' or *per accidens*, of the judgment that all bodies are divisible. (In connection with such inferences, recall that Kant, following Aristotelian logic, assumes subject-terms have existential import.)

81 This problem has been discussed by a number of commentators. Some of the difficulties it involves are vividly indicated in Bennett (1966, 81 ff.).

82 Nor does the little Kant says on this issue in, for example, Reflexion 5932 (Ak. 18, 391) really advance the discussion.

83 Inner-sense intuitions of other intuitions might be appealed to here. But because

inner-sense intuitions are not themselves subject to a still further form of intuition, yet are unity-of-apperception-governed and so known as accompanied by the *I think*, this would not show category application to the inner-sense intuitions themselves. See B157–58 note; A278 = B334; Kant to Herz, May 26, 1789 (Ak. 11, 51).

84 The issues here are very tricky and beyond the limits that I have set myself in this book. For some of the questions that arise, see, for example, Kemp Smith (1962, 311–12, 384–85, and 473–77); Wolff (1963, 166–74 and 191–202); Ameriks (1982a, Chapter 7); Allison (1983, Chapter 12); Guyer (1987, 373 ff.).

85 In the following remarks I ignore many complications, among them those of the sort raised in Parsons' (1964) discussion of the first *Critique* view of the possibility of experience.

86 The object known is of course so recognized in the form it takes as it is known. This point applies throughout the present discussion.

87 As noted in Chapter One, Section 2, Kant explicitly accepts the stronger, A51/B75–76 principle that our having knowledge requires our actually having an intuition of the object known. But in A225–26/B272–73 he allows the possibility of our knowing objects that we do not currently intuit. And, as noted below, he also argues explicitly for the point that we have no grounds to suppose any being besides us has any sort of intuition that allows that being to intuit the object known (as that object exists in itself or otherwise exists). Given that Kant urges this sort of point so often (and given also his A225–26/B272–73 claim), it seems that in the first *Critique* he does accept, at least implicitly, the overall general principle that any object known by us can in principle be intuited *by us or by some being* (and so on). As I suggest above in the main text, Kant's arguments that we have no grounds of the sort just noted then in effect allow him to derive from this overall, general principle the simpler result that any object known by us can in principle be intuited *by us* (and, as so intuited, can be known through a *de re*-like judgment). The stronger, A51/B75–76 principle then is itself best regarded as a further simplification, for expository purposes, of this last, simpler result.

88 See, for example, B-Deduction § 21, B138–39; § 23, B149; the Phenomena and Noumena chapter, A238–41/B297–300, B307–308, A252–53, A254/B310, and A256/B311; B72; and A770–71 = B798–99. The above requirement of grounds for such a supposition is the best interpretation of a complex set of Kantian claims about when a concept has meaning for us. These claims concern Kant's sense of 'has meaning' (*Bedeutung*, or occasionally *Sinn*) in the sense of 'has a relation to an object' – here note especially A241/B300. Despite some strong passages like A277–78 = B333–34, A636 = B664, and A696 = B724, these claims do not – or should not – concern the current sense of 'has meaning' that implies 'is comprehensible' and contrasts with 'has no sense.' After all, and given the conclusions of the Phenomena and Noumena chapter and the Transcendental Dialectic, when Kant says that without relation to sensible intuition the categories have no meaning (see A241/B300 or B306 ff.), he does not mean to assert that when the categories are used to *think* objects existing in themselves, they are then literally incomprehensible. Rather, he means that through the categories we then gain no cognitive relation to any objects, such a relation requiring the recognizability of the objects through sensible intuition. (See also A247–48/B304–305 and Kant's added handwritten note there – Nachträge CXXVII, in Schmidt, ed., 297; Kemp Smith trans., 265; Ak. 23, 48.)

89 A225–26/B273 speaks not just of such a lawful link but of 'being able in the series

of possible perceptions ... to make the transition from our actual perception to the thing in question' and of what we would perceive 'were our senses more refined.' See Parsons (1964).

90 That is, we might know some general, *de dicto*-like truths, such as 'all objects have grounds of existence,' that would hold true at the world of objects as they exist in themselves.

91 Inaugural Dissertation, § 10 (Ak. 2, 396; Kerferd and Walford trans., 60); note also § 1 and § 25.

92 See Inaugural Dissertation, § 25.

93 See A244/B302 and B302–303 note.

94 We saw in Chapter Nine that, if he tries to maintain his B-Deduction § 19 view that unity of apperception implies *H*'s thinking the (T*i*) thought that yields knowledge of an object distinct from all of *H*'s intuition-elements, Kant then faces the problem of allowing for *H*'s knowledge of merely subjective sequences of intuition-elements. And we saw that his best solution is to shift from the § 19 view to the Chapter Eight weakened (T*i*)-style result. Or else he should retain his basic § 17 position that unity of apperception implies *H*'s thinking the (T*i*) thought. But he should hold also that a separate, supplemental argument must be given to show that while in some cases that thought is about a distinct object, in other cases it may be about a subjective sequence of representations. I assume below that these Chapter Nine points are taken into account. (I also assume that, as noted in earlier chapters, an object distinct from all of *H*'s representations, taken separately or in sequence, can be an object treated, as on Kant's appearance theory, as identical to some categorically synthesized group of intuition-elements, as well as an object treated according to Kant's appearing theory.)

95 Strawson's well-known attempt on such issues (1966, 97–112) is fascinating but has been effectively criticized by, among others, Rorty (1970) (who offers a defense of his own, which has not convinced me) and Mackie (1980, 88–116). Mackie also criticizes suggestive arguments in Bennett (1966, Chapters 8 and 15). Of more recent discussions, Guyer (1987; see, for example, 224 ff., 253–59, 269–76, Chapter 13, and the Afterword), who rejects the Transcendental Deduction itself, makes an interesting attempt to argue from what is required for knowing the temporal relations of representations to knowledge of causally interacting objects distinct from those representations. However, Brueckner (1983, 1984) raises telling questions about this sort of attempt (and about other Kantian efforts), and while the attempt certainly deserves further discussion, I am currently quite pessimistic about its chances for success. Aquila's recent reflections on anticipation, retention, and self-consciousness seem not to attempt any sound argument from self-consciousness to knowledge of objects distinct from all our representations (see his 1989, 174–76).

96 I ignore the Chapter Three form-matter and regress issues and various complications. Other relevant texts include A103, A120–22, A124, A162–63/B203–204, B219, and A224/B271.

97 Recall from Chapter Three that Kant often does not bother to distinguish spatial parts from properties.

98 Note the A105, A106, B162, and A162–63/B203–204 texts cited above. I say 'specified' because Kant is not wholly clear about how the conceptual synthesis of the spatial-parts manifold occurs.

99 Kant's idea of the connection can be got from A168/B209 ff., which links the associated categories of reality and negation with intensive magnitude, and from

A143/B182–83. Helpful discussions exist in Wolff (1963, 232–38) and Guyer (1987, 196–205). See also Paton (1936, vol. II, Chapters 32 and 34).

[100] For the many further, and very obscure, details that I am passing over here, see A167–69/B209–11 and the authors cited in note 99. Because of the obscurities involved, I will not try to connect the present point about synthesis as yielding the object (with its intensive magnitude) with any example of a judgment about that object.

[101] It could of course be argued that at least sometimes (for example, in perception) that organization *is* given; that point would require further discussion.

[102] See, for example, Kemp Smith (1962, 331 and 414–17) and Rescher (1974). For criticism and a contrary position (with which I have some disagreement), see Thompson (1983).

[103] See Bxxvi, A255/B310–11, and various points in the Dialectic.

BIBLIOGRAPHY

Adickes, Erich. 1924. *Kant und das Ding an sich*. Berlin: Pan.

–. 1929. *Kants Lehre von der doppelten Affektion unseres Ich*. Tübingen: Mohr.

Allison, Henry E. 1983. *Kant's Transcendental Idealism*. New Haven, Conn., and London: Yale University Press, 1983.

–. 1987a. "Reflections on the B-Deduction." *Southern Journal of Philosophy* 25, suppl., 1–15.

–. 1987b. "Transcendental Idealism: The 'Two Aspect' View." In Bernard den Ouden and Marcia Moen, eds., *New Essays on Kant*. New York: Peter Lang, 155–78.

–. 1990. *Kant's Theory of Freedom*. Cambridge: Cambridge University Press.

Ameriks, Karl. 1978. "Kant's Transcendental Deduction as a Regressive Argument." *Kant-Studien* 69, 273–87.

–. 1982a. *Kant's Theory of Mind*. Oxford: Clarendon Press.

–. 1982b. "Recent Work on Kant's Theoretical Philosophy." *American Philosophical Quarterly* 19, 1–24.

Aquila, Richard E. 1983. *Representational Mind*. Bloomington: Indiana University Press.

–. 1989. *Matter in Mind*. Bloomington: Indiana University Press, 1989.

Aristotle. *Aristotle's Categories and De Interpretatione*. Trans. J. L. Ackrill. Oxford: Clarendon Press, 1963.

–. *Aristotle's De Anima*. Trans. D. W. Hamlyn. Oxford: Clarendon Press, 1968.

–. *Aristotle's Metaphysics*. Trans. Christopher Kirwan. Oxford: Clarendon Press, 1971.

–. *Aristotle's Posterior Analytics*. Trans. Jonathan Barnes. Oxford: Clarendon Press, 1975.

–. *The Basic Works of Aristotle*. Ed. Richard McKeon. New York: Random House, 1941.

Arnauld, Antoine, and Pierre Nicole. [1662] 1970. *La logique ou l'art de penser*. Reprint. Paris: Flammarion. Trans. as *The Art of Thinking*, by James Dickoff and Patricia James. Indianapolis: Bobbs-Merrill, 1964.

Barker, Stephen F. 1969. "Appearing and Appearances in Kant." In Lewis White Beck, ed., *Kant Studies Today*. LaSalle, Ill.: Open Court, 274–89.

Baum, Manfred. 1986. *Deduktion und Beweis in Kants Transzendentalphilosophie*. Königstein: Athenaeum.

–. 1987. "The B-Deduction and the Refutation of Idealism." *Southern Journal of Philosophy* 25, suppl., 89–107.

Beck, Lewis White. 1955. "Can Kant's Synthetic Judgments Be Made Analytic?" *Kant-Studien* 47, 168–81. Reprinted in Beck (1965), 74–91, and in Wolff, Ed. (1967), 3–22.

–. 1965. *Studies in the Philosophy of Kant*. Indianapolis: Bobbs-Merrill.

–. 1969. *Early German Philosophy*. Cambridge, Mass.: Harvard University Press.
–. 1978. *Essays on Kant and Hume*. New Haven, Conn., and London: Yale University Press.
–. 1989. "Two Ways of Reading Kant's Letter to Herz: Comments on Carl." In Eckart Förster, ed., *Kant's Transcendental Deductions*. Stanford, Calif.: Stanford University Press, 21–26.
Bennett, Jonathan. 1966. *Kant's Analytic*. Cambridge: Cambridge University Press.
–. 1974. *Kant's Dialectic*. Cambridge: Cambridge University Press.
Berkeley, George. *The Works of George Berkeley, Bishop of Cloyne*. Eds. A. A. Luce and T. E. Jessop. 9 vols. Edinburgh: Nelson, 1948–57.
Bird, Graham. 1962. *Kant's Theory of Knowledge*. London: Routledge & Kegan Paul.
Brittan, Gordon G., Jr. 1978. *Kant's Theory of Science*. Princeton, N.J.: Princeton University Press.
–. 1987. "The Reality of Reference: Comments on Carl Posy's 'Where Have All the Objects Gone?'" *Southern Journal of Philosophy* 25, suppl., 37–44.
Brouillet, Raymond. 1975. "Dieter Henrich et 'The Proof-Structure of Kant's Transcendental Deduction.' Réflexions critiques." *Dialogue* 14, 639–47.
Brueckner, Anthony L. 1983. "Transcendental Arguments I." *Noûs* 17, 551–75.
–. 1984. "Transcendental Arguments II." *Noûs* 18, 197–225.
Buroker, Jill. 1981. *Space and Incongruence*. Dordrecht: Reidel.
Carl, Wolfgang. 1989. "Kant's First Drafts of the Deduction of the Categories." In Eckart Förster, ed., *Kant's Transcendental Deductions*. Stanford, Calif.: Stanford University Press, 3–20.
Castañeda, Hector-Neri. 1966. "'He': A Study in the Logic of Self-Consciousness." *Ratio* 8, 130–57.
–. 1967. "Indicators and Quasi-indicators." *American Philosophical Quarterly* 4, 85–100.
–. 1968. "On the Logic of Attributions of Self-Knowledge to Others." *Journal of Philosophy* 55, 439–56.
–. 1989. "The Reflexivity of Self-Consciousness: Sameness/Identity, Data for Artificial Intelligence." *Philosophical Topics* 17, 27–58.
–. 1990. "The Role of Apperception in Kant's Transcendental Deduction of the Categories." *Noûs* 24, 141–57.
Chisholm, Roderick M. 1950. "The Theory of Appearing." In Max Black, ed., *Philosophical Analysis*. Englewood Cliffs, N.J.: Prentice-Hall, 97–112.
–. 1957. *Perceiving*. Ithaca, N.Y.: Cornell University Press.
–. 1981. *The First Person*. Minneapolis: University of Minnesota Press.
Descartes, René. *Oeuvres de Descartes*. Eds. Ch. Adam and P. Tannery. Paris: Vrin, 1957– .
–. *The Philosophical Writings of Descartes*. Vols. I and II. Trans. John Cottingham, Robert Stoothoff, and Dugald Murdoch. Cambridge: Cambridge University Press, 1985.
Ewing, A. C. 1950. *A Short Commentary on Kant's "Critique of Pure Reason."* 2d ed. London: Methuen.
Förster, Eckart. 1985. Review of Allison (1983). *Journal of Philosophy* 82, 734–38.

–, ed. 1989. *Kant's Transcendental Deductions*. Stanford, Calif.: Stanford University Press.

Friedman, Michael. 1985. "Kant's Theory of Geometry." *Philosophical Review* 94, 455–506.

–. 1990. "Kant on Concepts and Intuitions in the Mathematical Sciences." *Synthese* 84, 213–57.

Geach, Peter. 1957. *Mental Acts*. London: Routledge & Kegan Paul.

George, Rolf. 1981. "Kant's Sensationism." *Synthese* 47, 229–55.

Grayeff, Felix. 1970. *Kant's Theoretical Philosophy*. Trans. David Walford. Manchester: Manchester University Press.

Guyer, Paul. 1979. Review of Henrich (1976). *Journal of Philosophy* 76, 151–67.

–. 1987. *Kant and the Claims of Knowledge*. Cambridge: Cambridge University Press.

Harrison, Ross. 1982. "Transcendental Arguments and Idealism." In Godfrey Vesey, ed., *Idealism Past and Present*. Royal Institute of Philosophy Lecture Series 13. Cambridge: Cambridge University Press, 211–24.

Henrich, Dieter. 1968–69. "The Proof-Structure of Kant's Transcendental Deduction." *Review of Metaphysics* 22, 640–59. Reprinted in Walker (1982), 66–81.

–. 1976. *Identität und Objektivität*. Heidelberg: Carl Winter Universitätsverlag.

–. 1989. "Kant's Notion of a Deduction and the Methodological Background of the First *Critique*." In Eckart Förster, ed., *Kant's Transcendental Deductions*. Stanford, Calif.: Stanford University Press, 29–46.

Hintikka, Jaakko. 1962. *Knowledge and Belief*. Ithaca, N.Y.: Cornell University Press.

–. 1965. "Kant's 'New Method of Thought' and His Theory of Mathematics." *Ajatus* 27, 37–47.

–. 1967. "Kant on the Mathematical Method." *Monist* 51, 352–75.

–. 1969. "On Kant's Notion of Intuition (*Anschauung*)." In Terence Penelhum and J. J. MacIntosh, eds., *The First Critique*. Belmont, Calif.: Wadsworth, 38–53.

–. 1972. "Kantian Intuitions." *Inquiry* 15, 341–45.

–. 1973. *Logic, Language-Games, and Information*. Oxford: Oxford University Press.

–. 1974. *Knowledge and the Known*. Dordrecht: Reidel.

Howell, Robert. 1973. "Intuition, Synthesis, and Individuation in the *Critique of Pure Reason*." *Noûs* 7, 207–32.

–. 1979. "A Problem for Kant." In E. Saarinen, R. Hilpinen, I. Niiniluoto, and M. Provence Hintikka, eds., *Essays in Honour of Jaakko Hintikka*. Dordrecht: Reidel, 331–49.

–. 1981a. "Apperception and the 1787 Transcendental Deduction." *Synthese* 47, 385–448.

–. 1981b. "Kant's First-*Critique* Theory of the Transcendental Object." *Dialectica* 35, 85–125.

Hume, David. *Enquiries concerning the Human Understanding and concerning the Principles of Morals*. 2d ed. Ed. L. A. Selby-Bigge. Oxford: Clarendon Press, 1902.

–. *A Treatise of Human Nature*. Ed. L. A. Selby-Bigge. Oxford: Clarendon Press, 1888.

Kant, Immanuel. *Kant's gesammelte Schriften, herausgegeben von der Deutschen* [formerly *Königlichen Preussischen*] *Akademie der Wissenschaften*. 29 vols. Berlin: Reimer, later de Gruyter, 1902- . Cited as Ak.

–. *Critique of Judgment*. Trans. Werner S. Pluhar. Indianapolis: Hackett, 1987.

–. *Critique of Pure Reason*. Trans. Norman Kemp Smith. New York: St. Martin's Press, 1965.

–. *Kritik der reinen Vernunft*. Ed. Raymund Schmidt. Hamburg: Felix Meiner, 1956.

–. *Logic*. Trans. Robert S. Hartman and Wolfgang Schwarz. Indianapolis: Bobbs-Merrill, 1974.

–. *Metaphysical Foundations of Natural Science*. Trans. James Ellington. Indianapolis: Bobbs-Merrill, 1970.

–. *Philosophical Correspondence, 1759–99*. Ed. and trans. Arnulf Zweig. Chicago: University of Chicago Press, 1967.

–. *Prolegomena to Any Future Metaphysics*. Trans. Peter G. Lucas. Manchester: Manchester University Press, 1959.

–. *Selected Pre-Critical Writings*. Trans. G. B. Kerferd and D. E. Walford. Manchester: Manchester University Press, 1968.

–. *What Real Progress Has Metaphysics Made in Germany since the Time of Leibniz and Wolff?* Trans. Ted Humphrey. New York: Abaris Books, 1983.

Kaplan, David. 1989. "Demonstratives." Mimeographed. UCLA, 1977. Published in Joseph Almog, John Perry, Howard Wettstein, eds., *Themes from Kaplan*. New York and Oxford: Oxford University Press, 481–563.

Kemp Smith, Norman. 1962. *A Commentary to Kant's "Critique of Pure Reason."* 2d ed. New York: Humanities Press.

Kenny, Anthony. 1968. *Descartes*. New York: Random House.

Kitcher, Patricia. 1982. "Kant on Self-Identity." *Philosophical Review* 91, 41–72.

–. 1990. *Kant's Transcendental Psychology*. New York and Oxford: Oxford University Press.

Kitcher, Philip. 1975. "Kant and the Foundations of Mathematics." *Philosophical Review* 84, 23–50.

Leibniz, Gottfried Wilhelm. *Philosophical Papers and Letters*. 2d ed. Trans. and ed. Leroy E. Loemker. Dordrecht: Reidel, 1976.

–. *Die Philosophischen Schriften von G. W. Leibniz*. Vols. 1–7. Ed. C. I. Gerhardt. Berlin: Weidmann, 1875–90.

Lewis, Clarence Irving. 1929. *Mind and the World-Order*. New York: Scribner's.

Lewis, David. 1979. "Attitudes *De Dicto* and *De Se*." *Philosophical Review* 88, 513–43.

Locke, John. *An Essay concerning Human Understanding*. Ed. P. H. Nidditch. Oxford: Oxford University Press, 1975.

Mackie, J. L. 1980. *The Cement of the Universe*. Oxford: Clarendon Press.

Massey, G. J. 1986. "Kant's Theory of Propositional Attitudes." *Kant-Studien* 77, 423-40.

Meerbote, Ralf. 1980. Review of Allen W. Wood, *Kant's Rational Theology*, and I. Kant, *Lectures on Philosophical Theology*. *Philosophical Review* 89, 285–88.

–. 1986. "Introduction to Kant's 'Concerning Sensory Illusion and Poetic Fiction.'" In Lewis White Beck, ed., *Kant's Latin Writings*. New York: Peter Lang, 193–201.

–. 1987. "Apperception and Objectivity." *Southern Journal of Philosophy* 25, suppl., 115–30.

Mellin, G. S. A. 1797–1804. *Encyclopädisches Wörterbuch der Kritischen Philosophie*. Vols. 1–6. Züllichau and Leipzig: Frommann.

Melnick, Arthur. 1973. *Kant's Analogies of Experience*. Chicago and London: University of Chicago Press.

–. 1989. *Space, Time, and Thought in Kant*. Dordrecht: Kluwer.

Moore, George Edward. 1953. *Some Main Problems of Philosophy*. London: Allen & Unwin.

Moravcsik, J. M. E., ed. 1967. *Aristotle*. Garden City, N.Y.: Doubleday.

Nowotny, Viktor. 1981. "Die Struktur der Deduktion bei Kant." *Kant-Studien* 72, 270–79.

Parsons, Charles. 1964. "Infinity and Kant's Conception of the 'Possibility of Experience.'" *Philosophical Review* 73, 182–97. Reprinted with additional notes in Parsons (1983), 95–109.

–. 1969. "Kant's Philosophy of Arithmetic." In Sidney Morgenbesser, Patrick Suppes, and Morton White, eds., *Philosophy, Science and Method: Essays in Honor of Ernest Nagel*. New York: St. Martin's Press, 568–94. Reprinted with a new postscript in Parsons (1983), 110–49.

–. 1983. *Mathematics in Philosophy*. Ithaca, N.Y.: Cornell University Press.

Paton, H. J. 1936. *Kant's Metaphysic of Experience*. Vols. 1 and 2. London: Allen & Unwin.

Perry, John. 1977. "Frege on Demonstratives." *Philosophical Review* 86, 474–97.

–. 1979. "The Problem of the Essential Indexical." *Noûs* 13, 3–21.

Pippin, Robert. 1982. *Kant's Theory of Form*. New Haven, Conn., and London: Yale University Press.

Pistorius, H. A. [signed 'Sg.']. 1784. Review of Kant's *Prolegomena zu einer jeden künftigen Metaphysik*. *Allgemeine Deutsche Bibliothek* 59, 322–57.

Pollock, John. 1988. "My Brother, the Machine." *Noûs* 22, 173–211.

Posy, Carl. 1981. "The Language of Appearances and Things in Themselves." *Synthese* 47, 313–52.

–. 1983. "Dancing to the Antinomy: A Proposal for Transcendental Idealism." *American Philosophical Quarterly* 20, 81–94.

–. 1987. "Where Have All the Objects Gone?" *Southern Journal of Philosophy* 25, suppl., 17–36.

Powell, C. Thomas. 1990. *Kant's Theory of Self-Consciousness*. Oxford: Clarendon Press.

Prauss, Gerold. 1974. *Kant und das Problem der Dinge an sich*. Berlin: Bouvier.

Prichard, H. A. 1909. *Kant's Theory of Knowledge*. Oxford: Clarendon Press.

Rescher, Nicholas. 1974. "Noumenal Causality." In Lewis White Beck, ed., *Kant's Theory of Knowledge*. Dordrecht: Reidel, 175–83.

Robinson, Hoke. 1981. "Anschauung und Mannigfaltiges in der Transzendentalen Deduktion." *Kant-Studien* 72, 140–48. Trans. *Southern Journal of Philosophy* 22 (1984), 403–12.

–, ed. 1987. "Spindel Conference 1986: The B-Deduction." *Southern Journal of Philosophy* 25, suppl.

Rorty, Richard. 1970. "Strawson's Objectivity Argument." *Review of Metaphysics* 24, 207–44.

Schrader, George. 1958. "Kant's Theory of Concepts." *Kant-Studien* 49, 264–78. Reprinted in Wolff, Ed. (1967), 134–55.

Strawson, P. F. 1959. *Individuals*. London: Methuen.

–. 1966. *The Bounds of Sense*. London: Methuen.

Thompson, Manley. 1972–73. "Singular Terms and Intuitions in Kant's Epistemology." *Review of Metaphysics* 26, 314–43.

–. 1983. "Things in Themselves." *Proceedings and Addresses of the American Philosophical Association* 57, 33–48.

Vaihinger, Hans. 1922. *Commentar zu Kants Kritik der reinen Vernunft*. 2d ed. Vols. I and II. Stuttgart, Berlin, and Leipzig.

Vleeschauwer, H. J. de. 1934–37. *La déduction transcendantale dans l'oeuvre de Kant*. Vols. 1–3. Antwerp: De Sikkel.

–. 1962. *The Development of Kantian Thought*. Trans. A. R. C. Duncan. London: Nelson.

Wagner, Hans. 1980. "Der Argumentationsgang in Kants Deduktion der Kategorien." *Kant-Studien* 71, 352–66.

Walker, Ralph C. S. 1978. *Kant*. London, Henley, and Boston: Routledge & Kegan Paul.

–, ed. 1982. *Kant on Pure Reason*. Oxford: Oxford University Press.

Walsh, W. H. 1975. *Kant's Criticism of Metaphysics*. Edinburgh: Edinburgh University Press.

Weldon, T. D. 1958. *Kant's "Critique of Pure Reason."* 2d Ed. Oxford: Clarendon Press.

Wilson, Kirk Dallas. 1978. "Studies in the Formal Logic of Kant's Modal Functions of Judgment." *Kant-Studien* 69, 252–72.

Wilson, Margaret Dauler. 1978. *Descartes*. London, Henley, and Boston: Routledge & Kegan Paul.

Wolff, Robert Paul. 1960. "Hume's Theory of Mental Activity." *Philosophical Review* 49, 289–310.

–. 1963. *Kant's Theory of Mental Activity*. Cambridge, Mass.: Harvard University Press.

–, ed. 1967. *Kant: A Collection of Critical Essays*. Garden City, N.Y.: Doubleday.

INDEX

(See also the list of frequently cited displayed sentences on p. xiii.)